T0236230

# Lecture Notes in Computer Science 10163

Commenced Publication in 1973
Founding and Former Series Editors:
Gerhard Goos, Juris Hartmanis, and Jan van Leeuwen

More information about this series at http://www.springer.com/series/7412

Ana Fred · Maria De Marsico
Gabriella Sanniti di Baja (Eds.)

# Pattern Recognition Applications and Methods

5th International Conference, ICPRAM 2016
Rome, Italy, February 24–26, 2016
Revised Selected Papers

 Springer

*Editors*
Ana Fred
University of Lisbon
Lisbon
Portugal

Gabriella Sanniti di Baja
ICAR-CNR
Naples
Italy

Maria De Marsico
Sapienza University of Rome
Rome
Italy

ISSN 0302-9743                ISSN 1611-3349   (electronic)
Lecture Notes in Computer Science
ISBN 978-3-319-53374-2        ISBN 978-3-319-53375-9   (eBook)
DOI 10.1007/978-3-319-53375-9

Library of Congress Control Number: 2017930942

LNCS Sublibrary: SL6 – Image Processing, Computer Vision, Pattern Recognition, and Graphics

Printed on acid-free paper

This Springer imprint is published by Springer Nature
The registered company is Springer International Publishing AG
The registered company address is: Gewerbestrasse 11, 6330 Cham, Switzerland

# Preface

The present book includes extended and revised versions of a set of selected papers from the 5th International conference on Pattern Recognition Applications and Methods (ICPRAM 2016), held in Rome, Italy, during February 24–26, 2016. The purpose of the conference is to represent a major point of contact between researchers working along different research lines in the areas of pattern recognition, both from theoretical and application perspectives.

The papers represent the 10% most interesting and relevant part of those received for the conference. They were selected by the event chairs and their selection is based on a number of criteria that include the classifications and comments provided by the Program Committee members, the session chairs' assessment of presentation, and also the program chairs' global view of all papers included in the technical program. The authors of selected papers were then invited to submit a revised and extended version of their papers having a sufficient amount of innovative material, with respect to discussion of the proposed approaches, presentation of theoretical as well as operational details, and experiments.

We hope that the papers selected to be included in this book contribute to the understanding of relevant trends of current research on pattern recognition, especially in the areas covered by this book.

A first subdivision of the papers presented in this book is of course between papers related to methods and theoretical approaches, and those related to specific applications. The value of the former relies in providing support for possible implementation and improvement of applications and further field advancement. The latter are robust, reliable, and flexible in the measure they rely on solid methodological bases.

The first paper dealing with methods is "Experimental Evaluation of Graph Classification with Hadamard Code Graph Kernels" by Tetsuya Kataoka and Akihiro Inokuchi. Kernel methods are used to efficiently classify into the same class those graphs that have similar structures. The authors propose a novel graph kernel that can be used in this kind of operation that is based on the Hadamard code. The paper presents the Hadamard code kernel (HCK) and shortened HCK (SHCK), a version of HCK that compresses vertex labels in graphs. The performance and practicality of the proposed method are demonstrated in experiments that compare the computation time, scalability, and classification accuracy of HCK and SHCK with those of other approaches.

The paper "Document Clustering Games in Static and Dynamic Scenarios" by Rocco Tripodi and Marcello Pelillo proposes a game theoretic model for document clustering. Each document is represented as a player and each cluster as a strategy. The players receive a reward interacting with other players, according to the quality of the adopted strategy. Even in this case, the geometry of the data is modeled by a weighted graph that encodes the pairwise similarity among documents. Weights condition the chosen strategies. The system was evaluated using 13 document datasets with different settings.

In their paper "Criteria for Mixture-Model Clustering with Side-Information," Edith Grall-Maës and Duc Tung Dao consider mixture models using side-information, and use them for cluster analysis. Side-information gives the constraint that some data in a group originate from the same source. In this work the authors adapt three usual criteria, which are the Bayesian information criterion (BIC), the Akaike information criterion (AIC), and the entropy criterion (NEC), so that they can take into consideration the side-information.

The paper "Near-Boolean Optimization — A Continuous Approach to Set Packing and Partitioning" by Giovanni Rossi proposes to exploit near-Boolean functions to address the packing problem. Given a family of feasible subsets of the ground set, the packing problem is to find a largest subfamily of pairwise disjoint family members. The problem is first translated into a continuous version, with the objective function taking values on peculiar collections of points in a unit hypercube. Feasible solutions for the original combinatorial optimization problem are included in extremizers, and this allows for a gradient-based local search.

The book then presents the paper "Approximate Inference in Related Multi-output Gaussian Process Regression" by Ankit Chiplunkar, Emmanuel Rachelson, Michele Colombo, and Joseph Morlier. A relevant issue raised with Gaussian process regression is efficient inference when scaling up to large datasets. In this paper, the authors use approximate inference techniques upon multi-output kernels enforcing relationships between outputs. A multi-output kernel is a covariance function over correlated outputs. The main contribution of the paper is the application and validation of the proposed methodology on a dataset of real aircraft flight tests, achieved by also imposing knowledge of aircraft physics into the model.

Papers dealing with specific applications often start from considering specific new kinds of sensors. This is the case of the paper "An Online Data Validation Algorithm for Electronic Nose" by Mina Mirshahi, Vahid Partovi Nia, and Luc Adjengue. An electronic nose is one of the lesser known achievements in pattern recognition applications. An e-nose is a device that is trained to analyze the chemical components of an odor. It consists of an array of gas sensors for chemical detection, and a mechanism for pattern recognition to return the odor concentration. The latter defines the identifiability and perceivability of an odor. Specific impairment of the e-nose, and further environmental factors, e.g., wind, humidity, temperature, may introduce noise into the measurements, and affect recognition results. The paper proposes an online algorithm to evaluate the validity of sensor measurements during the sampling before using the data for pattern recognition phase.

One of the tasks included in image understanding is duplicate or near-duplicate retrieval. Among the local feature detectors and descriptors used for this task, the paper "Near-Duplicate Retrieval: A Benchmark Study of Modified SIFT Descriptors" by Afra'a Ahmad Alyosef and Andreas Nürnberger especially focuses on SIFT descriptors. The authors evaluate the accuracy and performance of variations of SIFT descriptors (reduced SIFT versions, RC-SIFT-64D, the original SIFT-128D) and SURF-64D, both using benchmarks of various sizes, and using one particular benchmark but extracting varying amounts of descriptors. Moreover, they also provide results of a comparative performance analysis using benchmarks generated by combining several image affine transformations.

Activity recognition is gaining a relevant role of its own in many application fields. The paper "Activity Recognition for Elderly Care by Evaluating Proximity to Objects and Human Skeleton Data" by Julia Richter, Christian Wiede, Enes Dayangac, Ahsan Shahenshah, and Gangolf Hirtz deals with remote control of activities of daily living (ADLs) of elderly for ambient-assisted living (AAL). The paper presents an algorithm that detects activities related to personal hygiene. The approach is based on the evaluation of pose information and a person's proximity to objects belonging to the typical equipment of bathrooms, such as sink, toilet, and shower. Moreover, a skeleton-based algorithm recognizes actions using a supervised learning model.

The paper "Real-Time Swimmer Tracking on Sparse Camera Array" by Paavo Nevalainen, M. Hashem Haghbayan, Antti Kauhanen, Jonne Pohjankukka, Mikko-Jussi Laakso, and Jukka Heikkonen deals with the analysis of swimming patterns from data captured from multiple cameras. This task is very important in a video-based athletics performance analysis. The presented real-time algorithm allows one to perform the planar projection of the image, fading the background to protect the intimacy of other swimmers, framing the swimmer at a specific swimming lane, and eliminating the redundant video stream from idle cameras. The generated video stream can be further analyzed. The geometric video transform accommodates a sparse camera array and enables geometric observations of swimmer silhouettes. The methodology allows for unknown camera positions and can be installed in many types of public swimming pools.

Recognition can also support higher-level tasks. In the paper "Fundamentals of Nonparametric Bayesian Line Detection" by Anne C. van Rossum, Hai Xiang Lin, Johan Dubbeldam, and H. Jaap van den Herik, Bayes approach is explored to solve the problem of line detection. In fact, line detection is a fundamental problem in the world of computer vision and a component step of higher-level recognition/detection activities.

The paper by Humberto Sossa and Hermilo Sánchez, "Computing the Number of Bubbles and Tunnels of a 3-D Binary Object," presents specific techniques to achieve separately the two results that can be used in a further analysis step.

An unusual application is presented in the paper "Raindrop Detection on a Windshield Based on Edge Ratio" by Junki Ishizuka and Kazunori Onoguchi. The method exploits data from an in-vehicle single camera. The rationale for this application is that raindrops on a windshield cause various types of bad influence for video-based automobile applications, such as pedestrian detection, lane detection and so on. For this reason, it is important to understand the state of the raindrop on a windshield for a driving safety support system or an automatic driving vehicle.

Last but not least, the paper "Comparative Analysis of PRID Algorithms Based on Results Ambiguity Evaluation" by V. Renò, A. Cardellicchio, T. Politi, C. Guaragnella, and T. D'Orazio deals with one of the faster-developing areas of pattern recognition that is most quickly developing at present, namely, biometric recognition. The re-identification of a subject among different cameras (Person Re-Identification or PRID) is a task that implicitly defines ambiguities, e.g., raising when images of two individuals dressed in a similar manner or with a comparable body shape are analyzed by a computer vision system. The authors propose an approach to find, exploit, and classify ambiguities among the results of PRID algorithms.

To conclude, we would like to thank all the authors for their contributions, and also the reviewers who helped ensure the quality of this publication. Finally, thanks are also due to the INSTICC staff that supported both the conference and the preparation of this book.

February 2016

Ana Fred
Maria De Marsico
Gabriella Sanniti di Baja

# Organization

## Conference Chair

Ana Fred                          Instituto de Telecomunicações/IST, Portugal

## Program Co-chairs

Maria De Marsico                  Sapienza Università di Roma, Italy
Gabriella Sanniti di Baja         ICAR-CNR, Italy

## Program Committee

Andrea F. Abate                   University of Salerno, Italy
Ashraf AbdelRaouf                 Misr International University MIU, Egypt
Rahib Abiyev                      Near East University, Turkey
Mayer Aladjem                     Ben-Gurion University of the Negev, Israel
Guillem Alenya                    Institut de Robòtica i Informàtica Industrial,
                                    CSIC-UPC, Spain
Luís Alexandre                    UBI/IT, Portugal
Francisco Martínez Álvarez        Pablo de Olavide University of Seville, Spain
Ioannis Arapakis                  Yahoo Labs Barcelona, Spain
Kevin Bailly                      ISIR Institut des Systèmes Intelligents et de Robotique,
                                    UPMC/CNRS, France
Emili Balaguer-Ballester          Bournemouth University, UK
Mohammed Bennamoun                School of Computer Science and Software
                                    Engineering, The University of Western Australia,
                                    Australia
Stefano Berretti                  University of Florence, Italy
Monica Bianchini                  University of Siena, Italy
Michael Biehl                     University of Groningen, The Netherlands
Isabelle Bloch                    Telecom ParisTech - CNRS LTCI, France
Mohamed-Rafik Bouguelia           Halmstad University, Sweden
Nizar Bouguila                    Concordia University, Canada
Francesca Bovolo                  Fondazione Bruno Kessler, Italy
Samuel Rota Bulò                  Fondazione Bruno Kessler, Italy
Javier Calpe                      Universitat de València, Spain
Francesco Camastra                University of Naples Parthenope, Italy
Virginio Cantoni                  Università di Pavia, Italy
Ramón A. Mollineda                Universitat Jaume I, Spain
  Cárdenas
Marco La Cascia                   Università degli Studi di Palermo, Italy

| | |
|---|---|
| Michal Haindl | Institute of Information Theory and Automation, Czech Republic |
| Barbara Hammer | Bielefeld University, Germany |
| Robert Harrison | Georgia State University, USA |
| Makoto Hasegawa | Tokyo Denki University, Japan |
| Mark Hasegawa-Johnson | University of Illinois at Urbana-Champaign, USA |
| Pablo Hennings-Yeomans | Sysomos, Canada |
| Laurent Heutte | Université de Rouen, France |
| Anders Heyden | Lund Institute of Technology/Lund University, Sweden |
| Kouichi Hirata | Kyushu Institute of Technology, Japan |
| Sean Holden | University of Cambridge, UK |
| Jose M. Iñesta | Universidad de Alicante, Spain |
| Yuji Iwahori | Chubu University, Japan |
| Sarangapani Jagannathan | Missouri University of Science and Technology, USA |
| Nursuriati Jamil | Universiti Teknologi MARA, Malaysia |
| Graeme A. Jones | Kingston University London, UK |
| Yasushi Kanazawa | Toyohashi University of Technology, Japan |
| Lisimachos Kondi | University of Ionnina, Greece |
| Mario Köppen | Kyushu Institute of Technology, Japan |
| Marco Körner | Technische Universität München, Germany |
| Walter Kosters | Universiteit Leiden, The Netherlands |
| Constantine Kotropoulos | Aristotle University of Thessaloniki, Greece |
| Sotiris Kotsiantis | University of Patras, Greece |
| Konstantinos Koutroumbas | National Observatory of Athens, Greece |
| Kidiyo Kpalma | INSA de Rennes, France |
| Marek Kretowski | Bialystok University of Technology, Poland |
| Adam Krzyzak | Concordia University, Canada |
| Piotr Kulczycki | Polish Academy of Sciences, Poland |
| Jaerock Kwon | Kettering University, USA |
| Shang-Hong Lai | National Tsing Hua University, Taiwan |
| Raffaella Lanzarotti | Università degli Studi di Milano, Italy |
| Nikolaos Laskaris | AUTH, Greece |
| Shi-wook Lee | National Institute of Advanced Industrial Science and Technology, Japan |
| Young-Koo Lee | Kyung Hee University, Republic of Korea |
| Qi Li | Western Kentucky University, USA |
| Aristidis Likas | University of Ioannina, Greece |
| Hantao Liu | University of Hull, UK |
| Gaelle Loosli | Clermont Université, France |
| Teresa Bernarda Ludermir | Federal University of Pernambuco, Brazil |
| Alessandra Lumini | Università di Bologna, Italy |
| Juan Luo | George Mason University, USA |
| Marco Maggini | University of Siena, Italy |
| Francesco Marcelloni | University of Pisa, Italy |
| Elena Marchiori | Radboud University, The Netherlands |
| Gian Luca Marcialis | Università degli Studi di Cagliari, Italy |

| | |
|---|---|
| Lorenza Saitta | Università degli Studi del Piemonte Orientale Amedeo Avogadro, Italy |
| Antonio-José Sánchez-Salmerón | Universitat Politecnica de Valencia, Spain |
| Carlo Sansone | University of Naples, Italy |
| K.C. Santosh | The University of South Dakota, USA |
| Atsushi Sato | NEC, Japan |
| Michele Scarpiniti | Sapienza University of Rome, Italy |
| Paul Scheunders | University of Antwerp, Belgium |
| Tanya Schmah | University of Toronto, Canada |
| Friedhelm Schwenker | University of Ulm, Germany |
| Ishwar Sethi | Oakland University, USA |
| Katsunari Shibata | Oita University, Japan |
| Lauro Snidaro | Università degli Studi di Udine, Italy |
| Bassel Solaiman | Institut Mines Telecom, France |
| Humberto Sossa | Instituto Politécnico Nacional-CIC, Mexico |
| Arcot Sowmya | University of New South Wales, Australia |
| Harry Strange | Aberystwyth University, UK |
| Mu-Chun Su | National Central University, Taiwan |
| Zhenan Sun | Institute of Automation, Chinese Academy of Sciences (CASIA), China |
| Alberto Taboada-Crispí | Universidad Central Marta Abreu de Las Villas, Cuba |
| Andrea Tagarelli | University of Calabria, Italy |
| Atsuhiro Takasu | National Institute of Informatics, Japan |
| Ichiro Takeuchi | Nagoya Institute of Technology, Japan |
| Oriol Ramos Terrades | Universitat Autònoma de Barcelona, Spain |
| Massimo Tistarelli | Università degli Studi di Sassari, Italy |
| Ricardo S. Torres | University of Campinas (UNICAMP), Brazil |
| Andrea Torsello | Università Ca' Foscari Venezia, Italy |
| Godfried Toussaint | New York University Abu Dhabi, United Arab Emirates |
| Kostas Triantafyllopoulos | University of Sheffield, UK |
| George Tsihrintzis | University of Piraeus, Greece |
| Olgierd Unold | Wroclaw University of Technology, Poland |
| Ernest Valveny | Universitat Autònoma de Barcelona, Spain |
| Antanas Verikas | Intelligent Systems Laboratory, Halmstad University, Sweden |
| M. Verleysen | Machine Learning Group, Université catholique de Louvain, Belgium |
| Markus Vincze | Technische Universität Wien, Austria |
| Panayiotis Vlamos | Ionian University, Greece |
| Asmir Vodencarevic | Schaeffler Technologies AG & Co. KG, Germany |
| Yvon Voisin | University of Burgundy, France |
| Toyohide Watanabe | Nagoya Industrial Science Research Institute, Japan |
| Jonathan Weber | Université de Haute-Alsace, France |
| Harry Wechsler | George Mason University, USA |

| Joost van de Weijer | Autonomous University of Barcelona, Spain |
| Laurent Wendling | LIPADE, France |
| Slawomir Wierzchon | Polish Academy of Sciences, Poland |
| Bing-Fei Wu | National Chiao Tung University, Taiwan |
| Xin-Shun Xu | Shandong University, China |
| Jing-Hao Xue | University College London, UK |
| Chan-Yun Yang | National Taipei University, Taiwan |
| Haiqin Yang | Chinese University of Hong Kong, Hong Kong, SAR China |
| Yusuf Yaslan | Istanbul Technical University, Turkey |
| Nicolas Younan | Mississippi State University, USA |
| Slawomir Zadrozny | Polish Academy of Sciences, Poland |
| Danuta Zakrzewska | Lodz University of Technology, Poland |
| Pavel Zemcik | Brno University of Technology, Czech Republic |
| Huiyu Zhou | Queen's University Belfast, UK |
| Jiayu Zhou | Michigan State University, USA |
| Reyer Zwiggelaar | Aberystwyth University, UK |

## Additional Reviewers

| Ranya Almohsen | West Virginia University, USA |
| Elhocine Boutellaa | Oulu University, Finland |
| Fabian Gieseke | Radboud University Nijmegen, The Netherlands |
| Sojeong Ha | POSTECH, Republic of Korea |
| Roberto Interdonato | DIMES, Unversità della Calabria, Italy |
| Jukka Komulainen | University of Oulu, Finland |
| Andrea Lagorio | University of Sassari, Italy |
| Marcello Vincenzo Tavano Lanas | Diego Portales University, Chile |
| Juho Lee | Pohang University of Science and Technology, Korea, Republic of |
| Yingying Liu | University of New South Wales, Australia |
| Pulina Luca | University of Sassari, Italy |
| Sebastian Sudholt | Technische Universität Dortmund, Germany |
| Suwon Suh | Pohang University of Science and Technology, Republic of Korea |
| Yang Wang | NICTA, Australia |

## Invited Speakers

| Arun Ross | Michigan State University, USA |
| Ludmila Kuncheva | Bangor University, UK |
| Fabio Roli | Università degli Studi di Cagliari, Italy |
| Tanja Schultz | Cognitive Systems Lab (CSL), University of Bremen, Germany |

# Contents

# Experimental Evaluation of Graph Classification with Hadamard Code Graph Kernels

Tetsuya Kataoka and Akihiro Inokuchi[✉]

School of Science and Technology, Kwansei Gakuin University,
2-1 Gakuen, Sanda, Hyogo, Japan
{TKataoka,inokuchi}@kwansei.ac.jp

**Abstract.** When mining information from a database comprising graphs, kernel methods are used to efficiently classify graphs that have similar structures into the same classes. Instances represented by graphs usually have similar properties if their graph representations have high structural similarity. The neighborhood hash kernel (NHK) and Weisfeiler–Lehman subtree kernel (WLSK) have previously been proposed as kernels that compute more quickly than the random-walk kernel; however, they each have drawbacks. NHK can produce hash collision and WLSK must sort vertex labels. We propose a novel graph kernel equivalent to NHK in terms of time and space complexities, and comparable to WLSK in terms of expressiveness. The proposed kernel is based on the Hadamard code. Labels assigned by our graph kernel follow a binomial distribution with zero mean. The expected value of a label is zero; thus, such labels do not require large memory. This allows the compression of vertex labels in graphs, as well as fast computation. This paper presents the Hadamard code kernel (HCK) and shortened HCK (SHCK), a version of HCK that compresses vertex labels in graphs. The performance and practicality of the proposed method are demonstrated in experiments that compare the computation time, scalability and classification accuracy of HCK and SHCK with those of NHK and WLSK for both artificial and real-world datasets. The effect of assigning initial labels is also investigated.

**Keywords:** Graph classification · Support vector machine · Graph kernel · Hadamard code

## 1 Introduction

A natural way of representing structured data is to use graphs [14]. As an example, the structural formula of a chemical compound is a graph, where each vertex corresponds to an atom in the compound and each edge corresponds to a bond between the two atoms therein. Using such graph representations, a new research field called graph mining has emerged from data mining, with the objective of mining information from a database consisting of graphs. With the potential to find meaningful information, graph mining has received much interest, and

© Springer International Publishing AG 2017
A. Fred et al. (Eds.): ICPRAM 2016, LNCS 10163, pp. 1–19, 2017.
DOI: 10.1007/978-3-319-53375-9_1

research in the field has grown rapidly in recent years. Furthermore, because the need for classifying graphs has strengthened in many real-world applications, such as the analysis of proteins in bioinformatics and chemical compounds in cheminformatics [11], graph classification has been widely researched worldwide. The main objective of graph classification is to classify graphs of similar structures into the same classes. This originates from the fact that instances represented by graphs usually have similar properties if their graph representations have high structural similarity.

Kernel methods such as the use of the support vector machine (SVM) are becoming increasingly popular because of their high performance in solving graph classification problems [8]. Most graph kernels are based on the decomposition of a graph into substructures and a feature vector containing counts of these substructures. Because the dimensionality of these feature vectors is typically high and this approach includes the subgraph isomorphism matching problem that is known to be NP-complete [6], kernels deliberately avoid the explicit computation of feature values and instead employ efficient procedures.

One representative graph kernel is the random-walk kernel (RWK) [8,10], which computes $k(g_i, g_j)$ in $O(|V(g)|^3)$ for graphs $g_i$ and $g_j$, where $|V(g)|$ is the number of vertices in $g_i$ and $g_j$. The kernel returns a high value if the random walk on the graph generates many sequences with the same labels for vertices and edges; i.e., the graphs are similar to each other. The neighborhood hash kernel (NHK) [7] and Weisfeiler–Lehman subtree kernel (WLSK) [9] are two other recently proposed kernels that compute $k(g_i, g_j)$ more quickly than RWK. NHK uses logical operations such as the exclusive OR on the label set of adjacent vertices, while WLSK uses a concatenation of label strings of the adjacent vertices to compute $k(g_i, g_j)$. The labels updated by repeating the hash or concatenation propagate the label information over the graph and uniquely represent the higher-order structures around the vertices beyond the vertex or edge level. An SVM with two graph kernels works well with benchmark data consisting of graphs.

The computation of NHK is efficient because it is a logical operation between fixed-length bit strings and does not require string sorting. However, its drawback is hash collision, which occurs when different induced subgraphs have identical hash values. Meanwhile, WLSK must sort the vertex labels, but it has high expressiveness because each vertex $v$ has a distribution of vertex labels within $i$ steps from $v$. To overcome the drawbacks of NHK and WLSK, in this paper, we propose a novel graph kernel that is equivalent to NHK in terms of time and space complexities and comparable to WLSK in terms of expressiveness. The graph kernel proposed in this paper is based on the Hadamard code [13]. The Hadamard code is used in spread spectrum-based communication technologies such as Code Division Multiple Access to spread message signals. Because the probability of occurrences of values of 1 and $-1$ are equivalent in each column of the Hadamard matrix except for the first column, labels assigned by our graph kernel follow a binomial distribution with zero mean under a certain assumption. Therefore, the expected value of the label is zero, and for such labels, a large memory space is not required. This characteristic is used to compress vertex labels in graphs, allowing the proposed graph kernel to be computed quickly.

Note that large portions of this paper were covered in our previous work [13]. Within the current work we demonstrate the performance and practicality of the proposed method in experiments that compare the computation time, scalability and classification accuracy of HCK and SHCK with those of NHK and WLSK for various artificial datasets. The effect of assigning initial labels for the graph kernels is also investigated.

The rest of this paper is organized as follows. Section 2 defines the graph classification problem and explains the framework of the existing graph kernels. Section 3 proposes the Hadamard code kernel (HCK), based on the Hadamard code, and another graph kernel called the shortened HCK (SHCK), which is a version of HCK that compresses vertex labels in graphs. Section 4 demonstrates the fundamental performance and practicality of the proposed method through experiments. Finally, we conclude the paper in Sect. 5.

## 2   Graph Kernels

### 2.1   Framework of Representative Graph Kernels

This paper tackles the classification problem of graphs. A graph is represented as $g = (V, E, \Sigma, \ell)$, where $V$ is a set of vertices, $E \subseteq V \times V$ is a set of edges, $\Sigma$ is a set of vertex labels, and $\ell : V \to \Sigma$ is a function that assigns a label to each vertex in the graph. Additionally, the set of vertices in graph $g$ is denoted by $V(g)$. Although we assume that only the vertices in the graphs have labels in this paper, the methods used in this paper can be applied to graphs where both the vertices and edges have labels. The vertices adjacent to vertex $v$ are represented as $N(v) = \{u \mid (v, u) \in E\}$. A sequence of vertices from $v$ to $u$ is called a path, and its step refers to the number of edges on that path. A path is described as being simple if and only if the path does not have repeating vertices. Paths in this paper are not always simple.

The graph classification problem is defined as follows. Given a set of $n$ training examples $D = \{(g_i, y_i)\}_{i=1}^{n}$, where each example is a pair consisting of a labeled graph $g_i$ and the class $y_i \in \{+1, -1\}$ to which it belongs, the objective is to learn a function $f$ that correctly predicts the classes of the test examples.

In this paper, graphs are classified by an SVM that uses graph kernels. Let $\Sigma$ and $c(g, \sigma)$ be $\{\sigma_1, \sigma_2, \cdots, \sigma_{|\Sigma|}\}$ and $c(g, \sigma) = |\{v \in V(g) \mid \ell(v) = \sigma\}|$, respectively. A function $\phi$ that converts a graph $g$ to a vector is defined as

$$\phi(g) = \big(c(g, \sigma_1), c(g, \sigma_2), \cdots, c(g, \sigma_{|\Sigma|})\big)^T.$$

Function $k'(g_i, g_j)$, defined as $\phi(g_i)^T \phi(g_j)$, is a semi-positive definite kernel. This function is calculated as

$$k'(g_i, g_j) = \phi(g_i)^T \phi(g_j) = \sum_{v_i \in V(g_i)} \sum_{v_j \in V(g_j)} \delta(\ell(v_i), \ell(v_j)), \tag{1}$$

where $\delta$ is the Kronecker delta. When $V(g_i)$ represents a set of vertices in $g_i$, $O(|V(g_i)| \times |V(g_j)|)$ is required to compute Eq. (1). However, Eq. (1) is solvable in $O(|V(g_i)| + |V(g_j)|)$ by using Algorithm 1 [7]. In Lines 1, a multiset of labels in the graph $g_i$ is sorted in ascending order by using the radix sort. This requires $O(|V(g_i)|)$. Similarly, a multiset of labels in the graph $g_i$ is also sorted in Line 2. In Lines 7 and 8, the $a$-th and $b$-th elements of the sorted labels are selected, respectively. Then, how many labels the graph have in common is counted in Line 10. This process is continued at most $|V(g_i)| + |V(g_j)|$ iterations.

---

**Algorithm 1.** Basic_Graph_Kernel.

---

**Data**: two graphs $g_i = (V_i, E_i, \Sigma, \ell_i)$ and $g_j = (V_j, E_j, \Sigma, \ell_j)$
**Result**: $k'(g_i, g_j)$
1  $V_i^{sort} \leftarrow$ Radix_Sort$(g_i, \ell_i)$;
2  $V_j^{sort} \leftarrow$ Radix_Sort$(g_j, \ell_j)$;
3  $\kappa \leftarrow 0$;
4  $a \leftarrow 1$;
5  $b \leftarrow 1$;
6  **for** $a \leq |V(g_i)| \wedge b \leq |V(g_j)|$ **do**
7  $\quad v_i \leftarrow V_i^{sort}[a]$;
8  $\quad v_j \leftarrow V_j^{sort}[b]$;
9  $\quad$ **if** $\ell_i(v_i) = \ell_j(v_j)$ **then**
10 $\quad\quad \kappa \leftarrow \kappa + 1$;
11 $\quad\quad a \leftarrow a + 1$;
12 $\quad\quad b \leftarrow b + 1$;
$\quad$ **else**
$\quad\quad$ **if** $\ell_i(v_i) < \ell_j(v_j)$ **then**
13 $\quad\quad\quad a \leftarrow a + 1$;
$\quad\quad$ **else**
14 $\quad\quad\quad b \leftarrow b + 1$;

15 **return** $\kappa$;

---

Given a $g^{(h)} = (V, E, \Sigma, \ell^{(h)})$, a procedure that converts $g^{(h)}$ to another graph $g^{(h+1)} = (V, E, \Sigma', \ell^{(h+1)})$ is called a relabel. Although relabel function $\ell^{(h+1)}$ is defined later in detail, the label of a $v$ in $g^{(h+1)}$ is defined using the labels of $v$ and $N(v)$ in $g^{(h)}$, and is denoted as $\ell^{(h+1)}(v) = r(v, N(v), \ell^{(h)})$. Let $\{g^{(0)}, g^{(1)}, \cdots, g^{(h)}\}$ be a series of graphs obtained by iteratively applying a relabel $h$ times, where $g^{(0)}$ is a graph contained in $D$. Given two graphs $g_i$ and $g_j$, a graph kernel is defined using $k'$ as

$$k(g_i, g_j) = k'(g_i^{(0)}, g_j^{(0)}) + k'(g_i^{(1)}, g_j^{(1)}) + \cdots + k'(g_i^{(h)}, g_j^{(h)}). \qquad (2)$$

Because $k$ is a summation of semi-positive definite kernels, $k$ is also semi-positive definite [3]. In addition, Eq. (2) is solvable in $O(h(|V(g_i)| + |V(g_j)|))$ according Algorithm 1.

Recently, various graph kernels have been applied to the graph classification problem. Representative graph kernels such as NHK and WLSK follow the above framework, where graphs contained in $D$ are iteratively relabeled. In these kernels, $\ell^{(h)}(v) = r(v, N(v), \ell^{(h-1)})$ characterizes a subgraph induced by the vertices that are reachable from $v$ within $h$ steps in $g^{(0)}$. Therefore, given $v_i \in V(g_i)$ and $v_j \in V(g_j)$, if subgraphs of the graphs induced by the vertices reachable from vertices $v_i$ and $v_j$ within $h$ steps are identical, the relabel assigns identical labels to them. Additionally, it is desirable for a graph kernel to fulfill the converse of this condition. However, it is not an easy task to design such a graph kernel.

We now review the representative graph kernels, NHK and WLSK.

**NHK:** Given a fixed-length bit string $\ell_1^{(0)}(v)$ of length $L$, $\ell_1^{(h)}(v)$ is defined as

$$\ell_1^{(h)}(v) = ROT(\ell_1^{(h-1)}(v)) \oplus \left( \bigoplus_{u \in N(v)} \ell_1^{(h-1)}(u) \right),$$

where $ROT$ is bit rotation to the left and $\oplus$ is the exclusive OR of the bit strings. NHK is efficient in terms of computation and space complexities because the relabel of NHK is computable in $O(L|N(v)|)$ for each vertex and its space complexity is $O(L)$.

Figure 1 shows an example of an NHK relabel and its detailed calculation for a vertex $v_2$, assuming that $L = 3$. First, $\ell_1^{(0)}(v_2) = \#011$ is rotated to return $\#110$. We then obtain $\#001$ using the exclusive OR of $\#110$, $\ell_1^{(0)}(v_1) = \#011$, $\ell_1^{(0)}(v_3) = \#001$, $\ell_1^{(0)}(v_4) = \#001$, and $\ell_1^{(0)}(v_5) = \#100$. In this computation, we do not require sorted bit strings because the exclusive OR is commutative. Three bits are required for $\ell_1^{(0)}(v_2)$ in this example, and $\ell_1^{(h)}(v_2)$ also requires three bits, even if $h$ is increased.

NHK has a drawback with respect to accidental hash collisions. For example, vertices $v_1$, $v_3$, and $v_4$ in $g^{(1)}$ in Fig. 1 have identical labels after the relabel. This is because $v_3$ and $v_4$ in $g^{(0)}$ have identical labels and the same number of adjacent vertices. However, despite the different labels and numbers of adjacent vertices of $v_1$ and $v_3$, these vertices have the same vertex labels in $g^{(1)}$, leading to low graph expressiveness and low classification accuracy.

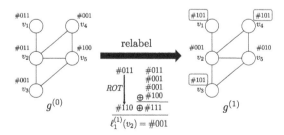

**Fig. 1.** Relabeling $g^{(0)}$ to $g^{(1)}$ in NHK.

We next describe WLSK, which is based on the Weisfeiler–Lehman algorithm, an algorithm that determines graph isomorphism.

**WLSK:** When $\ell_2^{(0)}(v)$ returns a string of characters, $\ell_2^{(h)}(v)$ is defined as

$$\ell_2^{(h)}(v) = \ell_2^{(h-1)}(v) \cdot \left( \bigodot_{u \in N(v)} \ell_2^{(h-1)}(u) \right),$$

where $\cdot$ and $\bigodot$ are string concatenation operators. Because concatenation is not commutative, $u$ is an iterator that obtains the vertices $N(v)$ adjacent to $v$ in alphabetical order. Because $\ell_2^{(h)}(v)$ has information on the distribution of labels for $h$ steps from $v$, it has high graph expressiveness.[1] If the labels are sorted using a bucket sort, the time complexity of WLSK is $O(|\Sigma||N(v)|)$ for each vertex.

Figure 2 shows an example of a relabel using WLSK. Vertices $v_1$, $v_2$, $v_3$, $v_4$, and $v_5$ in $g^{(0)}$ have labels $A$, $A$, $B$, $B$, and $C$, respectively. For each vertex, WLSK sorts the labels of the vertices adjacent to the vertex, and then concatenates these labels. In $g^{(1)}$, $v_3$ has the label BAC, meaning that $v_3$ has the label B in $g^{(0)}$ and two adjacent vertices whose labels are A and C.

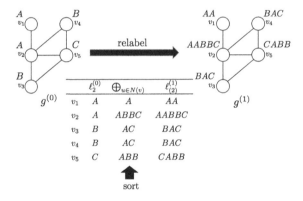

**Fig. 2.** Relabeling $g^{(0)}$ to $g^{(1)}$ in WLSK.

In addition to NHK and WLSK, we define the label aggregate kernel (LAK) to facilitate the understanding of the other kernels proposed in this paper.

**LAK:** In this kernel, $\ell_3^{(0)}(v)$ is a vector in $|\Sigma|$-dimensional space. In concrete terms, if a vertex in a graph has a label $\sigma_i$ among $\Sigma = \{\sigma_1, \sigma_2, \cdots, \sigma_{|\Sigma|}\}$, the $i$-th element in the vector is 1. Otherwise, it is zero. In LAK, $\ell_3^{(h)}(v)$ is defined as

$$\ell_3^{(h)}(v) = \ell_3^{(h-1)}(v) + \sum_{u \in N(v)} \ell_3^{(h-1)}(u).$$

---

[1] When $\ell_2^{(0)}(v)$ is a string of length 1, $\ell_2^{(1)}(v)$ is a string of length $|N(v)| + 1$. By replacing the later string with a new string of length 1, both the computation time and memory space required by WLSK are reduced.

The $i$-th element in $\ell_3^{(h)}(v)$ is the frequency of occurrence of character $\sigma_i$ in the string $\ell_2^{(h)}(v)$ concatenated by WLSK. Therefore, $\ell_3^{(h)}(v)$ has information on the distribution of labels within $h$ steps from $v$. Hence, LAK has high graph expressiveness. However, when $h$ is increased, the number of paths from $v$ that reach vertices labeled $\sigma_i$ increases exponentially. Thus, elements in $\ell_3^{(h)}(v)$ also increase exponentially. For example, if the average degree of vertices is $d$, there are $(d+1)^h$ vertices reachable from $v$ within $h$ steps. LAK thus requires a large amount of memory.

Figures 3 and 4 show an example of a relabel using LAK, assuming that $|\Sigma| = 3$. The vertex label of $v_5$ in $g^{(1)}$ is $(1, 2, 1)$, which means that there are one, two, and one vertices reachable from $v$ within one step that have labels $\sigma_1$, $\sigma_2$, and $\sigma_3$, respectively. Compared with relabeling $g^{(0)}$ to $g^{(1)}$, the additional number of values in $\ell_3^{(h)}(v)$ when relabeling $g^{(3)}$ to $g^{(4)}$ is large.

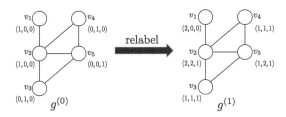

**Fig. 3.** Relabeling $g^{(0)}$ to $g^{(1)}$ in LAK.

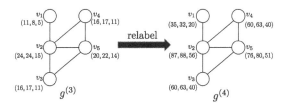

**Fig. 4.** Relabeling $g^{(3)}$ to $g^{(4)}$ in LAK.

## 2.2   Drawbacks of Existing Graph Kernels

We here summarize the characteristics of the above three graph kernels. NHK is efficient because its computation is a logical operation between fixed-length bit strings and does not require string sorting. However, its drawback is a tendency for hash collision, where different induced subgraphs have identical hash values. Meanwhile, WLSK requires vertex label sorting, but it has high expressiveness because $\ell_2^{(h)}(v)$ contains the distribution of the vertex labels within $h'$ steps

$(0 \leq h' \leq h)$ from $v$. LAK requires a large amount of memory to store vectors for high $h$ although it does not require label sorting. To overcome the drawbacks of NHK, WLSK and LAK, in this paper, we propose a novel graph kernel that is equivalent to NHK in terms of time and space complexities and equivalent to LAK in terms of expressiveness.

## 3    Graph Kernels Based on the Hadamard Code

In this section, we propose a novel graph kernel with the Hadamard code to overcome the aforementioned drawbacks. A Hadamard matrix is a square $(-1, 1)$ matrix in which any two row vectors are orthogonal, defined as

$$H_2 = \begin{pmatrix} 1 & 1 \\ 1 & -1 \end{pmatrix}, \quad H_{2^k} = \begin{pmatrix} H_{2^{k-1}} & H_{2^{k-1}} \\ H_{2^{k-1}} & -H_{2^{k-1}} \end{pmatrix}.$$

A Hadamard code is a row vector of the Hadamard matrix. Given a Hadamard matrix of order $2^k$, $2^k$ Hadamard codes having $2^k$ elements are generated from this matrix. Using the Hadamard codes, we propose HCK as follows.

**HCK:** Let $H$ be a Hadamard matrix of order $2^{\lceil \log_2 |\Sigma| \rceil}$ and $\ell_4^{(0)}(v)$ be a Hadamard code of order $|H|$. If a vertex $v$ has label $\sigma_i$, the $i$-th row in the Hadamard matrix of order $|H|$ is assigned to the vertex. $\ell_4^{(h)}(v)$ is then defined as

$$\ell_4^{(h)}(v) = \ell_4^{(h-1)}(v) + \sum_{u \in N(v)} \ell_4^{(h-1)}(u).$$

When $\ell_{\sigma_i}$ is a Hadamard code for a vertex label $\sigma_i$, $\ell_{\sigma_i}^T \ell_4^{(h)}(v)/|H|$ is the occurrence of $\sigma_i$ in a string $\ell_2^{(h)}(v)$ generated by WLSK. HCK therefore has the same expressiveness as LAK.

Figure 5 shows an example of a relabel using HCK. Each vertex $v$ in $g^{(1)}$ is represented as a vector produced by the summation of vectors for vertices adjacent to $v$ in $g^{(0)}$. Additionally, after the relabel, we obtain the distribution of the vertex labels within one step of $v$ according to

$$\frac{1}{|H|} H \ell_4^{(1)}(v_5) = \frac{1}{4} \begin{pmatrix} 1 & 1 & 1 & 1 \\ 1 & -1 & 1 & -1 \\ 1 & 1 & -1 & -1 \\ 1 & -1 & -1 & 1 \end{pmatrix} \begin{pmatrix} 4 \\ 0 \\ 2 \\ -2 \end{pmatrix} = \begin{pmatrix} 1 \\ 2 \\ 1 \\ 0 \end{pmatrix}.$$

In other words, there are one $\sigma_1$, two $\sigma_2$, and one $\sigma_3$ labels within one step of $v_5$. Furthermore, the result is equivalent to $\ell_3^{(1)}(v_5)$, as shown in Fig. 3. The reason why we divide $H \ell_4^{(h)}(v)$ by 4 is that the order of the Hadamard matrix is $|H| = 4$.

If each element in $\ell_4^{(h)}(v)$ is stored in four bytes (the commonly used size of integers in C, Java, and other languages), the space complexity of HCK is

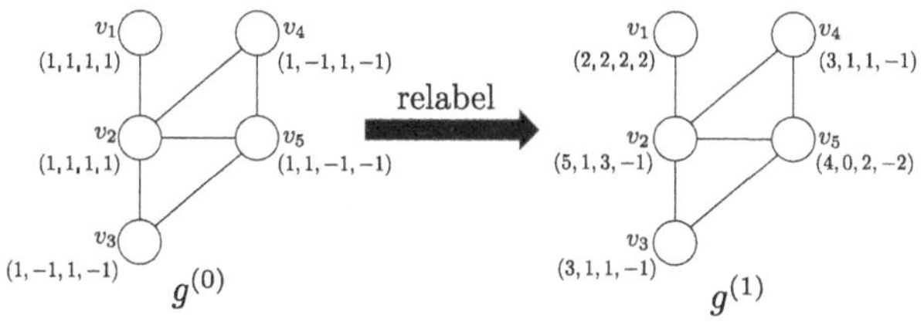

**Fig. 5.** Relabeling $g^{(0)}$ to $g^{(1)}$ in HCK.

equivalent to that of LAK. We therefore have not yet overcome the drawback of LAK. In this paper, we assume that each vertex label is assigned to a vertex with equal probability. The probabilities of occurrence of 1 and $-1$ are equivalent in each column of the Hadamard matrix except for the first column, and the $i$-th element $(1 < i \le |\Sigma|)$ in $\ell_4^{(h)}(v)$ follows a binomial distribution with zero mean under this assumption. Therefore, the expected value of the element in $\ell_4^{(h)}(v)$ is zero, and for the elements, a large memory space is not required. For example, Tables 1 and 2 present values of the $i$-th elements in $\ell_3^{(h)}(v_2)$ and $\ell_4^{(h)}(v_2)$, respectively, in a graph $g^{(h)}$, when $g^{(0)}$ (shown in Fig. 6) is relabeled iteratively $h$ times. Under the assumption of the vertex label probability, the expected value of all elements in $\ell_4^{(h)}(v_2)$ except for the first element becomes zero. The first element represents the number of paths from $v_2$ to the vertices reachable within $h$ steps. On the basis of this observation, we assign bit arrays of length $\rho$ in the $L$-bit array to the elements as follows.

**SHCK:** Similar to NHK, $\ell_5^{(0)}(v)$ is a fixed-length bit array of length $L$. The bit array is divided into $|H|$ fragments, one of which is a bit array of length $L - \rho(|H| - 1)$ and the rest are bit arrays of length $\rho$. The first fragment of length $L - \rho(|H| - 1)$ is assigned to store the first element of $\ell_4^{(0)}(v)$, the next

**Table 1.** Elements in a label in LAK.

| $h$ | Label |
|---|---|
| 0 | $\ell_3^{(0)}(v_2) = (0\ 1\ 0\ 0)$ |
| 1 | $\ell_3^{(1)}(v_2) = (1\ 1\ 1\ 0)$ |
| 2 | $\ell_3^{(2)}(v_2) = (2\ 3\ 2\ 2)$ |
| 3 | $\ell_3^{(3)}(v_2) = (7\ 7\ 7\ 6)$ |
| 4 | $\ell_3^{(4)}(v_2) = (20\ 21\ 20\ 20)$ |
| 5 | $\ell_3^{(5)}(v_2) = (61\ 61\ 61\ 60)$ |
| 6 | $\ell_3^{(6)}(v_2) = (182\ 183\ 182\ 182)$ |
| 7 | $\ell_3^{(7)}(v_2) = (547\ 547\ 547\ 546)$ |
| 8 | $\ell_3^{(8)}(v_2) = (1640\ 1641\ 1640\ 1640)$ |
| 9 | $\ell_3^{(9)}(v_2) = (4921\ 4921\ 4921\ 4920)$ |
| 10 | $\ell_3^{(10)}(v_2) = (14762\ 14763\ 14762\ 14762)$ |

**Table 2.** Elements in a label in HCK.

| $h$ | Label |
|---|---|
| 0 | $\ell_4^{(0)}(v_2) = (1\ -1\ -1\ 1)$ |
| 1 | $\ell_4^{(1)}(v_2) = (3\ -1\ -1\ -1)$ |
| 2 | $\ell_4^{(2)}(v_2) = (9\ -1\ -1\ 1)$ |
| 3 | $\ell_4^{(3)}(v_2) = (27\ -1\ -1\ -1)$ |
| 4 | $\ell_4^{(4)}(v_2) = (81\ -1\ -1\ 1)$ |
| 5 | $\ell_4^{(5)}(v_2) = (243\ -1\ -1\ -1)$ |
| 6 | $\ell_4^{(6)}(v_2) = (729\ -1\ -1\ 1)$ |
| 7 | $\ell_4^{(7)}(v_2) = (2187\ -1\ -1\ -1)$ |
| 8 | $\ell_4^{(8)}(v_2) = (6561\ -1\ -1\ 1)$ |
| 9 | $\ell_4^{(9)}(v_2) = (19683\ -1\ -1\ -1)$ |
| 10 | $\ell_4^{(10)}(v_2) = (59049\ -1\ -1\ 1)$ |

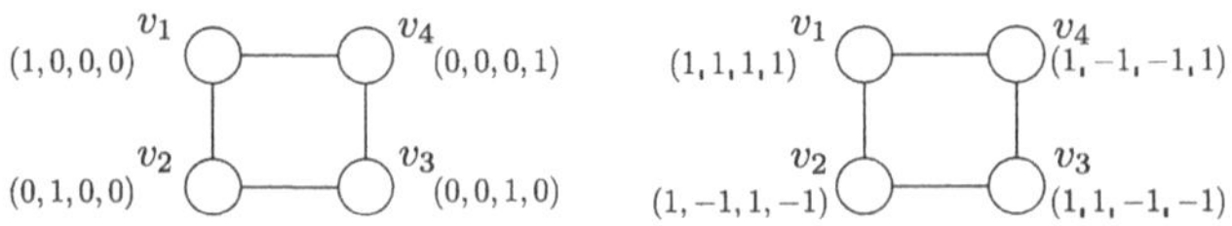

(a) A graph $g^{(0)}$ relabeled by LAK    (b) A graph $g^{(0)}$ relabeled by HCK

**Fig. 6.** Relabeled graphs.

fragment of length $\rho$ is assigned to store the second element, and so on. Here, $\rho$ is a positive integer fulfilling $\rho(|H| - 1) = \rho(2^{\lceil \log_2 |\Sigma| \rceil} - 1) \leq L$. Additionally, each element of $\boldsymbol{\ell}_4^{(0)}(v)$ is represented by its two's complement in $\ell_5^{(0)}(v)$ for the purpose of the following summation, which defines $\ell_5^{(h)}(v)$.

$$\ell_5^{(h)}(v) = \ell_5^{(h-1)}(v) + \sum_{u \in N(v)} \ell_5^{(h-1)}(u).$$

Because $\ell_5^{(h)}(v)$ is a fixed-length binary bit string and $\ell_5^{(h)}(v)$ is the summation of the values represented as bit strings, both the time and space complexities of SHCK are equivalent to those of NHK. Additionally, the expressiveness of SHCK is equivalent that of LAK, if overflow of the fixed-length bit array does not occur. The theoretical discussion on the overflow can be found in [13]. In the next section, we demonstrate that the proposed graph kernel, SHCK, has the ability to classify graphs with high accuracy.

## 4    Experimental Evaluation

The proposed method was implemented in Java. All experiments were done on an Intel Xeon X5670 2.93-GHz computer with 48-GB memory running Microsoft Windows 8. We compared the computation time and accuracy of the prediction performance of HCK and SHCK with those of NHK and WLSK. To learn from the kernel matrices generated by the above graph kernels, we used the LIBSVM package [2][2] using 10-fold cross validation.

### 4.1    Experiments on Artificial Datasets

We generated artificial datasets of graphs using the four parameters and their default values listed in Table 3. For each dataset, $|D|$ graphs, each with an average of $|V(g)|$ vertices, were generated. Two vertices in a graph were connected with probability $p$ of the existence of an edge, and one of $|\Sigma|$ labels was assigned to each vertex in the graph. In parallel with the dataset generation and using the same parameters $p$ and $|\Sigma|$, a graph pattern $g_s$ with 15 vertices was also generated for embedding in some of the $|D|$ graphs as common induced subgraphs.

---

[2] http://www.csie.ntu.edu.tw/~cjlin/libsvm/.

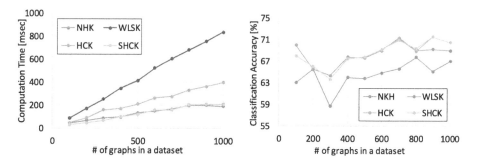

**Fig. 7.** Computation time for various $|D|$.

**Fig. 8.** Accuracy for various $|D|$.

**Table 3.** Default parameters for the data generation program.

|  | # of graphs in a dataset | Average number of vertices in graphs | Prob. of edge existence | # of vertex labels |
|---|---|---|---|---|
| Default values | $|D| = 200$ | $|V(g)| = 100$ | $p = 5\%$ | $|\Sigma| = 2$ |

$g_s$ was randomly embedded in half of the $|D|$ graphs in a dataset, and the class label 1 was assigned to the graphs containing $g_s$, while the class label $-1$ was assigned to the other graphs.

First, we varied only $|D|$ to generate various datasets with the other parameters set to their default values. The number of graphs in each dataset was varied from 100 to 1,000. Figures 7 and 8 show the computation time to relabel graphs in each dataset and classification accuracy, respectively, for the four graph kernels. In these experiments, $h$ and $\rho$ were set to 5 and 3, respectively, and the computation time does not contain time to generate kernel matrices from $g^{(0)}, g^{(1)}, \cdots, g^{(h)}$. As shown in Fig. 7, the computation time for all of the graph kernels is proportional to the number of graphs in a dataset, because the relabels in the kernels are applied to each graph independent to the other graphs in the dataset. As shown in Fig. 8, the classification accuracy increases when the number of graphs is increased. This is because the number of graphs in training datasets of SVMs also increase, when the number of graphs is increased. The classification accuracies for WLSK, HCK, and SHCK are superior to one for NHK, because they have high expressiveness. The classification accuracy for HCK is almost as equivalent as one for SHCK. Therefore, overflows in SHCK do not make much impact to the accuracy, which will be also shown in the other experiments.

Next, we varied only $|\Sigma|$ to generate various datasets with the other parameters set to their default values. Figures 9 and 10 show the computation time to relabel graphs in each dataset and classification accuracy, respectively, for the four graph kernels when the number of labels in each dataset was varied from 1 to 10. The computation time for NHK, WLSK, and SHCK is constant to the

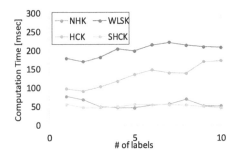

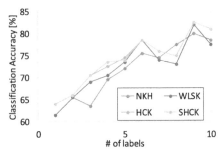

Fig. 9. Computation time for various $|\Sigma|$.

Fig. 10. Accuracy for various $|\Sigma|$.

number of labels in datasets, while one for HCK increases when the number of labels is increased. This is because HCK represents a label of each vertex in a graph by a $|\Sigma|$-dimensional vector and it sums up such $|\Sigma|$-dimensional vectors. On the other hand, NHK and SHCK represent the label by a fix-length bit string whose length is independent to the number of labels of graphs. The classification accuracy for all of the graph kernels increases when the number of labels in datasets, because labels which vertices in graphs have carry important information to classify graphs correctly. Especially in WLSK, HCK, and SHCK, since each vertex $v$ has a distribution of vertex labels within $i$ steps from $v$, their accuracies are superior to one for NHK.

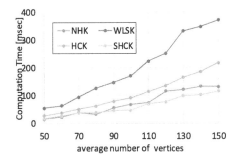

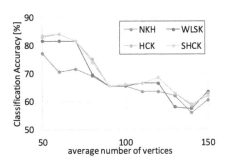

Fig. 11. Computation time for various $|V(g)|$.

Fig. 12. Accuracy for various $|V(g)|$.

We varied only $|V(g)|$ to generate various datasets with the other parameters set to their default values. Figures 11 and 12 show the computation time to relabel graphs in each dataset and classification accuracy, respectively, when the average number of vertices in graphs was varied from 50 to 150. The computation time for all of the graph kernels increases with the square of the average number $|V(g)|$ of vertices in graphs. This is because a new label for each vertex $v$ in a graph $g^{(h+1)}$ is obtained from labels of $v$ and its adjacent vertices $U$ in a graph

$g^{(h)}$ and the relabel for $g^{(h)}$ for any $h$ is computable in $O(|V(g)| \times |U|)$, where $|U|$ is $p \times |V(g)|$. Since graphs in most of real world datasets are sparse, the graph kernels in this paper are applicable enough to such sparse graphs even if the number of vertices in graphs is large. For example, chemical compounds which can be represented graphs where each vertex and edge correspond to an atom and edge in compounds, respectively, are sparse, since atoms in chemical compounds have two chemical bonds in average. The classification accuracy decreases, when the average number of vertices in graphs is increased, because the number of graphs whose class labels are $-1$ and which contain some small subgraphs of the embedded pattern $g_s$ increases when the average number of vertices in graphs is increased.

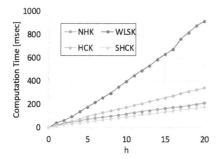

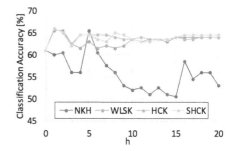

**Fig. 13.** Computation time for various $h$.

**Fig. 14.** Accuracy for various $h$.

We varied $h$ for the graph kernels. Figures 13 and 14 show the computation time to relabel graphs in each dataset and classification accuracy, respectively, for the four graph kernels in the experiments, when $h$ was varied from 1 to 20. The computation time for all of the graph kernels is proportional to $h$. When $h$ is increased, the graph kernel has information on a large subgraph induced by vertices reachable within $h$ steps from every vertex in a graph. However, the classification accuracy does not always increase, when $h$ is increased. We do not know an adequate value for $h$ in advance of training SVMs. One of ways for determining an adequate value for $h$ for a dataset is to divide the dataset into three portions, one of them is for the training dataset of an SVM, another is for the dataset to obtain an adequate value for $h$ and the SVM's parameters, and the other is for testing the SVM.

From these experimental results using artificial datasets, we confirmed that the proposed graph kernel SHCK is equivalent to NHK in terms of time complexity and equivalent to WLSK in terms of classification accuracy. The experimental results do not contain time for computing kernel matrices from $g^{(0)}, g^{(1)}, \cdots, g^{(h)}$ obtained by the graph kernels. A kernel matrix for a dataset $D$ is obtained in $O(h|D|^2|V(g)|)$, where $|V(g)|$ is the average number of vertices of graphs in $D$, because we need to run Algorithm 1 $h$ times for every pair of graphs in $D$.

## 4.2   Experiments on Real-World Graphs

To assess the practicability of our proposed method, we used five real-world datasets. The first dataset, MUTAG [4], contains information on 188 chemical compounds and their class labels. The class labels are binary values that indicate the mutagenicity of chemical compounds. The second dataset, ENZYMES, contains information on 600 proteins and their class labels. The class labels are one of six labels showing the six EC top-level classes [1,12]. The third dataset, D&D, contains information on 1178 protein structures, where each amino acid corresponds to a vertex and two vertices are connected by an edge if they are less than 6 Ångstroms apart [5]. The remaining datasets, NCI1 and NCI109, represent two balanced subsets of data sets of chemical compounds screened for activity against non-small cell lung cancer and ovarian cancer cell lines, respectively [15]. These datasets contain about 4000 chemical compounds, each of which has a class label among positive and negative. Each chemical compound is represented as an undirected graph where each vertex, edge, vertex label, and edge label corresponds to an atom, chemical bond, atom type, and bond type, respectively. Because we assume that only vertices in graphs have labels, the chemical graphs are converted following the literature [7]; i.e., an edge labeled with $\ell$ that is adjacent to vertices $v$ and $u$ in a chemical graph is replaced with a vertex labeled with $\ell$ that is adjacent to $v$ and $u$ with unlabeled edges. Table 4 summarizes the datasets.

**Table 4.** Summary of evaluation datasets.

|  | MUTAG | ENZYMES | D&D | NCI1 | NCI109 |
|---|---|---|---|---|---|
| Number of graphs $|D|$ | 188 | 600 | 1178 | 4110 | 4127 |
| Maximum graph size | 84 | 126 | 5748 | 349 | 349 |
| Average graph size | 53.9 | 32.6 | 284.3 | 94.5 | 93.9 |
| Number of labels $|\Sigma|$ | 12 | 3 | 82 | 40 | 41 |
| Number of classes (class distribution) | 2 (126,63) | 6 (100,100,100, 100,100,100) | 2 (487, 691) | 2 (2053, 2057) | 2 (2048, 2079) |
| Average degree of vertices | 2.1 | 3.8 | 5.0 | 2.7 | 2.7 |

Figures 15, 16, 17, 18, and 19 show the computation time required to obtain a graph $g^{(h)}$ from a graph $g^{(0)}$ in NHK, WLSK, HCK, and SHCK for various $h$ for the MUTAG, ENZYMES, D&D, NCI1, and NCI109 datasets, respectively. As shown in the figures, NHK and SHCK are faster than HCK, and much faster than WLSK. Additionally, the computation times of NHK, HCK, and SHCK increase

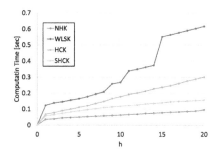

**Fig. 15.** Computation time for various $h$ (MUTAG).

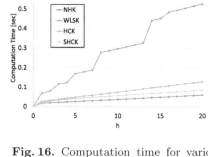

**Fig. 16.** Computation time for various $h$ (ENZYMES).

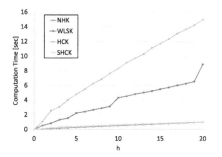

**Fig. 17.** Computation time for various $h$ (D&D).

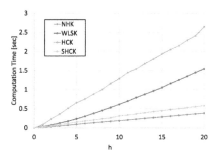

**Fig. 18.** Computation time for various $h$ (NCI1).

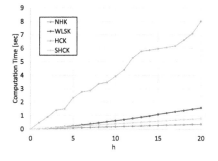

**Fig. 19.** Computation time for various $h$ (NCI109).

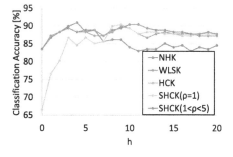

**Fig. 20.** Accuracy for various $h$ and $\rho$ (MUTAG).

linearly with $h$. The reason why WLSK requires such a long computation time is that WLSK must sort the labels of adjacent vertices and replace a string of length $|N(v)|+1$ with a string of length 1. This is especially true when $h = 11$ or 15 for the MUTAG dataset, $h = 8$ or 14 for the ENZYMES dataset, and $h = 10$ or 20 for the D&D dataset. In our implementation, this replacement is done with Java's HashMap class, where a string of length $|N(v)| + 1$ is the hash key and a string of length 1 is a value corresponding to that key. Although the average degree in the evaluated datasets is low, WLSK requires further computation time when the average degree of the data increases. HCK requires a long computation time for the D&D, NCI1, and NCI109 datasets, because there are many labels in the datasets and the computation time is proportional to the number of labels.

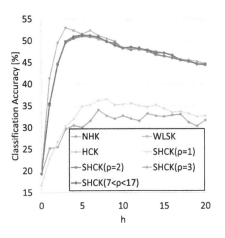

**Fig. 21.** Accuracy for various $h$ and $\rho$ (ENZYMES).

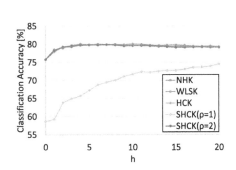

**Fig. 22.** Accuracy for various $h$ and $\rho$ (D&D).

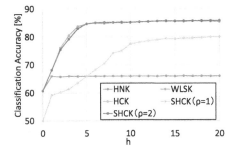

**Fig. 23.** Accuracy for various $h$ and $\rho$ (NCI1).

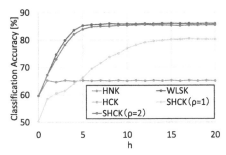

**Fig. 24.** Accuracy for various $h$ and $\rho$ (NCI109).

Figure 20 shows the classification accuracy of NHK, WLSK, HCK, and SHCK for various $h$ and $\rho$ in the case of the MUTAG dataset. The length of bit strings for NHK and SHCK was set to $L = 64$. The maximum accuracies for various $h$ are almost the same. When $h = 0$, the accuracy for SHCK ($\rho = 1$) is very low because a value of 1 or $-1$ (the values in the Hadamard matrix) cannot be stored as a two's complement consisting of one bit. The accuracy of HCK is exactly the same as that of SHCK ($1 < \rho < 5$), which means that although overflow may occur in SHCK, the kernel can assign identical vertex labels to the identical subgraphs induced by a vertex $v$ and the vertices within $h$ steps from $v$. Figure 21 shows the classification accuracy of NHK, WLSK, HCK, and SHCK for various $h$ and $\rho$ in the case of the ENZYMES dataset. WLSK is slightly more accurate than HCK and SHCK ($\rho = 2$, $\rho = 3$, and $7 < \rho < 17$), which are much more accurate than NHK and SHCK ($\rho = 1$). The performance of HCK is exactly the same as that of SHCK for high $\rho$ ($7 < \rho < 17$) and almost the same as that of SHCK for low $\rho$ ($\rho = 2$ and $\rho = 3$). The maximum accuracy of WLSK is 53.0%, while the maximum accuracies of HCK and SHCK ($\rho = 3$, 4, and $7 < \rho < 17$) are both 51.3%. WLSK is slightly more accurate than HCK, because $\ell_2^{(h)}(v)$ contains information on the distribution of labels at $h$ steps from $v$, while $\ell_4^{(h)}(v)$ contains information on the distribution of all labels within $h$ steps from $v$. Although the latter distribution can be obtained from the former distribution, the former distribution cannot be obtained from the latter distribution. Therefore, WLSK is more expressive than HCK and SHCK. When $\rho$ is increased to 16, the length of a bit string needed to store the first element of $\ell_4^{(h)}(v)$ is $L - \rho \times 2^{\lceil \log_2 |\Sigma| \rceil} = 64 - 16 \times 2^{\lceil \log_2 3 \rceil} = 0$. Even in this case, the accuracy of SHCK is equivalent to that of HCK, which means that the overflow of the first element of $\ell_4^{(h)}(v)$ has absolutely no effect on the classification accuracy.

Figure 22 shows the classification accuracy of NHK, WLSK, HCK, and SHCK for various $h$ and $\rho$ in the case of the D&D dataset. The lengths of bit strings of NHK and SHCK were set to $L = 256$. All accuracies except for that of SHCK ($\rho = 1$) are almost equivalent. In addition, Figs. 23 and 24 show the classification accuracy of NHK, WLSK, HCK, and SHCK for various $h$ and $\rho$ in the cases of the NCI1 and NCI109 datasets, respectively. All accuracies except those of NHK and SHCK ($\rho = 1$) are almost equivalent. The classification accuracy of NHK is low owing to hash collision, while the classification accuracys of SHCK ($\rho = 1$) is low because a value of 1 or $-1$ (the values in the Hadamard matrix) cannot be stored as a two's complement consisting of one bit. From the results of these experiments, we recommend setting as large a value as possible for $\rho$ to obtain high classification accuracy. To do so, we set $\rho = \lfloor \frac{L}{|\Sigma|} \rfloor$, where $L$ is the length of the bit array and $|\Sigma|$ is the number of labels in a dataset.

## 4.3   Effect of Assignment of Initial Labels

The proposed graph kernels HCK and SHCK randomly assign one of the Hadamard codes to each label. The assignment of Hadamard codes to labels affects the classification accuracy especially when there are few labels in a dataset. In this experiment, we demonstrate the relationship between classification accuracies and assignments of initial labels. Figure 25 shows the standard

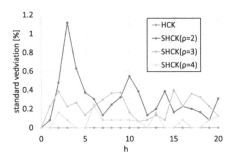

**Fig. 25.** Effect of assignment of initial labels for various $h$ and $\rho$ (ENZYMES).

deviations of five classification accuracies, each of which is obtained by 10-fold cross validation for the ENZYME dataset, which has the smallest number of labels among the five datasets. Figure 25 shows a high standard deviation for small $\rho$. In SHCK, a label $\ell_5^{(h)}(v)$ represents a subgraph induced by vertices reachable from $v$ within $h$ step. When there are many labels in a dataset, nonisomorphic subgraphs induced by vertices reachable from a vertex within $h$ steps are likely represented by distinct labels. Meanwhile, when there are few labels in a dataset, the nonisomorphic subgraphs may be represented by the same label. In addition, the small $\rho$ causes the overflow in $\ell_5^{(h)}(v)$, because a subgraph induced by vertices reachable from $v$ within $h$ is represented by an short bit array with length $\rho$ in $\ell_5^{(h)}(v)$. Whether nonisomonphic subgraphs are represented by the same labels varies according to the location where the overflow occurs, which affects the classification accuracy. Therefore, when $\rho$ is small, the standard deviation of classification accuracies of SHCK becomes high. However, it is possible to decrease the standard deviation by decreasing the value of $\rho$, because this reduces the possibility of overflow occurring. The standard deviation of the classification accuracy for HCK in which overflow does not occur is zero. The same experiment was conducted for the other datasets, and the standard deviation of the classification accuracy was zero because these datasets have enough labels.

## 5   Conclusion

In this paper, we proposed a novel graph kernel equivalent to NHK in terms of time and space complexities, and comparable to WLSK in terms of expressiveness. The proposed kernel is based on the Hadamard code. Labels assigned by our graph kernel follow a binomial distribution with zero mean. The expected value of a label is zero; thus, such labels do not require large memory. This allows the compression of vertex labels in graphs, as well as fast computation. We presented the Hadamard code kernel (HCK) and shortened HCK (SHCK), a version of HCK that compresses vertex labels in graphs. The fundamental performance and practicality of the proposed method were demonstrated in experiments that compare the computation time, scalability and classification accuracy of HCK and SHCK with those of NHK and WLSK for the artificial and real-world datasets.

# References

1. Borgwardt, K.M., Cheng, S.O., Schonauer, S., Vishwanathan, S.V.N., Smola, A.J., Kriegel, H.-P.: Protein function prediction via graph kernels. Bioinfomatics **21**(suppl 1), 47–56 (2005)
2. Chang, C.-C., Lin, C.-J.: LIBSVM: A Library for Support Vector Machines (2001). http://www.csie.ntu.edu.tw/cjlin/libsvm
3. Cristianini, N., Shawe-Taylor, J.: An Introduction to Support Vector Machines and Other Kernel-based Learning Methods. Cambridge University Press, Cambridge (2000)
4. Debnath, A.K., Lopez de Compadre, R.L., Debnath, G., Shusterman, A.J., Hansch, C.: Structure-activity relationship of mutagenic aromatic and heteroaromatic nitro compounds. Correlation with molecular orbital energies and hydrophobicity. J. Med. Chem. **34**, 786–797 (1991)
5. Dobson, P.D., Doig, A.J.: Distinguishing enzyme structures from non-enzymes without alignments. J. Mol. Biol. **330**(4), 771–783 (2003)
6. Garey, M.R., Johnson, D.S.: Computers and Intractability: A Guide to the Theory of NP-Completeness. W.H. Freeman, Gordonsville (1979)
7. Hido, S., Kashima, H.: A linear-time graph kernel. In: Proceedings of the IEEE International Conference on Data Mining (ICDM), pp. 179–188 (2009)
8. Kashima, H., Tsuda, K., Inokuchi, A.: Marginalized kernels between labeled graphs. In: Proceedings of the International Conference on Machine Learning (ICML), pp. 321–328 (2003)
9. Shervashidze, N., Schweitzer, P., van Leeuwen, E.J., Mehlhorn, K., Borgwardt, K.M.: Weisfeiler-Lehman graph kernels. J. Mach. Learn. Res. (JMLR) **12**, 2539–2561 (2011)
10. Schölkopf, B., Smola, A.J.: Learning with Kernels. MIT Press, Cambridge (2002)
11. Schölkopf, B., Tsuda, K., Vert, J.-P.: Kernel Methods in Computational Biology. MIT Press, Cambridge (2004)
12. Schomburg, I., Chang, A., Ebeling, C., Gremse, M., Heldt, C., Huhn, G., Schomburg, D.: BRENDA, the enzyme database: updates and major new developments. Nucleic Acids Res. **32D**, 431–433 (2004)
13. Kataoka, T., Inokuchi, A.: Hadamard code graph kernels for classifying graphs. In: Proceedings of the International Conference on Pattern Recognition Applications and Methods (ICPRAM), pp. 24–32 (2016)
14. Vinh, N.D., Inokuchi, A., Washio, T.: Graph classification based on optimizing graph spectra. In: Pfahringer, B., Holmes, G., Hoffmann, A. (eds.) DS 2010. LNCS (LNAI), vol. 6332, pp. 205–220. Springer, Heidelberg (2010). doi:10.1007/978-3-642-16184-1_15
15. Wale, N., Karypis, G.: Comparison of descriptor spaces for chemical compound retrieval and classification. In: Proceedings of the IEEE International Conference on Data Mining (ICDM), Hong Kong, pp. 678–689 (2006)

# Document Clustering Games
# in Static and Dynamic Scenarios

Rocco Tripodi[1]([⊠]) and Marcello Pelillo[1,2]

[1] ECLT, Ca' Foscari University, Ca' Minich, Venice, Italy
{rocco.tripodi,pelillo}@unive.it
[2] DAIS, Ca' Foscari University, Via Torino, Venice, Italy

**Abstract.** In this work we propose a game theoretic model for document clustering. Each document to be clustered is represented as a player and each cluster as a strategy. The players receive a reward interacting with other players that they try to maximize choosing their best strategies. The geometry of the data is modeled with a weighted graph that encodes the pairwise similarity among documents, so that similar players are constrained to choose similar strategies, updating their strategy preferences at each iteration of the games. We used different approaches to find the prototypical elements of the clusters and with this information we divided the players into two disjoint sets, one collecting players with a definite strategy and the other one collecting players that try to learn from others the correct strategy to play. The latter set of players can be considered as new data points that have to be clustered according to previous information. This representation is useful in scenarios in which the data are streamed continuously. The evaluation of the system was conducted on 13 document datasets using different settings. It shows that the proposed method performs well compared to different document clustering algorithms.

## 1 Introduction

Document clustering is a particular kind of clustering that involves textual data. It can be employed to organize tweets [24], news [4], novels [3] and medical documents [6]. It is a fundamental task in text mining and have different applications in document organization and language modeling [15].

State-of-the-art algorithms designed for this task are based on generative models [38], graph models [29,37] and matrix factorization techniques [20,35]. Generative models and topic models [5] aim at finding the underlying distribution that created the set of data objects, observing the sequences of objects and features. One problem with these approaches is the conditional-independence assumption that does not hold for textual data and in particular for streaming documents. In fact, streamed documents such as mails, tweets or news can be generated in response to past events, creating topics and stories that evolve over time.

CLUTO is a popular graph-based algorithm for document clustering [36]. It employs a graph to organize the documents and different criterion functions

© Springer International Publishing AG 2017
A. Fred et al. (Eds.): ICPRAM 2016, LNCS 10163, pp. 20–42, 2017.
DOI: 10.1007/978-3-319-53375-9_2

to partition this graph into a predefined number of clusters. The problem with partitional approaches is that these approaches require to know in advance the number of clusters into which the data points have to be divided. A problem that can be restrictive in real applications and in particular on streaming data.

Matrix factorization algorithms, such as Non-negative Matrix Factorization (NMF) [7,12], assume that words that occur together can represent the features that characterize a clusters. Ding et al. [7] demonstrated the equivalence between NMF and Probabilistic Latent Semantic Indexing, a popular technique for document clustering. Also with these approaches it is required to know in advance the number of clusters into which the data have to be organized.

A general problem, common to all these approaches, concerns the temporal dimension. In fact, for these approaches it is difficult to deal with streaming datasets. A non trivial problem, since in many real world applications documents are streamed continuously. This problem is due to the fact that these approaches operate on a dataset as a whole and need to be recomputed if the dataset changes. It can be relevant also in case of huge static datasets, because of scalability issues [1]. In these contexts an incremental algorithm would be preferable, since with this approach it is possible to cluster the data sequentially.

With our approach we try to overcome this problem. We cluster part of the data producing small clusters that at the beginning of the process can be considered as cluster representative. Then we cluster new instances according to this information. With our approach is also possible deal with situations in which the number of clusters is unknown, a common situation in real world applications. The clustering of new instances is defined as a game, in which there are labeled players (from an initial clustering), which always play the strategy associated to their cluster and unlabeled players that learn their strategy playing the games iteratively and obtaining a feedback from the strategy that their co-players are adopting.

In contrast to other stream clustering algorithm our approach is not based only on proximity relations, such as in methods based on partitioning representatives [2]. With these approaches the cluster membership of new data points is defined selecting the cluster of their closest representative. With our approach the cluster membership emerges dynamically from the interactions of the players and all the neighbors of a new data point contribute in different proportion to the final cluster assignment. It does not consider only local information to cluster new data points but find solutions that are globally consistent. In fact, if we consider only local information the cluster membership of a point in between two or more clusters could be arbitrary.

The rest of this contribution is organized as follows. In the next Section, we briefly introduce the basic concepts of classical game theory and evolutionary game theory that we used in our framework; for a more detailed analysis of these topics the reader is referred to [13,23,34]. Then we introduce the *dominant set* clustering algorithm [18,21] that we used in part of our experiments to find the initial clustering of the data. In Sect. 4 we describe our model and in the last section we present the evaluation of our approach in different scenarios. First we use it to cluster static datasets and then, in Sect. 5.6, we present the evaluation of our method on streaming data. This part extends our previous work [32] and demonstrates that the proposed framework can be used in different scenarios with good performances.

## 2    Game Theory

Game theory was introduced by Von Neumann and Morgenstern [33]. Their idea was to develop a mathematical framework able to model the essentials of decision making in interactive situations. In its *normal-form* representation, which is the one we use in this work, it consists of a finite set of players $I = \{1, \ldots, n\}$, a set of pure strategies, $S_i = \{s_1, \ldots, s_m\}$, and a utility function $u_i : S_1 \times \cdots \times S_n \to \mathbb{R}$ that associates strategies to payoffs; $n$ is the number of players and $m$ the number of pure strategies. The games are played among two different players and each of them have to select a strategy. The outcome of a game depends on the combination of strategies (strategy profile) played at the same time by the players involved in it, not just on the single strategy chosen by a player. For example we can consider the following payoff matrix (Table 1),

**Table 1.** The payoff matrix of the prisoner's dilemma game.

| $P_1 \backslash P_2$ | Strategy 1 | Strategy 2 |
|---|---|---|
| Strategy 1 | $-5, -5$ | $0, -6$ |
| Strategy 2 | $-6, 0$ | $-1, -1$ |

where, for example, player 1 get $-5$ when he chooses strategy 1 and player 2 chooses strategy 1. Furthermore, in *non-cooperative games* the players choose their strategies independently, considering what the other players can play and try to find the best strategy profile to employ in a game.

An important assumption in game theory is that the players try to maximize their utility in the games ($u_i$), selecting the strategies that can give the highest payoff, considering what strategies the other player can employ. The players try to find the strategies that are better than others regardless what the other player does. These strategies are called *strictly dominant* and can occur if and only if:

$$u(s_i^*, s_{-i}) > u_i(s_i, s_{-i}), \forall s_{-i} \in S_{-i} \tag{1}$$

where $s_{-i}$ denotes the strategy chosen by the other player(s).

The key concept of game theory is the Nash equilibrium that is used to predict the outcome of a strategic interaction. It can be defined as those strategy profiles in which no player has the incentive to unilaterally deviate from it, because there is no way to do increment the payoff. The strategies in a Nash equilibrium are best responses to all other strategies in the game, which means that they give the most favorable outcome for a player, given other players' strategies.

The players can play *mixed strategies*, which are probability distributions over pure strategies. In this context, the players select a strategy with a certain probability. A mixed strategy set can be defined as a vector $x = (x^1, \ldots, x^m)$, where $m$ is the number of pure strategies and each component $x^h$ denotes the probability that a particular player select its $h$th pure strategy. Each player has a strategy set that is defined as a standard simplex:

$$\Delta = \left\{ x \in \mathbb{R} : \sum_{h=1}^{m} x^h = 1, \text{ and } x^h \geq 0 \text{ for all } h \right\}. \tag{2}$$

A mixed strategy set corresponds to a point on the simplex $\delta$, whose corners represent pure strategies.

A strategy profile can be defined as a pair $(p, q)$ where $p \in \Delta_i$ and $q \in \Delta_j$. The payoff of this strategy profile is computed as:

$$u_i(p, q) = p \cdot A_i q, u_j(p, q) = q \cdot A_j p, \tag{3}$$

where $A_i$ and $A_j$ are the payoff matrices of player $i$ and $j$ respectively. The Nash equilibrium within this setting can be computed in the same way it is computed in pure strategies. In this case, it consists in a pair of mixed strategies such that each one is a best response to the other.

To overcome some limitations of traditional game theory, such as the hyper-rationality imposed on the players, a dynamic version of game theory was introduced. It was proposed by Smith and Price [26], as evolutionary game theory. Within this framework the games are not static and are played repeatedly. This reflect real life situations, in which the choices change according to past experience. Furthermore, players can change a behavior according to heuristics or social norms [28]. In this context, players make a choice that maximizes their payoffs, balancing cost against benefits [17].

From a machine learning perspective this process can be seen as an *inductive learning* process, in which agents play the games repeatedly and at each iteration of the system they update their beliefs on the strategy to take. The update is done considering what strategy has been effective and what has not in previous games. With this information, derived from the observation of the payoffs obtained by each strategy, the players can select the strategy with higher payoff.

The strategy space of each players is defined as a mixed strategy profile $x_i$ and the mixed strategy space of the game is given by the Cartesian product of all the players' strategy space:

$$\Theta = \times_{i \in I} \Delta_i. \tag{4}$$

The expected payoff of a strategy $e^h$ in a single game is calculated as in mixed strategies (see Eq. 3) but, in evolutionary game theory, the final payoff of each player is the sum of all the partial payoffs obtained during an iteration. The payoff corresponding to a single strategy is computed as:

$$u_i(e_i^h) = \sum_{j=1}^{n} (A_{ij} x_j)^h \tag{5}$$

and the average payoff is:

$$u_i(x) = \sum_{j=1}^{n} x_i^T A_{ij} x_j, \tag{6}$$

where $n$ is the number of players with whom player $i$ play the games and $A_{ij}$ is the payoff matrix among player $i$ and $j$. At each iteration a player can update his strategy space according to the payoffs gained during the games, it allocates more probability on the strategies with high payoff, until an equilibrium is reached, a situation in which it is not possible to obtain higher payoffs.

To find the Nash equilibrium of the system it is common to use the replicator dynamic equation [30],

$$\dot{x} = [u(e^h) - u(x)] \cdot x^h . \forall h \in x. \tag{7}$$

This equation allows better than average strategies to increase at each iteration. It can be used to analyze frequency-dependent selection processes [16], furthermore, the fixed points of Eq. 7 correspond to Nash equilibria [34]. We used the discrete time version of the replicator dynamic equation for the experiments of this work.

$$x^h(t+1) = x^h(t) \frac{u(e^h)}{u(x)} \ \forall h \in x(t). \tag{8}$$

The players update their strategies at each time step $t$ considering the strategic environment in which they are playing.

The complexity of each step of the replicator dynamics is quadratic but there are more efficient dynamics that can be used, such as the *infection and immunization* dynamics that has a linear-time/space complexity per step and it is known to be as accurate as the replicator dynamics [22].

## 3   Dominant Set Clustering

*Dominant set* is a graph based clustering algorithm that generalizes the notion of maximal clique from unweighted undirected graphs to edge-weighted graphs [18, 21]. With this algorithm it is possible to extract compact structures from a graph in an efficient way. Furthermore, it can be used on symmetric and asymmetric similarity graphs and does not require any parameter. With this framework it is possible to obtain measures of clusters cohesiveness and to evaluate the strength of participation of a vertex to a cluster. It models the well-accepted definition of a cluster, which states that a cluster should have high internal homogeneity and that there should be high inhomogeneity between the objects in the cluster and those outside [10].

The extraction of compact structures from graphs that reflect these two conditions, is given by the following quadratic form:

$$f(x) = x^T A x. \tag{9}$$

where $A$ is a similarity graph and $x$ is a probability vector, whose components indicate the participation of each node of the graph to a cluster. In this context, the clustering task corresponds to the task of finding a vector $x$ that maximizes $f$ and this can be done with the following program:

$$\begin{aligned} \text{maximize } & f(x) \\ \text{subject to } & x \quad \in \quad \Delta. \end{aligned} \tag{10}$$

where $\Delta$ represents the standard simplex. A (local) solution of program (10) corresponds to a maximally cohesive structure in the graph [10].

The solution of program (10) can be found using the discrete time version of the replicator dynamic equation, computed as follows,

$$x(t+1) = x \frac{Ax}{x^T Ax}, \tag{11}$$

where $x$ represent the strategy space at time $t$.

The clusters are extracted sequentially from the graph using a peel-off strategy to remove the objects belonging to the extracted cluster, until there are no more objects to cluster or some predefined criteria are satisfied.

## 4 Document Clustering Games

In this section we present step by step our approach to document clustering. First we describe how the documents are represented and how we prepare the data and structure them using a weighted graph. Then we pass to the preliminary clustering in order to divide the data points in two disjoint sets of labeled and unlabeled players. With this information we can initialize the strategy space of the players and run the dynamics of the system that lead to the final clustering of the data.

### 4.1 Document Representation

The documents of a datasets are processed with a *bag-of-words* (BoW) model. With this method each document is represented as a vector indexed according to the frequency of the words in it. To do this it is necessary to construct the vocabulary of the text collection that is composed by the set of unique words in the corpus. BoW represents a corpus with a *document-term matrix*. It consists in a $N \times T$ matrix $M$, where $N$ is the number of documents in the corpus and $T$ the number of words in the vocabulary. The words are considered as the features of the documents. Each element of the matrix $M$ indicates the frequency of a word in a document.

The BoW representation can lead to a high dimensional space, since the vocabulary size increases as sample size increases. Furthermore, it does not incorporate semantic information treating homonyms as the same feature. These problems can result in bad representations of the data and for this reason there where introduced different approaches to balance the importance of the features and also to reduce their number, focusing only on the most relevant.

An approach to weigh the importance of a feature is the *term frequency - inverse document frequency* (tf-idf) method [15]. This technique takes as input a document-term matrix $M$ and update it with the following equation,

$$tf\text{-}idf(d, t) = tf(d, t) \cdot log \frac{D}{df(d, t)}, \tag{12}$$

where $df(d, t)$ is the number of documents that contain word $t$. Then the vectors are normalized to balance the importance of each feature.

Latent Semantic Analysis (LSA) is a technique used to infer semantic information [11] from a *document-term matrix*, reducing the number of features. Semantic information is obtained projecting the documents into a *semantic space*, where the relatedness of two terms is computed considering the number of times they appear in a similar context. Single Value Decomposition (SVD) is used to create an approximation of the *document-term matrix* or *tf-idf matrix*. It decomposes a matrix $M$ in:

$$M = U \Sigma V^T, \tag{13}$$

where $\Sigma$ is a diagonal matrix with the same dimensions of $M$ and $U$ and $V$ are two orthogonal matrices. The dimensions of the feature space is reduced to $k$, taking into account only the first $k$ dimensions of the matrices in Eq. (13).

## 4.2   Data Preparation

This new representation of the data is used to compute the pairwise similarity among documents and to construct the proximity matrix $W$, using the cosine distance as metric,

$$cos \theta \frac{v_i \cdot v_j}{||v_i|| ||v_j||} \tag{14}$$

where the nominator is the intersection of the words in the vectors that represent two documents and $||v||$ is the norm of the vectors, which is calculated as: $\sqrt{\sum_{i=1}^{n} w_i^2}$.

## 4.3   Graph Construction

$W$ can be used to represent a text collection as a graph $G$, whose nodes represent documents and edges are weighted according to the information stored in $W$. Since, the cosine distance acts as a linear kernel, considering only information between vectors under the same dimension, it is common to smooth the data using a kernel function and transforming the proximity matrix $W$ into a similarity matrix $S$ [25]. It can also transform a set of complex and nonlinearly separable patterns into linearly separable patterns [9]. For this task we used the Gaussian kernel,

$$s_{ij} = exp \left\{ -\frac{w_{ij}^2}{\sigma^2} \right\} \tag{15}$$

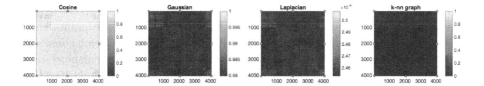

**Fig. 1.** Different data representations for a dataset with 5 classes of different sizes.

where $w_{ij}$ is the dissimilarity among pattern $i$ and $j$ and $\sigma$ is a positive real number that determines the kernel width. This parameter is calculated experimentally, since it is not possible to know in advance the nature of the data and the clustering separability indices [19]. The data representation on the graph can be improved using graph Laplacian techniques. These techniques are able to decrease the weights of the edges between different clusters making them more distant. The normalized graph Laplacian is computed as $L = D^{-1/2}SD^{-1/2}$, where $D$ is the degree matrix of $S$.

Another technique that can be used to better represent the data is sparsification, that consists in reducing the number of nodes in the graph, focusing only on the most important. This refinement is aimed at modeling the local neighborhood relationships among nodes and can be done with two different methods, the $\epsilon$-neighborhood technique, which maintains only the edges which have a value higher than a predetermined threshold, $\epsilon$; and the $k$-nearest neighbor technique, which maintains only the highest $k$ values. It results in a similarity matrix that can be used as the adjacency matrix of a weighted graph $G$.

The effect of the processes described above is presented in Fig. 1. Near the main diagonal of the matrices it is possible to recognize some blocks which represent clusters. The values of those points are low in the cosine matrix, since it encodes the proximity of the points. Then the matrix is transformed into a similarity matrix by the Gaussian kernel, in fact, the values of the points near the main diagonal in this representation are high. It is possible to note that some noise was removed with the Laplacian matrix. The points far from the diagonal appear now clearer and the blocks are more compact. Finally the $k$-nn matrix remove many nodes from the representation, giving a clear picture of the clusters.

We used the Laplacian matrix $L$ for the experiments with the *dominant set*, since it requires that the similarity values among the elements of a cluster are very close to each other. Graph $G$ is used to run the clustering games, since this framework does not need a dense graph to cluster the data points.

## 4.4   Clustering

We use the *dominant set* algorithm to extract the prototypical elements of each cluster with two different settings, one in which we give as input the number of clusters to extract and the other without this information. In the fist case we extract the first $K$ clusters from a dataset and then run the document clustering games to cluster the remaining clusters. This situation can be interpreted as the case in which there are some labeled points in the data and new points have to be clustered according to this evidence. In the second case we run *dominant set* recursively to extract small clusters and then use the document clustering games to cluster the clusters, merging them according to their similarity. The similarity among two clusters $C_i$ and $C_j$ is computed as:

$$sim(C_i, C_j) = \frac{\sum_{r \in C_i} \sum_{t \in C_j} s_{rt}}{|C_i| + |C_j|} \tag{16}$$

We conducted also experiments in which we simulated the streaming data process. This is done dividing a dataset in random folds and clustering the dataset iteratively adding a fold at time to measure if the performances of the system are constant. In this case we used a fold (8% of the data) as initial clustering.

### 4.5    Strategy Space Implementation

The clustering phase serves as preliminary phase to partition the data into two disjoint sets, one containing clustered objects and the other unclustered. Clustered objects supply information to unclustered nodes in the graph. We initialized the strategy space of the player in these two sets as follows,

$$x_i^h = \begin{cases} K^{-1}, & \text{if node } i \text{ is unclustred.} \\ 1, & \text{if node } i \text{ is in cluster } h, \end{cases} \tag{17}$$

where $K$ is the number of clusters to extract and $K^{-1}$ ensures that the constraints required by a game theoretic framework are met (see Eq. (2)).

### 4.6    Clustering Games

We assume that each player $i \in I$ that participates in the games is a document and that each strategy $s \in S_i$ is a particular cluster. The players can choose a determined strategy from their strategy space that is initialized as described in previous section and can be considered as a mixed strategy space (see Sect. 2). The games are played among two similar players, $i$ and $j$. The payoff matrix among two players $i$ and $j$ is defined as an identity matrix of rank $K$, $A_{ij}$.

This choice is motivated by the fact that in this context all the players have the same number of strategies and in the studied contexts the number of clusters of each dataset is low. In works in which there are many interacting classes it is possible to use a similarity function to construct the payoff matrix, as described in [31].

The best choice for two similar players is to be clustered in the same cluster, this is imposed with the entry $A_{ij} = 1, i = j$. This kind of game is called *imitation game* because the players try to learn their strategy observing the choices of their co-players. For this reason the payoff function of each player is additively separable and is computed as described in Sect. 2. Specifically, in the case of clustering games there are labeled and unlabeled players that, as proposed in [8], can be divided in two disjoint sets, $I_l$ and $I_u$. We have $K$ disjoint subsets, $I_l = \{I_{l|1}, \ldots, I_{l|K}\}$, where each subset denotes the players that always play their $h$th pure strategy.

Only unlabeled players play the games, because they have to decide their best strategy (cluster membership). This strategy is selected taking into account the similarity that a player share with other players and the choices of these players. Labeled players act as bias over the choices of unlabeled players because they always play a defined strategy and unlabeled players influence each other. The players adapt to the strategic environment, gradually adjusting their preferences over strategies [23]. Once equilibrium is reached, the cluster of each player $i$, corresponds to the strategy, with the highest value.

The payoffs of the games are calculated with Eqs. 5 and 6, which in this case, with labeled and unlabeled players, can be defined as,

$$u_i(e_i^h) = \sum_{j \in I_u} (g_{ij} A_{ij} x_j)^h + \sum_{h=1}^{K} \sum_{j \in I_{l|h}} (g_{ij} A_{ij})^h \tag{18}$$

and,

$$u_i(x) = \sum_{j \in I_u} x_i^T g_{ij} A_{ij} x_j + \sum_{k=1}^{K} \sum_{j \in I_{l|h}} x_i^T (g_{ij} A_{ij})^h. \tag{19}$$

where the first part of the equations calculates the payoffs that each player obtains from unclustered players and the second part computes the payoffs obtained from labeled players. The Nash equilibria of the system are calculated the replicator dynamics Eq. 8.

## 5   Experimental Setup

The performances of the systems are measured using the accuracy (AC) and the normalized mutual information (NMI). AC is calculated with the following equation,

$$AC = \frac{\sum_{i=1}^{n} \delta(\alpha_i, map(l_i))}{n}, \tag{20}$$

where $n$ denotes the total number of documents in the dataset and $\delta(x, y)$ is equal to 1 if $x$ and $y$ are clustered in the same cluster. The function $map(L_i)$ maps each cluster label $l_i$ to the equivalent label in the benchmark, aligning the labeling provided by the benchmark and those obtained with our clustering algorithm. It is done using the Kuhn-Munkres algorithm [14]. The NMI was introduced by Strehl and Ghosh [27] and indicates the level of agreement between the clustering $C$ provided by the ground truth and the clustering $C'$ produced by a clustering algorithm. This measure takes into account also the partitioning similarities of the two clustering and not just the number of correctly clustered objects. The mutual information (MI) between the two clusterings is computed with the following equation,

$$MI(C, C') = \sum_{c_i \in C, c'_j \in C'} p(c_i, c'_j) \cdot log_2 \frac{p(c_i, c'_j)}{p(c_i) \cdot p(c'_j)}, \tag{21}$$

where $p(c_i)$ and $p(c'_i)$ are the probabilities that a document belongs to cluster $c_i$ and $c'_i$, respectively; $p(c_i, c'_i)$ is the probability that the selected document belongs to $c_i$ as well as $c'_i$ at the same time. The MI information is then normalized with the following equation,

$$NMI(C, C') = \frac{MI(C, C')}{max(H(C), H(C'))} \tag{22}$$

where $H(C)$ and $H(C')$ are the entropies of $C$ and $C'$, respectively, This measure ranges from 0 to 1. It is equal to 1 when the two clustering are identical and it is equal to 0 if the two sets are independent. We run each experiment 50 times and present the mean results with standard deviation ($\pm$).

**Table 2.** Datasets description.

| Data | $n_d$ | $n_v$ | K | $n_c$ | Balance |
|------|-------|-------|---|-------|---------|
| NG17-19 | 2998 | 15810 | 3 | 999 | 0.998 |
| classic | 7094 | 41681 | 4 | 1774 | 0.323 |
| k1b | 2340 | 21819 | 6 | 390 | 0.043 |
| hitech | 2301 | 10800 | 6 | 384 | 0.192 |
| reviews | 4069 | 18483 | 5 | 814 | 0.098 |
| sports | 8580 | 14870 | 7 | 1226 | 0.036 |
| la1 | 3204 | 31472 | 6 | 534 | 0.290 |
| la12 | 6279 | 31472 | 6 | 1047 | 0.282 |
| la2 | 3075 | 31472 | 6 | 513 | 0.274 |
| tr11 | 414 | 6424 | 9 | 46 | 0.046 |
| tr23 | 204 | 5831 | 6 | 34 | 0.066 |
| tr41 | 878 | 7453 | 10 | 88 | 0.037 |
| tr45 | 690 | 8261 | 10 | 69 | 0.088 |

We evaluated our model on the same datasets[1] used in [38]. In that work it has been conducted an extensive comparison of different document clustering algorithms. The description of these datasets is shown in Table 2. The authors used 13 datasets (described in Table 2). The datasets have different sizes ($n_d$), from 204 documents (tr23) to 8580 (*sports*). The number of classes ($K$) is also

---

[1] http://www.shi-zhong.com/software/docdata.zip.

different and ranges from 3 to 10. Another important feature of the datasets is the size of the vocabulary ($n_w$) of each dataset that ranges from 5832 (*tr23*) to 41681 (*classic*) and is function of the number of documents in the dataset, their size and the number of different topics in it, that can be considered as clusters. The datasets are also described with $n_c$ and *Balance*. $n_c$ indicates the average number of documents per cluster and *Balance* is the ratio among the size of the smallest cluster and that of the largest.

## 5.1  Basic Experiments

We present in this Section an experiment in which all the features of each dataset are used, constructing the graphs as described in Sect. 4. We first used *dominant set* to extract the prototypical elements of each cluster and then we applied our approach to cluster the remaining data points.

The results of this series of experiments are presented in Table 3. They can be used as point of comparison for our next experiments, in which we used different settings. From the analysis of the table it is not possible to find a stable pattern. The results range from NMI .27 on the *hitech*, to NMI .67 on *k1b*. The reason of this instability is due to the representation of the datasets that in some cases is not appropriate to describe the data.

**Table 3.** Results as AC and NMI, with the entire feature space.

|  | NG17-19 | classic | k1b | hitech | review | sports | la1 | la12 | la2 | tr11 | tr23 | tr41 | tr45 |
|---|---|---|---|---|---|---|---|---|---|---|---|---|---|
| AC | .56±.0 | .66±.07 | .82±.0 | .44±.0 | .81±.0 | .69±.0 | .49±.04 | .57±.02 | .54±.0 | .68±.02 | .44±.01 | .64±.07 | .64±.02 |
| NMI | .42±.0 | .56±.22 | .66±.0 | .27±.0 | .59±.0 | .62±.0 | .45±.04 | .46±.01 | .46±.01 | .63±.02 | .38±.0 | .53±.06 | .59±.01 |

An example of the graphical representation of the two datasets mentioned above is presented in Fig. 2, where the similarity matrices constructed for *k1b* and *hitech* are shown. We can see that the representation of *hitech* does not show a clear structure near the main diagonal, to the contrary, it is possible to recognize a block structures on the graphs representing *k1b*.

## 5.2  Experiments with Feature Selection

In this section we present an experiment in which we conducted feature selection to see if it is possible to reduce the noise introduced by determined features. To do this, we decided to apply to the corpora a basic frequency selection heuristic that eliminates the features that occur more (or less) often than a determined thresholds. In this study were kept only the words that occur more than once.

This basic reduction leads to a more compact feature space, which is easier to handle. Words that appear very few times in the corpus can be special characters or miss-spelled words and for this reason can be eliminated. The number of features of the reduced datasets are shown in Table 4. From the table, we can see that the reduction is significant for 5 of the datasets used, with a reduction

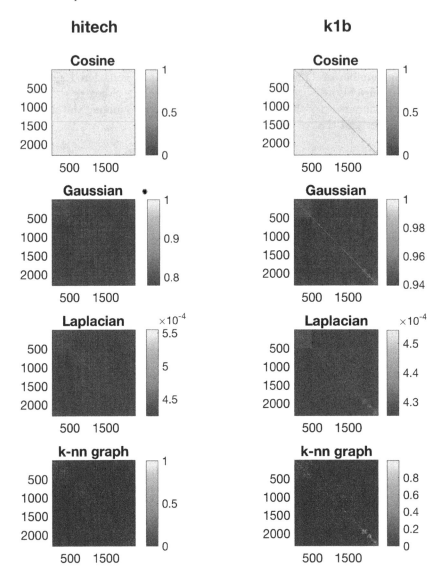

**Fig. 2.** Different representations for the datasets *hitech* and *k1b*.

of 82% for *classic*. The datasets that are not listed in the table were not affected by this process.

In Table 5 we present the results obtained employing feature selection. This technique can be considered a good choice to reduce the size of the datasets and the computational cost, but in this case does not seem to have a big impact on the performances of the algorithm. In fact, the improvements in the performance of the algorithm are not substantial. There is an improvement of 1%,

**Table 4.** Number of features for each dataset before and after feature selection.

| | classic | k1b | la1 | la12 | la2 |
|---|---|---|---|---|---|
| pre | 41681 | 21819 | 31472 | 31472 | 31472 |
| post | 7616 | 10411 | 13195 | 17741 | 12432 |
| % | 0.82 | 0.52 | 0.58 | 0.44 | 0.6 |

**Table 5.** Mean results as AC and NMI, with frequency selection.

| | classic | k1b | la1 | la12 | la2 |
|---|---|---|---|---|---|
| AC | .67 ± .0 | .79 ± .0 | .56 ± .11 | .56 ± .03 | .57 ± .0 |
| NMI | .57 ± .0 | .67 ± .0 | .47 ± .12 | .44 ± .01 | .47 ± .0 |

in terms of $NMI$, in four datasets over five and in one case we obtained lower results. This could be due to the fact that we do not know exactly what features have been removed, because this information is not provided with the datasets. It is possible that the reduction has removed some important (discriminative) word, compromising the representation of the data and the computation of the similarities. Also for this reason we did not use any other frequency selection technique.

### 5.3 Experiments with LSA

In this Section we used LSA (see Sect. 4.1) to reduce the number of features that describe the data. The evaluation was conducted using different numbers of features to describe each dataset, ranging from 10 to 400. This operation is required because there is no agreement on the correct number of features to extract for a determined dataset, for this reason this value has to be calculate experimentally.

The results of this evaluation are shown in two different tables, Table 6 indicates the results as NMI and Table 7 indicates the results as AC for each dataset and number of features. The performances of the algorithm measured as NMI are similar on average (excluding the case of $n_v$ with 10 features), but there is no agreement on different datasets. In fact, different data representations affect heavily the performances on datasets such as $NG17\text{-}19$, where the performances ranges from .27 to .46. This phenomenon is due to the fact that each dataset has different characteristics, as shown in Table 2 and that their representation require an appropriate semantic space. With $n_v = 250$ we obtained the higher results on average, both in terms of NMI and AC.

The results with the representation provided by LSA show how this technique is effective in terms of performances. In fact, it is possible to achieve higher results than using the entire feature space or with the frequency selection technique. The improvements are substantial and in many cases are 10% higher. Furthermore, with this new representation it is easier to handle the data.

### 5.4 Comparison with State-of-the-Art Algorithms

The results of the evaluation of the document clustering games are shown in Tables 8 and 9 (third column, DCG). We compared the best results obtained

**Table 6.** NMI results for all the datasets. Each column indicates the results obtained with a reduced version of the feature space using LSA.

| Data\$n_v$ | 10 | 50 | 100 | 150 | 200 | 250 | 300 | 350 | 400 |
|---|---|---|---|---|---|---|---|---|---|
| NG17-19 | .27 | .37 | **.46** | .26 | .35 | .37 | .36 | .37 | .37 |
| classic | .53 | .63 | .71 | .73 | **.76** | .74 | .72 | .72 | .69 |
| k1b | **.68** | .61 | .58 | .62 | .63 | .63 | .62 | .61 | .62 |
| hitech | **.29** | .28 | .25 | .26 | .28 | .27 | .27 | .26 | .26 |
| reviews | **.60** | .59 | .59 | .59 | .59 | .59 | .58 | .58 | .58 |
| sports | .62 | .63 | **.69** | .67 | .66 | .66 | .66 | .64 | .62 |
| la1 | .49 | .53 | .58 | .58 | .58 | .57 | **.59** | .57 | **.59** |
| la12 | .48 | .52 | .52 | .52 | .53 | **.56** | .54 | .55 | .54 |
| la2 | .53 | .56 | .58 | .58 | .58 | .58 | **.59** | .58 | .58 |
| tr11 | .69 | .65 | .67 | .68 | **.71** | .70 | .70 | .69 | .70 |
| tr23 | .42 | **.48** | .41 | .39 | .41 | .40 | .41 | .40 | .41 |
| tr41 | .65 | .75 | .72 | .69 | .71 | .74 | **.76** | .69 | .75 |
| tr45 | .65 | **.70** | .67 | .69 | .69 | .68 | .68 | .67 | .69 |
| avg. | .53 | .56 | **.57** | .56 | **.57** | **.57** | **.57** | .56 | **.57** |

**Table 7.** AC results for all the datasets. Each column indicates the results obtained with a reduced version of the feature space using LSA.

| Data\$n_v$ | 10 | 50 | 100 | 150 | 200 | 250 | 300 | 350 | 400 |
|---|---|---|---|---|---|---|---|---|---|
| NG17-19 | .61 | **.63** | .56 | .57 | .51 | .51 | .51 | .51 | .51 |
| classic | .64 | .76 | .87 | .88 | **.91** | .88 | .85 | .84 | .80 |
| k1b | .72 | .55 | .58 | .73 | **.75** | **.75** | .73 | .70 | .73 |
| hitech | **.48** | .36 | .42 | .41 | .47 | .46 | .41 | .43 | .42 |
| reviews | **.73** | .72 | .69 | .69 | .69 | .71 | .71 | .71 | .71 |
| sports | .62 | .61 | **.71** | .69 | .68 | .68 | .68 | .68 | .61 |
| la1 | .59 | .64 | .72 | .70 | **.73** | .72 | **.73** | .72 | **.73** |
| la12 | .63 | .63 | .62 | .62 | .63 | **.67** | .64 | **.67** | .65 |
| la2 | **.69** | .66 | .60 | .60 | .61 | .60 | .65 | .60 | .60 |
| tr11 | .69 | .66 | .69 | .70 | **.72** | .71 | .71 | .71 | .71 |
| tr23 | .44 | **.51** | .43 | .42 | .43 | .43 | .43 | .43 | .43 |
| tr41 | .60 | .76 | .68 | .68 | .65 | .75 | **.77** | .67 | **.77** |
| tr45 | .57 | **.69** | .66 | .68 | .67 | .67 | .67 | .67 | .67 |
| avg. | .62 | .63 | .63 | .64 | .65 | **.66** | .65 | .64 | .64 |

with the document clustering games approach and the best results indicated in [38] and in [20]. In the first article it was conducted an extensive evaluation of different generative and discriminative models, specifically tailored for document clustering and two graph-based approaches, CLUTO and a bipartite spectral co-clustering method. In this evaluation the results are reported as NMI and graphical approaches obtained better performances than generative. In the second article were evaluated different NMF approaches to document clustering, on the same datasets, here the results are reported as AC.

From Table 8 it is possible to see that the results of the document clustering games are higher than those of state-of-the-art algorithms on ten datasets out of thirteen. On the remaining three datasets we obtained the same results on two datasets and a lower result in one. On *classic*, *tr23* and tr26 the improvement of our approach is substantial, with results higher than 5%. Form Table 9 we can see that our approach performs substantially better that NMF on all the datasets.

### 5.5 Experiments with No Cluster Number

In this section we present the experiments conducted with our system in a context in which the number of clusters to extract from the dataset is not used. It has been tested the ability of *dominant set* to find natural clusters and the performances that can be obtained in this context by the document clustering games. We first run *dominant set* to discover many small clusters, setting the parameter of the gaussian kernel with a small value ($\sigma = 0.1$), then these clusters

**Table 8.** Results as NMI of generative models and graph partitioning algorithm (*Best*) compared to our approach with and without $k$.

| Data | $DCG_{noK}$ | $DCG$ | $Best$ |
|---|---|---|---|
| NG17-19 | .39 ± .0 | **.46** ± .0 | .46 ± .01 |
| classic | .71 ± .0 | **.76** ± .0 | .71 ± .06 |
| k1b | **.73** ± .02 | .68 ± .02 | .67 ± .04 |
| hitech | **.35** ± .01 | .29 ± .02 | .33 ± .01 |
| reviews | .57 ± .01 | **.60** ± .01 | .56 ± .09 |
| sports | .67 ± .0 | **.69** ± .0 | .67 ± .01 |
| la1 | .53 ± .0 | **.59** ± .0 | .58 ± .02 |
| la12 | .52 ± .0 | **.56** ± .0 | .56 ± .01 |
| la2 | .53 ± .0 | **.59** ± .0 | .56 ± .01 |
| tr11 | **.72** ± .0 | .71 ± .0 | .68 ± .02 |
| tr23 | **.57** ± .02 | .48 ± .03 | .43 ± .02 |
| tr41 | .70 ± .01 | **.76** ± .06 | .69 ± .02 |
| tr45 | **.70** ± .02 | **.70** ± .03 | .68 ± .05 |

**Table 9.** Results as AC of nonnegative matrix factorization algorithms (*Best*) compared to our approach with and without $k$.

| Data | $DCG_{noK}$ | $DCG$ | $Best$ |
|---|---|---|---|
| NG17-19 | .59 ± .0 | **.63** ± .0 | – |
| classic | .80 ± .0 | **.91** ± .0 | .59 ± .07 |
| k1b | **.86** ± .02 | .75 ± .03 | .79 ± .0 |
| hitech | **.52** ± .01 | .48 ± .02 | .48 ± .04 |
| reviews | .64 ± .01 | **.73** ± .01 | .69 ± .07 |
| sports | **.78** ± .0 | .71 ± .0 | .50 ± .07 |
| la1 | .63 ± .0 | **.73** ± .0 | .66 ± .0 |
| la12 | .59 ± .0 | **.67** ± .0 | – |
| la2 | .55 ± .0 | **.69** ± .0 | .53 ± .0 |
| tr11 | **.74** ± .0 | .72 ± .0 | .53 ± .05 |
| tr23 | **.52** ± .02 | .51 ± .05 | .43 ± .06 |
| tr41 | .75 ± .01 | **.77** ± .08 | .53 ± .06 |
| tr45 | **.71** ± .01 | .69 ± .04 | .54 ± .06 |

are re-clustered as described in Sect. 4.4 constructing a graph that encodes their pairwise similarity (see Eq. 16).

The evaluation of this model was conducted on the same datasets used in previous experiments and the results are shown in Tables 8 and 9 (second column, $DCG_{noK}$). From these tables we can see that this new formulation of the clustering games performs well in many datasets. In fact, in datasets such as *k1b*, *hitech*, *tr11* and *tr23* it has results higher than those obtained in previous experiments. This can be explained by the fact that with this formulation the number of clustered points used by our framework is higher that in the previous experiments. Furthermore, this new technique is able to extract clusters of any shape. In fact, as we can see in Fig. 3, datasets such as *la1* and *la2* present a more compact cluster structure, whereas in datasets such as *k1b* and *hitech* the clusters structure is loose[2].

The performances of the system can be improved with this setting when it is able to extract the exact number of natural clusters from the graph. To the contrary, when it is not able to predict this number, the performances decrease drastically. This phenomenon can explain why this approach performs poorly in some datasets. In fact, in datasets such as, *NG17-19*, *la1*, *la12* and *l2* the system performs poorly compared to our previous experiments. In many cases this happens because during the clustering phase we extract more clusters than expected. The results as NMI of our system are higher than those of related algorithms on 8 over 13 datasets, even if $k$ is not given as input. Also the results as AC are good, in fact on 9 datasets over 11 we obtained better performances.

---

[2] The dataset have been visualized using t-SNE to reduce the features to 3d.

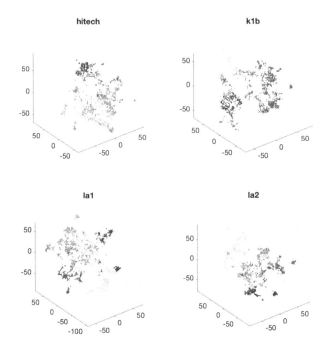

**Fig. 3.** Representation of the datasets *hitech*, *k1b*, *la1* and *la2*.

## 5.6    Experiments on Streaming Data

In this section we present the evaluation of our apporoach on streaming datasets. For this task we used the same datasets used in previous experiments but this time we divided each of them in 12 random folds. In this way we simulated the data streaming process, clustering the data iterativelly. We performed the experiments 15 times to not bias our test sets. For each experiment we selected a random fold as initial clustering and performed 11 runs of the algorithm, each time including a new fold in the test set. Previous clusterings are used to drive the choices of new data points to specific clusters, making the final clustering coherent.

The results of this evaluation are presented in Tables 10 and 11 as NMI and AC, respectively. From the tables we can see that the performances of the system are stable over time. In fact, we can see that in 9 datasets over 13, the different among the results as NMI with the entire dataset (12 folds) and those with 2 folds is 2%. The results as AC are even better. In fact, with the entire dataset the performances are stable and in two cases higher (*la2* and *tr45*). The latter behavior can be explained considering the fact that the algorithm exploit contextual information and in many cases having more information to use leads to better solutions. We can see that just in one case we have a drop of 5% in performances, comparing the results in fold 2 with those in fold 12. The most negative results have been achieved on small datasets, this because in these

**Table 10.** Results as NMI for all the datasets. Each column indicates the results obtained including the corresponding fold in the test set.

| Data | 2 | 3 | 4 | 5 | 6 | 7 | 8 | 9 | 10 | 11 | 12 |
|---|---|---|---|---|---|---|---|---|---|---|---|
| ng17-19 | .57 ± .07 | .55 ± .05 | .55 ± .04 | .55 ± .03 | .55 ± .03 | .55 ± .03 | .55 ± .03 | .55 ± .03 | .55 ± .03 | .55 ± .03 | .55 ± .03 |
| classic | .81 ± .02 | .81 ± .02 | .81 ± .02 | .81 ± .01 | .81 ± .01 | .81 ± .01 | .81 ± .01 | .81 ± .01 | .81 ± .01 | .81 ± .01 | .81 ± .01 |
| k1b | .85 ± .04 | .83 ± .03 | .83 ± .02 | .83 ± .02 | .83 ± .02 | .83 ± .01 | .83 ± .01 | .83 ± .01 | .83 ± .02 | .83 ± .01 | .83 ± .01 |
| hitech | .38 ± .04 | .34 ± .04 | .34 ± .03 | .33 ± .03 | .33 ± .02 | .32 ± .02 | .32 ± .02 | .32 ± .02 | .32 ± .02 | .32 ± .02 | .32 ± .02 |
| reviews | .77 ± .03 | .75 ± .02 | .75 ± .02 | .74 ± .02 | .74 ± .01 | .74 ± .02 | .74 ± .02 | .74 ± .01 | .74 ± .02 | .74 ± .01 | .74 ± .01 |
| sports | .86 ± .02 | .85 ± .02 | .84 ± .02 | .84 ± .01 | .84 ± .01 | .83 ± .01 | .83 ± .01 | .83 ± .01 | .83 ± .01 | .83 ± .01 | .83 ± .01 |
| la1 | .65 ± .05 | .63 ± .04 | .63 ± .04 | .63 ± .03 | .64 ± .02 | .64 ± .02 | .63 ± .02 | .63 ± .02 | .63 ± .02 | .63 ± .02 | .63 ± .02 |
| la12 | .68 ± .03 | .67 ± .02 | .66 ± .01 | .67 ± .01 | .66 ± .01 | .66 ± .01 | .66 ± .01 | .66 ± .01 | .66 ± .01 | .66 ± .01 | .66 ± .01 |
| la2 | .68 ± .03 | .67 ± .02 | .67 ± .02 | .67 ± .02 | .66 ± .02 | .66 ± .01 | .66 ± .01 | .67 ± .01 | .67 ± .01 | .67 ± .01 | .67 ± .02 |
| tr11 | .69 ± 10 | .64 ± .09 | .61 ± 10 | .58 ± .08 | .56 ± .08 | .56 ± .07 | .55 ± .07 | .54 ± .07 | .54 ± .07 | .54 ± .07 | .54 ± .07 |
| tr23 | .66 ± 11 | .57 ± 10 | .52 ± .08 | .50 ± .09 | .50 ± .08 | .49 ± .08 | .48 ± .09 | .48 ± .09 | .47 ± .08 | .46 ± .08 | .45 ± .08 |
| tr41 | .86 ± .05 | .84 ± .05 | .83 ± .04 | .83 ± .04 | .83 ± .03 | .82 ± .03 | .82 ± .03 | .82 ± .03 | .82 ± .03 | .82 ± .03 | .81 ± .03 |
| tr45 | .79 ± .04 | .76 ± .04 | .76 ± .04 | .75 ± .04 | .74 ± .04 | .74 ± .04 | .73 ± .04 | .73 ± .03 | .73 ± .03 | .73 ± .04 | .73 ± .04 |

**Table 11.** Results as AC for all the datasets. Each column indicates the results obtained including the corresponding fold in the test set.

| Data | 2 | 3 | 4 | 5 | 6 | 7 | 8 | 9 | 10 | 11 | 12 |
|---|---|---|---|---|---|---|---|---|---|---|---|
| ng17-19 | .85 ± .03 | .84 ± .03 | .84 ± .02 | .84 ± .01 | .84 ± .02 | .84 ± .01 | .84 ± .01 | .84 ± .01 | .84 ± .01 | .84 ± .01 | .84 ± .01 |
| classic | .94 ± .01 | .94 ± .01 | .94 ± .01 | .94 ± .01 | .94 ± .00 | .94 ± .00 | .94 ± .00 | .94 ± .00 | .94 ± .00 | .94 ± .00 | .94 ± .00 |
| k1b | .94 ± .02 | .94 ± .01 | .94 ± .01 | .94 ± .01 | .95 ± .01 | .95 ± .01 | .95 ± .01 | .95 ± .01 | .94 ± .01 | .94 ± .01 | .94 ± .01 |
| hitech | .61 ± .04 | .61 ± .03 | .61 ± .03 | .61 ± .03 | .60 ± .02 | .60 ± .02 | .60 ± .02 | .60 ± .02 | .60 ± .02 | .60 ± .02 | .60 ± .02 |
| reviews | .92 ± .01 | .91 ± .01 | .91 ± .01 | .91 ± .01 | .91 ± .01 | .91 ± .01 | .91 ± .01 | .91 ± .01 | .91 ± .01 | .91 ± .01 | .91 ± .01 |
| sports | .95 ± .01 | .95 ± .01 | .95 ± .01 | .95 ± .01 | .95 ± .01 | .95 ± .01 | .95 ± .01 | .95 ± .01 | .95 ± .01 | .95 ± .01 | .95 ± .01 |
| la1 | .82 ± .03 | .82 ± .03 | .82 ± .02 | .82 ± .02 | .82 ± .02 | .82 ± .02 | .82 ± .02 | .82 ± .02 | .82 ± .01 | .82 ± .01 | .82 ± .01 |
| la12 | .85 ± .02 | .84 ± .01 | .84 ± .01 | .84 ± .01 | .84 ± .00 | .84 ± .01 | .84 ± .00 | .84 ± .01 | .84 ± .01 | .84 ± .01 | .84 ± .00 |
| la2 | .83 ± .02 | .84 ± .01 | .84 ± .01 | .84 ± .01 | .84 ± .01 | .84 ± .01 | .84 ± .01 | .84 ± .01 | .84 ± .01 | .84 ± .01 | .84 ± .01 |
| tr11 | .72 ± .07 | .72 ± .08 | .71 ± .08 | .70 ± .07 | .69 ± .07 | .69 ± .06 | .69 ± .06 | .69 ± .06 | .69 ± .06 | .69 ± .06 | .69 ± .06 |
| tr23 | .73 ± .08 | .71 ± .08 | .69 ± .08 | .69 ± .07 | .69 ± .07 | .68 ± .07 | .68 ± .07 | .68 ± .07 | .68 ± .07 | .68 ± .07 | .68 ± .07 |
| tr41 | .90 ± .04 | .90 ± .03 | .90 ± .03 | .90 ± .03 | .90 ± .02 | .90 ± .02 | .90 ± .02 | .90 ± .02 | .90 ± .02 | .90 ± .02 | .90 ± .02 |
| tr45 | .80 ± .04 | .81 ± .04 | .82 ± .04 | .82 ± .04 | .82 ± .04 | .82 ± .03 | .82 ± .04 | .82 ± .03 | .82 ± .03 | .82 ± .04 | .82 ± .04 |

cases the clusters are small and unbalanced. In particular dealing with clusters of very different sizes makes the $k$-nn algorithm, used to sparsify the graph, not useful. In fact, the resulting structure allow the elements of small clusters to have connections with elements belonging to other clusters. In these cases the dynamics of our system converge to solutions in which small clusters are absorbed by bigger ones. This because the elements belonging to small clusters are likely to receive influence from the elements belonging to large clusters if $k$ is larger than the cardinality of the small clusters. This phenomenon can be seen in Fig. 4, where we compare the clustering results of our method against the ground truth, on $k1b$. We can see that the orange cluster disappears in fold 2 and that this error is propagated on the other folds. The other clusters are partitioned correctly.

If we compare the results in this Section with the results proposed in Sect. 5.4 we can see that with this approach we can have a bust in performances. In fact, in all datasets, except one (tr11) the results are higher both in terms of NMI and AC. We can see that using just few labeled points allows our approach to

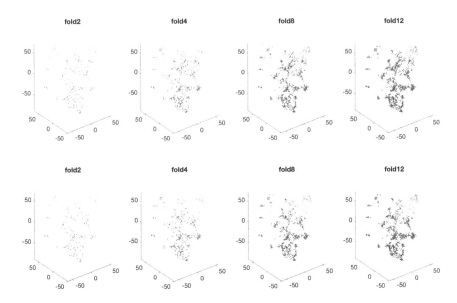

**Fig. 4.** Visualizations of the results on *k1b* on different folds. The upper row shows the ground truth and the lower row shows the results of our approach.

substantially improve its performances. Furthermore we see that these performance are stable over time and that the standard deviation is very low in all experiments, ≤0.11 for NMI and ≤0.8 for AC.

**Comparison with k-NN.** We conducted the same experiment described in previous Section to compare the performances of our method with the k-nearest neighbor (k-NN) algorithm. We used k-NN to classify iteratively the folds of each dataset treating the data in same way of previous experiments and setting $k = 1$. Experimentally we noticed that this value achieve the best performances. Higher values have very low NMI, leading to situations in which small clusters are merged in bigger ones.

The results of this evaluation are shown in Tables 12 and 13 as NMI and AC, respectively. From these tables we can see that the performances of k-NN are not stable and tend to increase at each step. We can notice that the results in fold 2 in many cases are doubled in fold 12, this behaviour demonstrate that this algorithm requires many data to achieve good classification performances. To the contrary with our approach it is possible to obtain stable performances in each fold.

The performances of k-NN are very low compared with our approaches. In particular, we can see that it does not perform well in the first seven folds. This can be explained considering that it classify new instances taking into account only local information (the information on the class membership of its nearest neighbour), without considering any other source of information and without imposing any coherence constraint using contextual information.

Form Tables 12 and 13 we can see that the results of k-NN in fold 12 (entire dataset) are almost always lower that those obtained with our method, both in terms of NMI and AC. In fact, just in two cases k-NN obtain equal and higher results, in *tr11* and *tr23* if we consider the NMI. If we consider the results as AC we can see that in two datasets k-NN has the same performances of our method (*NG17-19* and *tr11*) and that it has higher performances on *hitech* (+1%).

## 6  Conclusions

With this work we explored new methods for document clustering based on game theory and consistent labeling principles. We have conducted an extensive series of experiments to test the approach on different scenarios. We have also evaluated the system with different implementations and compared the results with state-of-the-art algorithms.

**Table 12.** Results as NMI for all the datasets using k-NN. Each column indicates the results obtained including the corresponding fold in the test set.

| Data | 2 | 3 | 4 | 5 | 6 | 7 | 8 | 9 | 10 | 11 | 12 |
|---|---|---|---|---|---|---|---|---|---|---|---|
| ng3sim | .25 ± .03 | .31 ± .03 | .36 ± .03 | .40 ± .02 | .43 ± .02 | .46 ± .01 | .48 ± .01 | .49 ± .01 | .51 ± .01 | .52 ± .01 | .53 ± .01 |
| classic | .31 ± .02 | .39 ± .02 | .44 ± .02 | .49 ± .02 | .52 ± .01 | .55 ± .01 | .58 ± .01 | .60 ± .01 | .62 ± .01 | .63 ± .01 | .64 ± .01 |
| k1b | .32 ± .04 | .38 ± .03 | .44 ± .02 | .49 ± .02 | .53 ± .02 | .57 ± .02 | .60 ± .01 | .62 ± .01 | .64 ± .01 | .66 ± .01 | .67 ± .01 |
| hitech | .17 ± .03 | .18 ± .02 | .20 ± .02 | .21 ± .02 | .23 ± .02 | .24 ± .01 | .26 ± .01 | .27 ± .01 | .28 ± .01 | .29 ± .01 | .29 ± .01 |
| reviews | .35 ± .03 | .41 ± .03 | .46 ± .02 | .50 ± .02 | .53 ± .02 | .55 ± .01 | .57 ± .01 | .59 ± .01 | .60 ± .01 | .61 ± .01 | .62 ± .01 |
| sports | .48 ± .02 | .56 ± .02 | .62 ± .01 | .66 ± .01 | .69 ± .01 | .71 ± .01 | .73 ± .01 | .75 ± .01 | .76 ± .01 | .77 ± .00 | .78 ± .00 |
| la1 | .31 ± .02 | .35 ± .02 | .39 ± .02 | .42 ± .02 | .44 ± .02 | .46 ± .02 | .48 ± .01 | .50 ± .01 | .51 ± .01 | .52 ± .01 | .53 ± .01 |
| la12 | .32 ± .02 | .37 ± .02 | .41 ± .01 | .45 ± .01 | .48 ± .01 | .50 ± .01 | .52 ± .01 | .53 ± .01 | .55 ± .01 | .56 ± .01 | .57 ± .01 |
| la2 | .33 ± .03 | .37 ± .03 | .41 ± .02 | .44 ± .02 | .47 ± .01 | .49 ± .01 | .51 ± .01 | .53 ± .01 | .54 ± .01 | .55 ± .01 | .56 ± .01 |
| tr11 | .36 ± .07 | .38 ± .04 | .40 ± .04 | .43 ± .04 | .45 ± .03 | .47 ± .03 | .49 ± .03 | .50 ± .02 | .52 ± .02 | .53 ± .02 | .54 ± .02 |
| tr23 | .34 ± .12 | .35 ± .09 | .39 ± .06 | .40 ± .06 | .41 ± .07 | .44 ± .06 | .46 ± .06 | .47 ± .05 | .49 ± .04 | .50 ± .04 | .52 ± .04 |
| tr41 | .41 ± .05 | .47 ± .04 | .51 ± .03 | .55 ± .03 | .59 ± .02 | .61 ± .02 | .63 ± .02 | .65 ± .02 | .67 ± .02 | .68 ± .02 | .70 ± .01 |
| tr45 | .46 ± .05 | .48 ± .05 | .52 ± .04 | .55 ± .03 | .57 ± .02 | .60 ± .02 | .62 ± .02 | .63 ± .02 | .64 ± .02 | .65 ± .01 | .66 ± .01 |

**Table 13.** Results as AC for all the datasets using k-NN. Each column indicates the results obtained including the corresponding fold in the test set.

| Data | 2 | 3 | 4 | 5 | 6 | 7 | 8 | 9 | 10 | 11 | 12 |
|---|---|---|---|---|---|---|---|---|---|---|---|
| ng3sim | .60 ± .02 | .67 ± .02 | .72 ± .01 | .76 ± .01 | .78 ± .01 | .80 ± .01 | .81 ± .01 | .82 ± .01 | .83 ± .01 | .84 ± .01 | .84 ± .00 |
| classic | .59 ± .02 | .68 ± .01 | .73 ± .01 | .77 ± .01 | .80 ± .01 | .82 ± .01 | .84 ± .01 | .85 ± .00 | .86 ± .00 | .87 ± .00 | .87 ± .00 |
| k1b | .53 ± .04 | .62 ± .02 | .69 ± .02 | .74 ± .01 | .78 ± .01 | .81 ± .01 | .83 ± .01 | .84 ± .01 | .86 ± .01 | .87 ± .01 | .88 ± .01 |
| hitech | .40 ± .03 | .44 ± .02 | .48 ± .02 | .51 ± .02 | .53 ± .01 | .55 ± .01 | .57 ± .01 | .58 ± .01 | .59 ± .01 | .60 ± .01 | .61 ± .01 |
| reviews | .58 ± .03 | .66 ± .02 | .72 ± .01 | .76 ± .01 | .78 ± .01 | .81 ± .01 | .82 ± .01 | .83 ± .00 | .84 ± .00 | .85 ± .00 | .86 ± .00 |
| sports | .72 ± .01 | .79 ± .01 | .83 ± .01 | .86 ± .01 | .88 ± .00 | .89 ± .00 | .90 ± .00 | .91 ± .00 | .92 ± .00 | .92 ± .00 | .93 ± .00 |
| la1 | .46 ± .02 | .55 ± .01 | .61 ± .01 | .66 ± .01 | .69 ± .01 | .71 ± .01 | .73 ± .01 | .74 ± .01 | .76 ± .01 | .77 ± .01 | .78 ± .01 |
| la12 | .49 ± .01 | .58 ± .01 | .64 ± .01 | .68 ± .01 | .72 ± .01 | .74 ± .01 | .76 ± .01 | .77 ± .01 | .78 ± .01 | .79 ± .01 | .80 ± .00 |
| la2 | .49 ± .03 | .58 ± .02 | .64 ± .02 | .68 ± .01 | .71 ± .01 | .73 ± .01 | .75 ± .01 | .76 ± .01 | .78 ± .01 | .78 ± .00 | .79 ± .00 |
| tr11 | .42 ± .05 | .43 ± .04 | .46 ± .04 | .50 ± .04 | .55 ± .03 | .58 ± .03 | .61 ± .02 | .63 ± .02 | .66 ± .02 | .67 ± .02 | .69 ± .02 |
| tr23 | .49 ± .07 | .49 ± .05 | .54 ± .04 | .59 ± .04 | .63 ± .04 | .66 ± .04 | .69 ± .04 | .71 ± .03 | .73 ± .03 | .75 ± .03 | .76 ± .03 |
| tr41 | .45 ± .05 | .50 ± .03 | .55 ± .02 | .62 ± .02 | .67 ± .02 | .71 ± .01 | .73 ± .01 | .76 ± .01 | .78 ± .01 | .80 ± .01 | .81 ± .01 |
| tr45 | .50 ± .04 | .56 ± .04 | .62 ± .03 | .67 ± .02 | .71 ± .02 | .74 ± .01 | .76 ± .01 | .78 ± .01 | .79 ± .01 | .80 ± .01 | .81 ± .01 |

Our method can be considered as a continuation of graph based approaches but it combines together the partition of the graph and the propagation of the information across the network. With this method we used the structural information about the graph and then we employed evolutionary dynamics to find the best labeling of the data points. The application of a game theoretic framework is able to exploit relational and contextual information and guarantees that the final labeling of the data is consistent.

The system has demonstrated to perform well compared with state-of-the-art system and to be extremely flexible. In fact, it has been tested with different features, with and without the information about the number of clusters to extract and on static and dynamic context. Furthermore, it is not difficult to implement new graph similarity measure and new dynamics to improve its performances or to adapt to new contexts.

The experiments without the use of $K$, where the algorithm collects together highly similar points and then merges the resulting groups, demonstrated how it is able to extract clusters of any size without the definition of any centroid. The experiments on streaming data demonstrated that our approach can be used to cluster data dynamically. In fact, the performances of the system does not change much when the test set is enriched with new instances to cluster. This is an appealing feature, since it makes the framework flexible and not computationally expensive. On this scenario it was demonstrated that the use of contextual information helps the clustering task. In fact, using the k-NN algorithm on streaming data produces lower and not stable results.

As future work we are planning to apply this framework to other kind of data and also to use it in the context of *big data*, where, in many cases, it is necessary to deal with datasets that do not fit in memory and have to be divided in different parts in order to be clustered or classified.

**Acknowledgements.** This work was partly supported by the Samsung Global Research Outreach Program.

# References

1. Aggarwal, C.C.: A survey of stream clustering algorithms. In: Data Clustering: Algorithms and Applications, pp. 231–258 (2013)
2. Aggarwal, C.C.: Data Streams: Models and Algorithms. Springer, Heidelberg (2007)
3. Ardanuy, M.C., Sporleder, C.: Structure-based clustering of novels. In: EACL 2014, pp. 31–39 (2014)
4. Bharat, K., Curtiss, M., Schmitt, M.: Methods and apparatus for clustering news content. US Patent 7,568,148, 28 July 2009
5. Blei, D.M., Ng, A.Y., Jordan, M.I.: Latent Dirichlet allocation. J. Mach. Learn. Res. **3**, 993–1022 (2003). http://dl.acm.org/citation.cfm?id=944919.944937
6. Dhillon, I.S.: Co-clustering documents and words using bipartite spectral graph partitioning. In: Proceedings of the Seventh ACM SIGKDD International Conference on Knowledge Discovery and Data Mining, pp. 269–274. ACM (2001)

7. Ding, C., Li, T., Peng, W.: Nonnegative matrix factorization and probabilistic latent semantic indexing: equivalence chi-square statistic, and a hybrid method. In: Proceedings of the National Conference on Artificial Intelligence, vol. 21, p. 342. AAAI Press/MIT Press, Menlo Park/Cambridge, London (1999, 2006)
8. Erdem, A., Pelillo, M.: Graph transduction as a noncooperative game. Neural Comput. **24**(3), 700–723 (2012)
9. Haykin, S., Network, N.: A comprehensive foundation. Neural Netw. **2**, 1–3 (2004)
10. Jain, A.K., Dubes, R.C.: Algorithms for clustering data. Prentice-Hall, Inc., Upper Saddle River (1988)
11. Landauer, T.K., Foltz, P.W., Laham, D.: An introduction to latent semantic analysis. Discourse Process. **25**(2–3), 259–284 (1998)
12. Lee, D.D., Seung, H.S.: Learning the parts of objects by non-negative matrix factorization. Nature **401**(6755), 788–791 (1999)
13. Leyton-Brown, K., Shoham, Y.: Essentials of game theory: a concise multidisciplinary introduction. Synth. Lect. Artif. Intell. Mach. Learn. **2**(1), 1–88 (2008)
14. Lovasz, L.: Matching Theory (North-Holland Mathematics Studies) (1986)
15. Manning, C.D., Raghavan, P., Schütze, H., et al.: Introduction to Information Retrieval, vol. 1. Cambridge University Press, Cambridge (2008)
16. Nowak, M.A., Sigmund, K.: Evolutionary dynamics of biological games. Science **303**(5659), 793–799 (2004)
17. Okasha, S., Binmore, K.: Evolution and Rationality: Decisions, Co-operation and Strategic Behaviour. Cambridge University Press, Cambridge (2012)
18. Pavan, M., Pelillo, M.: Dominant sets and pairwise clustering. IEEE Trans. Pattern Anal. Mach. Intell. **29**(1), 167–172 (2007)
19. Peterson, A.D.: A separability index for clustering and classification problems with applications to cluster merging and systematic evaluation of clustering algorithms (2011)
20. Pompili, F., Gillis, N., Absil, P.A., Glineur, F.: Two algorithms for orthogonal nonnegative matrix factorization with application to clustering. Neurocomputing **141**, 15–25 (2014)
21. Rota Bulò, S., Pelillo, M.: A game-theoretic approach to hypergraph clustering. IEEE Trans. Pattern Anal. Mach. Intell. **35**(6), 1312–1327 (2013)
22. Rota Buló, S., Pelillo, M., Bomze, I.M.: Graph-based quadratic optimization: a fast evolutionary approach. Comput. Vis. Image Underst. **115**(7), 984–995 (2011)
23. Sandholm, W.H.: Population Games and Evolutionary Dynamics. MIT Press, Cambridge (2010)
24. Sankaranarayanan, J., Samet, H., Teitler, B.E., Lieberman, M.D., Sperling, J.: TwitterStand: news in tweets. In: Proceedings of the 17th ACM SIGSPATIAL International Conference on Advances in Geographic Information Systems, pp. 42–51. ACM (2009)
25. Shawe-Taylor, J., Cristianini, N.: Kernel Methods for Pattern Analysis. Cambridge University Press, Cambridge (2004)
26. Smith, J.M., Price, G.: The logic of animal conflict. Nature **246**, 15 (1973)
27. Strehl, A., Ghosh, J.: Cluster ensembles—a knowledge reuse framework for combining multiple partitions. J. Mach. Learn. Res. **3**, 583–617 (2003)
28. Szabó, G., Fath, G.: Evolutionary games on graphs. Phys. Rep. **446**(4), 97–216 (2007)
29. Tagarelli, A., Karypis, G.: Document clustering: the next frontier. In: Data Clustering: Algorithms and Applications, p. 305 (2013)
30. Taylor, P.D., Jonker, L.B.: Evolutionary stable strategies and game dynamics. Math. Biosci. **40**(1), 145–156 (1978)

31. Tripodi, R., Pelillo, M.: A game-theoretic approach to word sense disambiguation. Comput. Linguist. (in press)
32. Tripodi, R., Pelillo, M.: Document clustering games. In: Proceedings of the 5th International Conference on Pattern Recognition Applications and Methods, pp. 109–118 (2016)
33. Von Neumann, J., Morgenstern, O.: Theory of Games and Economic Behavior (60th Anniversary Commemorative Edition). Princeton University Press, Princeton (1944)
34. Weibull, J.W.: Evolutionary Game Theory. MIT Press, Cambridge (1997)
35. Xu, W., Liu, X., Gong, Y.: Document clustering based on non-negative matrix factorization. In: Proceedings of the 26th Annual International ACM SIGIR Conference on Research and Development in Informaion Retrieval, pp. 267–273. ACM (2003)
36. Zhao, Y., Karypis, G.: Empirical and theoretical comparisons of selected criterion functions for document clustering. Mach. Learn. **55**(3), 311–331 (2004)
37. Zhao, Y., Karypis, G., Fayyad, U.: Hierarchical clustering algorithms for document datasets. Data Min. Knowl. Discov. **10**(2), 141–168 (2005)
38. Zhong, S., Ghosh, J.: Generative model-based document clustering: a comparative study. Knowl. Inf. Syst. **8**(3), 374–384 (2005)

# Criteria for Mixture-Model Clustering
# with Side-Information

Edith Grall-Maës[(✉)] and Duc Tung Dao

ICD - LM2S - UMR 6281 CNRS - Troyes University of Technology, Troyes, France
edith.grall@utt.fr

**Abstract.** The estimation of mixture models is a well-known approach for cluster analysis and several criteria have been proposed to select the number of clusters. In this paper, we consider mixture models using side-information, which gives the constraint that some data in a group originate from the same source. Then the usual criteria are not suitable. An EM (Expectation-Maximization) algorithm has been previously developed to jointly allow the determination of the model parameters and the data labelling, for a given number of clusters. In this work we adapt three usual criteria, which are the bayesian information criterion (BIC), the Akaike information criterion (AIC), and the entropy criterion (NEC), so that they can take into consideration the side-information. One simulated problem and two real data sets have been used to show the relevance of the modified criterion versions and compare the criteria. The efficiency of both the EM algorithm and the criteria, for selecting the right number of clusters while getting a good clustering, is in relation with the amount of side-information. Side-information being mainly useful when the clusters overlap, the best criterion is the modified BIC.

## 1   Introduction

The goal of clustering is to determine a partition rule of data such that observations in the same cluster are similar to each other. A probabilistic approach to clustering is based on mixture models [4]. It is considered that the sample is composed of a finite number of sub-populations which are all distributed according to an a priori parameterized probability density function [11]. If the parameter set of the mixture is known, then the unknown labels can be estimated and a partition of the sample is obtained. Thus the clustering result depends on the number of sub-populations and on the estimated parameters given this number. In the context of clustering via finite mixture models the problem of choosing a particular number of clusters is important.

In this work, the clustering problem consists in a semi-supervised learning using side-information, which is given as constraints between data points. These constraints determine whether points were generated by the same source. Spatiotemporal data for which the statistical properties do not depend on the time is a good example of such a context. Since all the measures originating from the same position in space are realization of the same random variable, it is

© Springer International Publishing AG 2017
A. Fred et al. (Eds.): ICPRAM 2016, LNCS 10163, pp. 43–59, 2017.
DOI: 10.1007/978-3-319-53375-9_3

known that they belong to a same cluster and a constraint has to be imposed. An example is the temperature in a given town in a given month. The measured temperature for different years can be considered as different realizations of the same random variable. Then a group is formed by the values for a given town and for the different years. Using the temperature measures in different towns, the aim of the clustering problem is to group similar towns, considering the values of the different years for one town as one group. It has to be noticed that the number of samples in a group is not fixed. Clustering of time series is another example of clustering data with side-information. All points in a series have to be assigned to the same cluster and thus they make a group.

The mixture models using side-information have been previously introduced. In [14] an algorithm has been proposed for determining the parameters of a Gaussian mixture model when the number of components in the mixture is a priori known. In [9] the clustering of processes has been proposed in the case of any mixture model approach specifically with the issue of data partition. However, the main difficulty as in classical clustering problems is to choose the number of clusters since it is usually not known a priori.

In this paper, criteria are proposed for assessing the number of clusters in a mixture-model clustering approach with side-information. The side-information defines constraints, grouping some points. Because usual criteria do not consider the side-information in the data they are not suitable. We propose to modify three well-known criteria and to compare the efficiency of the adapted version with the original version of the criteria. We also compare each criterion to each others.

This paper is organized as follows. Section 2 introduces notations and describes the method for determining jointly the parameters of the mixture model and the cluster labels, with the constraint that some points are in the same group, for the case of a given a priori number of clusters. Section 3 is devoted to the criteria. To take into consideration the side-information, adjustments have been made to the Bayesian information criterion (BIC), the entropy criterion (NEC), and the Akaike information criterion (AIC). The results using one example on simulated data and two examples on real data are reported in Sect. 4. A comparison with the classical criteria demonstrates the relevance of the modified versions. The efficiency of the criteria are compared and the influence of the group size is discussed. We conclude the paper in Sect. 5.

## 2    Clustering with Side-Information

### 2.1    Data Description

The objects to be classified are considered as a sample composed of $N$ observations $\mathcal{X} = \{s_n\}_{n=1..N}$. Each observation $s_n$ is a group $s_n = \{x_i^n\}_{i=1..|s_n|}$ which is a small subset of $|s_n|$ independent data points that are known to belong to a single unknown class. An example of a data set with 9 observations is given on Fig. 1.

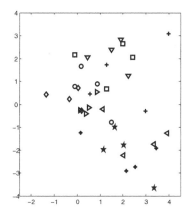

**Fig. 1.** An example of data set with side-information: there are 9 groups, each one being composed of 4 points.

It is considered that the sample is composed of $K$ sub-populations which are all distributed according to an a priori statistical parameterized law. The latent cluster labels are given by $\mathcal{Z} = \{z_n\}_{n=1..N}$ where $z_n = k$ means that the $n^{th}$ realization originates from the $k^{th}$ cluster, which is modeled by the statistical law with parameter $\theta_k$. The $|s_n|$ points $x_i^n$ in the group $n$ are assigned to the same cluster $z_n$. Thus the constraints between data points are observed.

In summary, the observation set and the cluster label set in the case of data with side-information are respectively given by:

$$\mathcal{X} = \{s_n\}_{n=1..N} \text{ with } s_n = \{x_i^n\}_{i=1..|s_n|} \tag{1}$$

and

$$\mathcal{Z} = \{z_n\}_{n=1..N}. \tag{2}$$

Let's define $\mathcal{X}'$ the same observation set with the removal of the side-information. $\mathcal{X}'$ is composed of $N'$ points with $N' = \sum_{n=1}^{N} |s_n|$:

$$\mathcal{X}' = \{x_i^n\}_{i=1..|s_n|, n=1..N} \tag{3}$$

and the cluster label set $\mathcal{Z}'$ is:

$$\mathcal{Z}' = \{z_i^n\}_{i=1..|s_n|, n=1..N} \tag{4}$$

in which $z_i^n$ is not constrained to be equal to $z_j^n$.

## 2.2 Mixture Model

The mixture model approach to clustering [12] assumes that data are from a mixture of a number $K$ of clusters in some proportions. The density function of a sample $s$ writes as:

$$f(s|\boldsymbol{\theta}_K) = \sum_{k=1}^{K} \alpha_k f_k(s|\theta_k)$$

where $f_k(s|\theta_k)$ is the density function of the component $k$, $\alpha_k$ is the probability that a sample belongs to class $k$: $\alpha_k = P(Z = k)$, $\theta_k$ is the model parameter value for the class $k$, and $\boldsymbol{\theta}_K$ is the is parameter set $\boldsymbol{\theta}_K = \{\theta_k, \alpha_k\}_{k=1..K}$.

The maximum likelihood approach to the mixture problem given a data set and a number of clusters consists of determining the parameters that maximizes the log-likelihood:

$$L_{\mathcal{X}}(\boldsymbol{\theta}_K) = \sum_{n=1}^{N} \log f(s_n|\boldsymbol{\theta}_K)$$

in which $\mathcal{X}$ is the data set, $K$ is the number of clusters and $\boldsymbol{\theta}_K$ is the parameter set to be determined.

Due to the constraint given by the side-information, and the independence of points within a group, we get

$$f(s_n|\boldsymbol{\theta}_K) = \sum_{k=1}^{K} \alpha_k f_k(s_n|\theta_k) = \sum_{k=1}^{K} \alpha_k \prod_{i=1}^{|s_n|} f_k(x_i^n|\theta_k). \tag{5}$$

Then

$$L_{\mathcal{X}}(\boldsymbol{\theta}_K) = \sum_{n=1}^{N} \log \left( \sum_{k=1}^{K} \alpha_k \prod_{i=1}^{|s_n|} f_k(x_i^n|\theta_k) \right). \tag{6}$$

This expression can be compared to the log-likelihood for the same observation set with the removal of the side-information

$$
\begin{aligned}
L_{\mathcal{X}'}(\boldsymbol{\theta'}_K) &= \sum_{n=1}^{N} \sum_{i=1}^{|s_n|} \log f(x_i^n|\boldsymbol{\theta'}_K) \\
&= \sum_{n=1}^{N} \sum_{i=1}^{|s_n|} \log \left( \sum_{k=1}^{K} \alpha_k' f_k(x_i^n|\theta_k') \right)
\end{aligned}
\tag{7}
$$

in which $\mathcal{X}'$ is the data set and $\boldsymbol{\theta'}_K = \{\theta_k', \alpha_k'\}_{k=1..K}$ is the parameter set.

A common approach for optimizing the parameters of a mixture model is the expectation-maximization (EM) algorithm [3]. This is an iterative method that produces a set of parameters that locally maximizes the log-likelihood of a given sample, starting from an arbitrary set of parameters.

A modified EM algorithm has been proposed in [9]. Its aim is to take into consideration the side-information, as in [14], and to get a hard partition, as in [4]. It repeats an estimation step (E step), a classification step, and a maximization step (M step). The algorithm is the following one.

- Initialize the parameter set $\boldsymbol{\theta}_K^{(0)}$
- Repeat from $m = 1$ until convergence
  - E step: compute the posteriori probability $c_{nk}^{(m)}$ that the $n^{th}$ group originates from the $k^{th}$ cluster according to

$$c_{nk}^{(m)} = p(Z_n = k|s_n, \boldsymbol{\theta}_K^{(m-1)}) = \frac{\alpha_k^{(m-1)} \prod_{i=1}^{|s_n|} f_k(x_i^n|\theta_k^{(m-1)})}{\sum_{r=1}^{K} \alpha_r^{(m-1)} \prod_{i=1}^{|s_n|} f_r(x_i^n|\theta_r^{(m-1)})} \tag{8}$$

- determine the cluster label set $\mathbf{Z}^{(m)}$: choose $z_n^{(m)} = k$ corresponding to the largest value $c_{nk}^{(m)}$
- M step: determine the parameters $\theta_k^{(m)}$ that maximizes the log-likelihood for each class $k$ and compute the new value of $\alpha_k^{(m)}$ for each $k$.

This algorithm provides an optimal parameter set $\boldsymbol{\theta}_K^*$ for a given number $K$. We define $L^*(K)$ as the corresponding log-likelihood

$$L^*(K) = \max_{\boldsymbol{\theta}_K} L_{\mathcal{X}}(\boldsymbol{\theta}_K) = L_{\mathcal{X}}(\boldsymbol{\theta}_K^*) \tag{9}$$

In the case without side-information using the initial set $\mathcal{X}'$, we get a similar optimal log-likelihood $L'^*(K)$. The maximum likelihood increases with the model complexity as reminded in [7]. Then the maximized log-likelihood $L'^*(K)$ is an increasing function of $K$, and the maximized log-likelihood cannot be used as a selection criterion for choosing the number $K$.

In the case of data with side-information, the maximized log-likelihood $L^*(K)$ is also an increasing function of $K$, and thus it is not adapted as a selection criterion for choosing the number of clusters.

## 3 Criteria

The aim of a criterion is to select a model in a set. The criterion should measure the model's suitability which balances the model fit and the model complexity depending on the number of parameters in the model.

In the case of data without side-information, there are several information criteria that help to support the selection of a particular model or clustering structure. However, specific criteria may be more suitable for particular applications [7].

In this paper we have chosen three usual criteria and have adapted them to data with side-information. The bases of the criteria are briefly described below and their modified version are given.

### 3.1 Criterion BIC

The criterion BIC [13] is a widely used likelihood criterion penalized by the number of parameters in the model. It is based on the selection of a model from a set of candidate models $\{M_m\}_{m=1,\dots,M}$ by maximizing the posterior probability:

$$P(M_m|\mathcal{X}) = \frac{P(\mathcal{X}|M_m)P(M_m)}{P(\mathcal{X})}$$

where $M_m$ corresponds to a model of dimension $r_m$ parameterized by $\phi_m$ in the space $\Phi_m$.

Assuming that $P(M_1) = P(M_m), \forall m = 1, \ldots, M$, the aim is to maximize $P(\mathcal{X}|M_m)$ which can be determined from the integration of the joint distribution:

$$P(\mathcal{X}|M_m) = \int_{\Phi_m} P(\mathcal{X}, \phi_m|M_m) \, d\phi_m$$

$$= \int_{\Phi_m} P(\mathcal{X}|\phi_m, M_m) P(\phi_m|M_m) \, d\phi_m$$

An approximation of this integral can be obtained by using the Laplace approximation [10]. Neglecting error terms, it is shown that

$$\log P(\mathcal{X}|M_m) \approx \log P(\mathcal{X}|\widehat{\phi_m}, M_m) - \frac{r_m}{2} \log(N)$$

The criterion BIC is derived from this expression. It corresponds to the approximation of $\log P(\mathcal{X}|M_m)$ and then is given by

$$BIC(K) = -2 \, L'^*(K) + r \, \log(N')$$

where $L'$ and $N'$ are respectively the log-likelihood and the number of points in the set $\mathcal{X}'$ defined by relation (3) for the case of no side-information, and $r$ is the number of free parameters. The model, i.e. the number of clusters, to be selected has to minimize this criterion.

For the case of data with side-information, described by the set $\mathcal{X}$ given by relation (1), the criterion has to be adapted. We suggest to use the modified BIC (BICm) criterion:

$$BICm(K) = -2 \, L^*(K) + r \, \log(N). \tag{10}$$

It has to be noticed that the criterion does not depend directly on the total number of points $\sum_{n=1}^{N} |s_n|$ but depends on the number of groups $N$. The maximum log-likelihood $L^*(K)$, which takes into account the positive constraints is computed differently to classical criterion.

### 3.2   Criterion AIC

Another frequently used information criterion to select a model from a set is the criterion AIC proposed in [1]. This criterion is based on the minimization of the Kullback-Leibler distance between the model and the truth $M_0$:

$$d_{KL}(M_0, M_i) = \int_{-\infty}^{+\infty} P(\mathcal{X}|M_0) \log(\mathcal{X}|M_0) - \int_{-\infty}^{+\infty} P(\mathcal{X}|M_0) \log P(\mathcal{X}|M_i).$$

The criterion AIC to be minimized in the case of data without side-information takes the form:

$$AIC(K) = -2 \, L'^*(K) + 2 \, r$$

For taking into account that some points arise from the same source, we propose to replace $L'^*(K)$ by $L^*(K)$. Then the modified AIC (AICm) is given by:

$$AICm(K) = -2\,L^*(K) + 2\,r$$

$$= -2 \sum_{n=1}^{N} \log \left( \sum_{k=1}^{K} \alpha_k \prod_{i=1}^{|s_n|} f_k(x_i^n|\theta_k) \right) + 2\,r \qquad (11)$$

### 3.3   Criterion NEC

The normalized entropy criterion (NEC) was proposed in [5] to estimate the number of clusters arising from a mixture model. It is derived from a relation linking the likelihood and the classification likelihood of a mixture.

In the case of a data set without side-information $\mathcal{X}'$, the criterion to be minimized in order to assess the number of clusters of the mixture components is defined as:

$$NEC(K) = \frac{E'^*(K)}{L'^*(K) - L'^*(1)} \qquad (12)$$

where $E'^*(K)$ denotes the entropy term which measures the overlap of the mixture components.

Because this criterion suffers of limitation to decide between one and more than one clusters, a procedure that consists in setting $NEC(1) = 1$ was proposed in [2].

In the case of data set with side-information, the computation of the terms of entropy and of log-likelihood has to be modified. Since $\sum_{k=1}^{K} c_{nk} = 1$, we can rewrite $L_\mathcal{X}(\boldsymbol{\theta}_K)$ given by relation (6) as:

$$L_\mathcal{X}(\boldsymbol{\theta}_K) = \sum_{n=1}^{N} \sum_{k=1}^{K} c_{nk} \log \sum_{r=1}^{K} \alpha_r \prod_{i=1}^{|s_n|} f_r(x_i^n|\theta_r).$$

The expression of $c_{nk}$ for data with side-information is given by:

$$c_{nk} = \frac{\alpha_k \prod_{i=1}^{|s_n|} f_k(x_i^n|\theta_k)}{\sum_{r=1}^{K} \alpha_r \prod_{i=1}^{|s_n|} f_r(x_i^n|\theta_r)}$$

Thus the log-likelihood can be expressed as:

$$L_\mathcal{X}(\boldsymbol{\theta}_K) = \sum_{k=1}^{K} \sum_{n=1}^{N} \log \frac{\alpha_k \prod_{i=1}^{|s_n|} f_k(x_i^n|\theta_k)}{c_{nk}}$$

and can be rewritten as

$$L_\mathcal{X}(\boldsymbol{\theta}_K) = C_\mathcal{X}(\boldsymbol{\theta}_K) + E_\mathcal{X}(\boldsymbol{\theta}_K)$$

with

$$C_{\mathcal{X}}(\boldsymbol{\theta}_K) = \sum_{k=1}^{K} \sum_{n=1}^{N} c_{nk} \log \left( \alpha_k \prod_{i=1}^{|s_n|} f_k(x_i^n | \theta_k) \right)$$

and

$$E_{\mathcal{X}}(\boldsymbol{\theta}_K) = - \sum_{k=1}^{K} \sum_{n=1}^{N} c_{nk} \log c_{nk}.$$

Thus we propose to use the criterion given by:

$$NECm(K) = \frac{E^*(K)}{L^*(K) - L^*(1)} \tag{13}$$

where $E^*(K) = E_{\mathcal{X}}(\boldsymbol{\theta}_K^*)$ and $L^*(K)$ is given by (9).

## 4    Results

The performances of the three criteria have been assessed using simulated problems, which are composed of Gaussian distributions, and two real problems using iris data and climatic data.

First of all, the relevance of the adapted version of the three criteria is shown. Then the performances of the criteria are compared in different situations in order to select the best one.

### 4.1    Comparison of the Original and Adapted Versions of the Three Criteria

In order to compare the original criteria with the modified version, we have used a mixture of four two dimensional Gaussian components. The observation data have been generated with the following parameter values

$$m_1 = [-1, -1], m_2 = [-0.5, 1], m_3 = [1, -1], m_4 = [0.5, 1]$$

$$\Sigma_1 = \Sigma_2 = \Sigma_3 = \Sigma_3 = \sigma^2 I$$

$N = 40$ (10 groups per cluster) and $|s_n| = 30 \; \forall n$. Consequently the total number of points is equal to 1200.

Examples of data sets are given on Fig. 2, with a variance equal to 1.2, 0.7, 0.3 and 0.03. Let note that on this figure the groups are not shown and then one can only see the points.

The data set was modeled by a Gaussian mixture. Thus the model fits the data when the number of clusters is equal to 4. For a given number of clusters $K$, the parameters of the Gaussian mixture were estimated according to the algorithm described in Sect. 2.2. The parameters to be estimated are the proportions of each class, the mean vector, and the variance-covariance matrix, in dimension 2. Thus the total number of parameters is equal to $6K - 1$. Then the value of each criterion was computed using the initial expression and the modified expression.

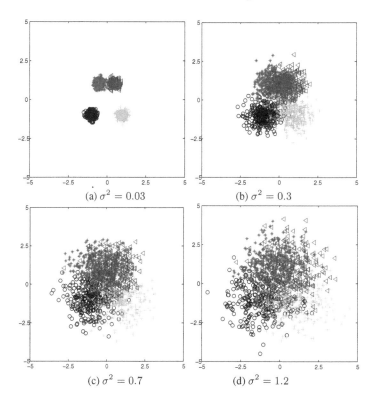

**Fig. 2.** Data sets composed of four Gaussian components, with a variance equal to 0.03, 0.3, 0.7 and 1.2.

The number of clusters was selected for each of the six criteria (BIC, AIC, NEC, BICm, AICm, NECm). This experiment has been repeated 200 times in order to estimate the percent frequency of choosing a K-component mixture for each criterion. This experiment has been done for 6 different values of $\sigma^2$: 0.01, 0.03, 0.1, 0.3, 0.7, 1.2. The results are reported in Table 1.

Comparing the criteria BIC and BICm, it can be observed that BIC allows to select the correct number of clusters only when the clusters do not overlap ($\sigma^2 < 0.3$), whereas BICm is efficient even when the clusters are not obviously separated. Comparing the criteria AIC and AICm, it can be observed that when the clusters do not overlap AIC is slightly better than AICm, but when the clusters are not obviously separated AICm remains efficient contrary to AIC. Comparing the criteria NEC and NECm, NEC selects only one class when the clusters are not well separated whereas NECm selects 2 or 3 classes whereas the true value is 4. Thus NECm performs better than NEC.

This experiment shows that the modified criteria are better suited to the data with side-information than the original criteria.

Comparing the three modified criterion, one observes that the best modified criterion is BICm and the worst results are obtained with NECm. With BICm

**Table 1.** Percent frequencies of choosing K clusters.

| $\sigma^2$ | K | BIC | AIC | NEC | BICm | AICm | NECm |
|---|---|---|---|---|---|---|---|
| 0.01 | 1 | 0 | 0 | 0 | 0 | 0 | 0 |
|  | 2 | 0 | 0 | 0 | 0 | 0 | 0 |
|  | 3 | 0 | 0 | 1 | 0 | 0 | 0 |
|  | 4 | 100 | 100 | 99 | 100 | 82 | 100 |
|  | 5 | 0 | 0 | 0 | 0 | 16 | 0 |
|  | 6 | 0 | 0 | 0 | 0 | 2 | 0 |
| 0.03 | 1 | 0 | 0 | 0 | 0 | 0 | 0 |
|  | 2 | 0 | 0 | 66 | 0 | 0 | 0.5 |
|  | 3 | 0 | 0 | 0 | 0 | 0 | 1 |
|  | 4 | 100 | 100 | 34 | 100 | 80.5 | 98.5 |
|  | 5 | 0 | 0 | 0 | 0 | 17 | 0 |
|  | 6 | 0 | 0 | 0 | 0 | 2.5 | 0 |
| 0.1 | 1 | 0 | 0 | 0 | 0 | 0 | 0 |
|  | 2 | 0 | 0 | 92.5 | 0 | 0 | 96.5 |
|  | 3 | 0 | 0 | 0 | 0 | 0 | 0 |
|  | 4 | 100 | 100 | 7.5 | 99 | 80 | 3.5 |
|  | 5 | 0 | 0 | 0 | 1 | 16.5 | 0 |
|  | 6 | 0 | 0 | 0 | 0 | 3.5 | 0 |
| 0.3 | 1 | 0 | 0 | 35.5 | 0 | 0 | 0 |
|  | 2 | 0 | 0 | 41.5 | 0 | 0 | 75.5 |
|  | 3 | 32 | 25 | 22.5 | 0 | 0 | 22 |
|  | 4 | 68 | 75 | 0.5 | 99.5 | 78.5 | 2.5 |
|  | 5 | 0 | 0 | 0 | 0.5 | 17.5 | 0 |
|  | 6 | 0 | 0 | 0 | 0 | 4 | 0 |
| 0.7 | 1 | 2.5 | 0 | 100 | 0 | 0 | 0 |
|  | 2 | 36.5 | 0.5 | 0 | 0 | 0 | 26 |
|  | 3 | 60.5 | 98.5 | 0 | 0 | 0 | 74 |
|  | 4 | 0.5 | 1 | 0 | 99.5 | 78.5 | 0 |
|  | 5 | 0 | 0 | 0 | 0.5 | 17.5 | 0 |
|  | 6 | 0 | 0 | 0 | 0 | 4 | 0 |
| 1 | 1 | 76 | 0 | 98 | 0 | 0 | 0 |
|  | 2 | 23.5 | 22.5 | 2 | 0 | 0 | 34 |
|  | 3 | 0.5 | 77.5 | 0 | 0 | 0 | 66 |
|  | 4 | 0 | 0 | 0 | 99.5 | 84 | 0 |
|  | 5 | 0 | 0 | 0 | 0.5 | 15 | 0 |
|  | 6 | 0 | 0 | 0 | 0 | 1 | 0 |
| 1.2 | 1 | 94.5 | 9.5 | 98 | 0 | 0 | 0 |
|  | 2 | 5.5 | 36.5 | 2 | 0 | 0 | 35.5 |
|  | 3 | 0 | 54 | 0 | 0 | 0 | 64.5 |
|  | 4 | 0 | 0 | 0 | 99.5 | 83 | 0 |
|  | 5 | 0 | 0 | 0 | 0.5 | 16 | 0 |
|  | 6 | 0 | 0 | 0 | 0 | 1 | 0 |

over 99% success is obtained in all cases. AICm has a slight tendency to over-estimate the number of clusters, while NECm has a tendency to underestimate. This conclusion is in accordance with the results given in [7] obtained in case of the classical criteria for mixture-model clustering without side-information. It is also mentioned that for normal distributions mixtures, BIC performs very well. In [5], it is mentioned that NEC is efficient when some cluster structure exists. Since the cluster structure is not obvious in this experiment, this criterion has low performance.

### 4.2   Influence of the Amount of Side-Information

The efficiency of the modified criteria has been studied in relation with the amount of side-information and also the overlapping of the clusters, which depends directly on the number of points in the groups. For that a mixture of three two dimensional Gaussian components has been used, with the following parameter values

$$m_1 = [0,0], m_2 = [a,a], m_3 = [a,-a] \quad \text{and} \quad \Sigma_1 = \Sigma_2 = \Sigma_3 = I.$$

The number of points can vary with the group but for the study we set the same number for all groups. We have used 4 different values for the triplet $(a, N, |s_n|)$. The total number of points given by the product $N|s_n|$ was always equal to 750. Thus when the number of groups $(N)$ decreases, the number of points in each group $(|s_n|)$ increases and the "side-information" increases. An example of data is given on Fig. 3 for $a = 2, N = 150, |s_n| = 5$.

A Gaussian mixture was used as the model. For a given number of clusters $K$, the $6K - 1$ parameters of the Gaussian mixture were estimated according to the algorithm described in Sect. 2.2. The number of clusters was determined using each criterion. This experiment has been repeated 200 times in order to estimate the percent frequency of choosing a K-component mixture for each criterion. The

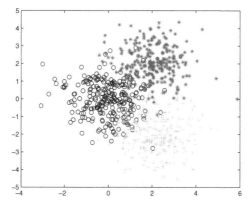

**Fig. 3.** An example of a mixture of three Gaussian components.

**Table 2.** Percent frequencies of choosing K clusters.

| | | K | BICm | AICm | NECm |
|---|---|---|---|---|---|
| $m_1 = [0,0]$ | $N = 150$ | 1 | 0 | 0 | 0 |
| $m_2 = [2,2]$ | $|s_n| = 5$ | 2 | 0 | 0 | 40 |
| $m_3 = [2,-2]$ | | 3 | **97** | **79** | **59** |
| | | 4 | 3 | 17 | 1 |
| | | 5 | 0 | 4 | 0 |
| $m_1 = [0,0]$ | $N = 15$ | 1 | 0 | 0 | 0 |
| $m_2 = [2,2]$ | $|s_n| = 50$ | 2 | 0 | 0 | 32 |
| $m_3 = [2,-2]$ | | 3 | **99** | **95** | **68** |
| | | 4 | 1 | 5 | 0 |
| | | 5 | 0 | 0 | 0 |
| $m_1 = [0,0]$ | $N = 150$ | 1 | 0 | 0 | 0 |
| $m_2 = [1,1]$ | $|s_n| = 5$ | 2 | 0 | 0 | **78** |
| $m_3 = [1,-1]$ | | 3 | **96** | **78** | 17 |
| | | 4 | 4 | 15 | 0 |
| | | 5 | 0 | 7 | 5 |
| $m_1 = [0,0]$ | $N = 15$ | 1 | 0 | 0 | 0 |
| $m_2 = [1,1]$ | $|s_n| = 50$ | 2 | 0 | 0 | 33 |
| $m_3 = [1,-1]$ | | 3 | **97** | **94** | **67** |
| | | 4 | 3 | 6 | 0 |
| | | 5 | 0 | 0 | 0 |

results for the four cases and for each of the criteria BICm, AICm and NECm are given in Table 2.

The results confirm the conclusions obtained in the previous section: the best results are obtained with BICm and that holds for any amount of side-information. Comparing the four cases, when the side-information increases, or when the cluster overlapping decreases, it is easier to determine the right number of clusters.

### 4.3   Iris Data Set

The Iris flower data set is a multivariate data set, available on the web (archive.ics.uci.edu). This is a well-known database to be found in the pattern recognition literature (for example [6]). Four features were measured from each sample: the length and the width of the sepals and petals, in centimetres. The data set contains 3 classes of 50 instances each, where each class refers to a type of iris plant. The data set only contains two clusters with rather obvious separation. One of the clusters contains Iris setosa, while the other cluster contains both Iris virginica and Iris versicolor and is not obviously separable.

Side-information has been added to the original data. Some groups in each class have been randomly built, given a number of groups per class. Four values have been used: 50, 25, 16, 10 and 6, corresponding respectively to $|s_n|$

approximatively equal to 1, 2, 3, 5, 8. The case with 50 groups corresponds to the case without side-information (only one observation par class).

For the clustering, a Gaussian mixture was used as the model. In order to reduce the number of parameters to be estimated we have considered that the random variables are independent. Thus, for a given number of clusters equal to $K$, the number of parameters $r$ is equal to $9K - 1$, corresponding to $4K$ for the mean vectors, $4K$ for the mean standard deviations, and $K - 1$ for the cluster probabilities. The number of clusters was selected using each of the three proposed criteria (BICm, AICm, and NECm). The experiment has been repeated 20 times for each value of $K$ for estimating the percent frequency of choosing $K$ clusters. The classification results for a realization with 10 groups are reported on Fig. 4. The probability of correct classification in the case of 3 clusters has also been estimated for studying the influence of the number of groups on the classification rate. The estimated value of this probability and the mean of the selected number of clusters are given in Table 3. The results of the percent frequency of choosing a K-component mixture are given in Table 4.

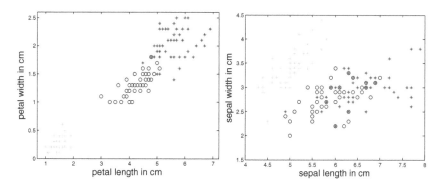

**Fig. 4.** Classification results of iris, in the case of 3 clusters and 10 groups in each cluster.

The results in Table 3 show that the probability of correct classification is quite good when the number of clusters is known. However the results in Table 4

**Table 3.** Mean of the selected number of clusters using the 3 criteria (BICm, AICm, and NECm) and estimated correct classification probability ($P_C$).

| Number of groups per class | AICm | BICm | NECm | $P_C$ |
|---|---|---|---|---|
| 50 | 6 | 3 | 2 | 90.67 |
| 25 | 5.6 | 3.55 | 2.05 | 98.2 |
| 16 | 4.75 | 3.5 | 2.05 | 98.97 |
| 10 | 4.15 | 3.2 | 2 | 100 |
| 6 | 3.3 | 3.15 | 2.05 | 100 |

**Table 4.** Percent frequencies of choosing K clusters for different numbers of groups.

|            | K | BICm | AICm | NECm |
|------------|---|------|------|------|
| N = 25 * 3 | 2 | 0    | 0    | 95   |
|            | 3 | 65   | 0    | 5    |
|            | 4 | 20   | 5    | 0    |
|            | 5 | 10   | 30   | 0    |
|            | 6 | 5    | 65   | 0    |
| N = 16 * 3 | 2 | 0    | 0    | 95   |
|            | 3 | 65   | 20   | 5    |
|            | 4 | 25   | 20   | 0    |
|            | 5 | 5    | 25   | 0    |
|            | 6 | 5    | 35   | 0    |
| N = 10 * 3 | 2 | 0    | 0    | 100  |
|            | 3 | 80   | 25   | 0    |
|            | 4 | 20   | 50   | 0    |
|            | 5 | 0    | 10   | 0    |
|            | 6 | 0    | 15   | 0    |
| N = 6 * 3  | 2 | 0    | 0    | 95   |
|            | 3 | 90   | 75   | 5    |
|            | 4 | 5    | 20   | 0    |
|            | 5 | 5    | 5    | 0    |
|            | 6 | 0    | 0    | 0    |

show that the determination of the right number of clusters is quite difficult. As previously, the best results are obtained with the criterion BICm whatever the value of $K$. The results with the criteria AICm and NECm are not good. Because the clusters Iris virginica and Iris versicolor and are not obviously separable, NECm gives almost always two clusters.

## 4.4   Climatic Data

Climatic data in France are available on the public website donneespubliques.meteofrance.fr. We have used the average of the daily maximum and minimum temperatures and the cumulative rainfall amounts for the months of January and July in 109 available towns. Then each measure for each town is a realization of a random variable of dimension 6. We have used the data for the years 2012 to 2015, and have assumed that the random variable for a given town has the same distribution within all years. Then we can consider that 4 realizations are available for each random variable. It also means that the number of points for each town (group) is equal to 4. We have used a Gaussian mixture model for the clustering algorithm and have considered that the random variables are indepen-

**Table 5.** Mixture model parameters for the climatic data in the case of 6 clusters.

| | Cluster number → | 1 | 2 | 3 | 4 | 5 | 6 |
|---|---|---|---|---|---|---|---|
| Mean | Temp min Jan | 2.76 | 2.39 | 3.74 | 4.69 | 4.89 | 0.18 |
| | Temp max Jan | 9.90 | 7.84 | 11.33 | 9.63 | 13.44 | 6.27 |
| | Rainfall Jan | 79.01 | 58.74 | 152.48 | 101.23 | 74.94 | 78.54 |
| | Temp min July | 16.57 | 14.38 | 16.25 | 14.35 | 19.66 | 14.41 |
| | Temp max July | 29.32 | 25.90 | 26.70 | 22.63 | 30.22 | 26.48 |
| | Rainfall July | 52.89 | 62.12 | 53.38 | 52.39 | 20.03 | 88.32 |
| Standard deviation | Temp min Jan | 1.38 | 1.40 | 2.13 | 1.54 | 1.75 | 1.64 |
| | Temp max Jan | 1.53 | 1.62 | 1.11 | 1.39 | 1.06 | 1.59 |
| | Rainfall Jan | 49.26 | 22.85 | 72.86 | 49.25 | 64.92 | 37.40 |
| | Temp min July | 1.71 | 1.37 | 1.66 | 1.42 | 1.58 | 1.67 |
| | Temp max July | 2.36 | 2.20 | 2.10 | 2.27 | 1.96 | 2.74 |
| | Rainfall July | 40.01 | 37.74 | 35.26 | 17.73 | 24.90 | 58.67 |
| Probability | | 0.14 | 0.32 | 0.08 | 0.15 | 0.10 | 0.21 |

dent in order to limit the number of parameters to be estimated. Thus, for a given number of clusters equal to $K$, the number of parameters $r$ is equal to $13K - 1$.

The selected number of clusters with the criterion BICm is 6, although the criterion values for $K$ equal to 5, 6 and 7 are very similar.

The values of the mixture model parameters are reported in Table 5. The class labels in respect to the longitude and latitude are reported on Fig. 5. We can see different towns with climates which are rather semi-continental (1), continental (2), maritime and warm (3), maritime (4), mediterranean (5), and mountainous (6). In comparison with the results in [8], it can be observed that the different clusters appear more homogeneous, thanks to a larger number of variables (6 instead of 4) and to a larger number of points in each group (4 instead of 3).

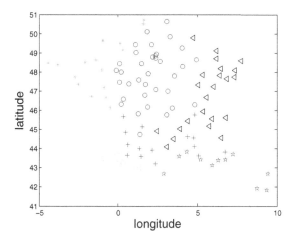

**Fig. 5.** Classification of climates in France.

## 5    Conclusion

This paper addresses the problem of assessing the number of clusters in a mixture model for data with the constraint that some points arise from the same source. To select the number of clusters, usual criteria are not suitable because they do not consider the side-information in the data. Thus we have proposed suitable criteria which are based on usual criteria.

The Bayesian information criterion (BIC), the Akaike information criterion (AIC) and the entropy criterion (NEC) have been modified, for taking into account the side-information. Instead of using the total number of points, the criteria use the number of groups, and instead of using the log-likelihood of the points, they use the log-likelihood of the groups.

The relevance of the modified versions of the criteria in respect to the original versions has been shown through a simulated problem. The performance of the proposed criteria in relation with the number of points per group has been studied using simulated data and using the Iris data set. According to the obtained results, we conclude that BICm is the best performing criterion. NECm is a criterion which has a tendency to underestimate the number of clusters and AICm tends to overestimate the cluster number. These results are similar with former results on the original criteria, in the case without side-information [7]. The side-information is relevant for overlapped clusters. On the contrary NEC is suitable for well separated components. Thus the modified criterion NECm was not efficient for the experimental data. Finally the EM algorithm and the modified criterion BICm were used for the classification of climatic data.

The proposed approach for the clustering with side-information allows to deal with a variable number of points within each group. For example, in the Iris problem the number of groups in each cluster could be different, in the climatic problem the number of years for each town could be different.

This clustering approach can be used to only cluster available data. It can also be used as a learning stage, for afterwards classifying new groups. A new group could be classified according to the log-likelihood, and thus associated to a sub-population described by the parameters of the corresponding mixture component. Thus it is important to estimate correctly both the number of clusters and the parameters of the component model. Then it would be necessary to define a criterion that takes into account these two items to assess an estimated mixture model.

## References

1. Akaike, H.: A new look at the statistical model identification. IEEE Trans. Autom. Control **19**(6), 716–723 (1974)
2. Biernacki, C., Celeux, G., Govaert, G.: An improvement of the nec criterion for assessing the number of clusters in a mixture model. Pattern Recogn. Lett. **20**(3), 267–272 (1999)
3. Celeux, G., Govaert, G.: A classification EM algorithm for clustering and two stochastic versions. Comput. Stat. Data Anal. **14**(3), 315–332 (1992)

4. Celeux, G., Govaert, G.: Gaussian parcimonious clustering models. Pattern Recogn. **28**, 781–793 (1995)
5. Celeux, G., Soromenho, G.: An entropy criterion for assessing the number of clusters in a mixture model. J. Classif. **13**(2), 195–212 (1996)
6. Duda, R.O., Hart, P.E.: Pattern Classification and Scene Analysis. Wiley, Hoboken (1973)
7. Fonseca, J.R., Cardoso, M.G.: Mixture-model cluster analysis using information theoretical criteria. Intell. Data Anal. **11**(1), 155–173 (2007)
8. Grall-Maës, E., Dao, D.: Assessing the number of clusters in a mixture model with side-information. In: Proceedings of 5th International Conference on Pattern Recognition Applications and Methods (ICPRAM 2016), Rome, Italy, pp. 41–47, 24–26 February 2016
9. Grall-Maës, E.: Spatial stochastic process clustering using a local a posteriori probability. In: Proceedings of IEEE International Workshop on Machine Learning for Signal Processing (MLSP 2014), Reims, France, 21–24 September 2014
10. Lebarbier, E., Mary-Huard, T.: Le critère BIC: fondements théoriques et interprétation. Research report, INRIA (2006)
11. McLachlan, G., Peel, D.: Finite Mixture Models. Wiley, Hoboken (2000)
12. McLachlan, G., Basford, K.: Mixture models. Inference and applications to clustering. In: Statistics: Textbooks and Monographs, vol. 1. Dekker, New York (1988)
13. Schwarz, G.: Estimating the dimension of a model. Ann. Stat. **6**, 461–464 (1978)
14. Shental, N., Bar-Hillel, A., Hertz, T., Weinshall, D.: Computing Gaussian mixture models with EM using side-information. In: Proceedings of 20th International Conference on Machine Learning. Citeseer (2003)

# Near-Boolean Optimization: A Continuous Approach to Set Packing and Partitioning

Giovanni Rossi[(✉)]

Department of Computer Science and Engineering - DISI,
University of Bologna, via Mura Anteo Zamboni 7, 40126 Bologna, Italy
giovanni.rossi6@unibo.it

**Abstract.** Near-Boolean functions essentially associate with every partition of a finite ground set the sum of the real values taken on blocks by an underlying set function. Given a family of feasible subsets of the ground set, the packing problem is to find a largest subfamily of pairwise disjoint family members. Through a multilinear polynomial whose variables are indexed by the elements of the ground set and correspond each to a unit membership distribution over those feasible subsets where their index is included, the problem is translated into a continuous version with the objective function taking values on peculiar collections of points in a unit hypercube. Extremizers are shown to include feasible solutions for the original combinatorial optimization problem, allowing for a gradient-based local search. Least-squares approximations with bounded degree and coalition formation games are also finally discussed.

**Keywords:** Pseudo-Boolean function · Möbius inversion · Set packing · Partition function · Polynomial multilinear extension · Local search

## 1 Introduction

Boolean functions and pseudo-Boolean optimization methods are key analytical tools in a variety of theoretical and applicative scenarios [1–3]. Pseudo-Boolean functions are in fact set functions, taking real values over the subsets of a finite set, and their polynomial multilinear extension, or MLE for short, allows to deal with several discrete optimization problems in a continuous setting. This paper[1] proposes a novel approach enabling to include among such problems those where the objective function takes real values over collections of pairwise disjoint subsets. In particular, these collections are evaluated by summing the values taken on their elements by a set function, and the corresponding combinatorial optimization problems are set packing and set partitioning.

---

[1] This is an extended version of a work [4] presented at the 5th Int. Conf. on Pattern Recognition Applications and Methods - ICPRAM 2016. I am grateful to the editors for their kind suggestions that significantly improved the original manuscript. Responsibility for any remaining error rests with the author.

A. Fred et al. (Eds.): ICPRAM 2016, LNCS 10163, pp. 60–87, 2017.
DOI: 10.1007/978-3-319-53375-9_4

For a ground set $N = \{1, \ldots, n\}$ and a family $\mathcal{F} \subseteq 2^N = \{A : A \subseteq N\}$ of feasible subsets, the standard packing problem is to find a largest subfamily $\mathcal{F}^* \subseteq \mathcal{F}$ whose members are pairwise disjoint, i.e. $A \cap B = \emptyset$ for all $A, B \in \mathcal{F}^*$. In addition, a set function $w : \mathcal{F} \to \mathbb{R}_+$ may assign weights to feasible subsets, in which case optimality attains at those such subfamilies $\mathcal{F}^*$ with maximum weight $W(\mathcal{F}^*) = \sum_{A \in \mathcal{F}^*} w(A)$. If $w(A) = 1$ for all $A \in \mathcal{F}$, then of course largest subfamilies have maximum weight. Accordingly, the proposed approach relies on using the MLE of set function $w$ in order to evaluate families of fuzzy feasible subsets. In this way, the search for locally optimal subfamilies $\mathcal{F}^*$ can be undertaken in a continuous domain.

In combinatorial optimization [5], set packing is very important from both theoretical and applicative perspectives. The problem considered in computational complexity is to approximate optimal solutions within a provable bounded factor by means of efficient algorithms. In particular, the concern is mostly with non-approximability results for $k$-set packing [6], where every feasible subset has size no greater than some $k \ll n$ (and unit weight as above). Also, if feasible subsets have all size $k = 2$, then set packing reduces to maximal matching for a graph with $N$ as vertex set, in which case there is an algorithm with polynomial running time whose output is an exact solution [7]. More generally, $k$-set packing with $k > 2$ is equivalent to vertex colouring in hypergraphs [8], and in the $k$-uniform and $d$-regular case every feasible subset $A \in \mathcal{F}$ has size $|A| = k$, while elements $i \in N$ of the ground set are each contained in $d > 1$ feasible subsets, i.e. $|\{A : i \in A \in \mathcal{F}\}| = d$.

As for applications, combinatorial auctions provide a main example: elements $i \in N$ are items to be sold with the objective to maximize the revenue, accepting bids on any subset or bundle (or combination) $A \in 2^N$. If the maximum received bid on each bundle is regarded as a weight, then optimization may be dealt with as a maximum-weight set packing problem [9]. Since the search space is exponentially large, time constraints often lead to use heuristics (rather than approximate algorithms, i.e. without worst-case guarantees) or simultaneous ascending auctions, where each item is sold independently and bids may be updated over a predetermined time period [10].

Throughout this work, set packing is approached by expanding traditional pseudo-Boolean optimization methods [1] into a novel near-Boolean one. Specifically, while the MLE of set functions is commonly employed to enlarge the domain of each of the $n$ variables from $\{0, 1\}$ to $[0, 1]$, the proposed technique relies on $n$ variables ranging each in the $2^{n-1}$-set of extreme points of a unit simplex, and uses the MLE to include the continuum provided by the whole simplex. As usual, elements of the ground set or integers $i \in N$ are the indices of the $n$ variables. On the other hand, the $n$ involved $2^{n-1} - 1$-dimensional unit simplices $\Delta_i, i \in N$ have their $2^{n-1}$ extreme points indexed by those (feasible) subsets where each $i \in N$ is included. The resulting near-Boolean function takes values on the $n$-fold product $\times_{i \in N} ex(\Delta_i)$, where $ex(\Delta_i)$ is the $2^{n-1}$-set of extreme points of $\Delta_i$, while its MLE evaluates collections of fuzzy subsets of $N$ or, equivalently, fuzzy subfamilies of (feasible) subsets. Such a MLE is precisely the objective function to be maximized through a gradient-based local search.

It can be mentioned that maximum-weight set packing may be tackled via constrained maximization of pseudo-Boolean function $v : \{0,1\}^{|\mathcal{F}|} \to \mathbb{R}_+$ with $v\left(x_{A_1}, \ldots, x_{A_{|\mathcal{F}|}}\right) = \sum_{1 \leq k \leq |\mathcal{F}|} x_{A_k} w(A_k)$ s.t. $A_k \cap A_l \neq \emptyset \Rightarrow x_{A_k} + x_{A_l} \leq 1$ for all $1 \leq k < l \leq |\mathcal{F}|$, where $x_A \in \{0,1\}$ for all $A \in \mathcal{F} = \{A_1, \ldots A_{|\mathcal{F}|}\}$. In fact, a suitable $|\mathcal{F}| \times |\mathcal{F}|$-matrix $M$ allows to replace $v$ with $xMx \simeq v$, where $x = (x_{A_1}, \ldots, x_{A_{|\mathcal{F}|}})$. Then, a constrained maximizer $x^*$ can be found by means of an heuristic, the corresponding solution being $\mathcal{F}^* = \{A : x_A^* = 1\}$. Such an approach [11] is very different from what is proposed here, in that $v$ is an objective function of $|\mathcal{F}|$ constrained Boolean variables, while the expanded MLE described above is a function of $n$ unconstrained near-Boolean variables.

The following section comprehensively frames the case where $\mathcal{F} = 2^N$, applying to set partitioning. Section 3 shows next that by introducing the empty set $\emptyset$ and all singletons $\{i\} \in 2^N, i \in N$ into the family $\mathcal{F} \subset 2^N$ of feasible subsets (with null weights $w(\emptyset) = 0 = w(\{i\})$ if $\{i\} \notin \mathcal{F}$) the proposed method also applies to set packing. The gradient-based local search for set packing is detailed in Sect. 4, where a cost function $c : \mathcal{F}^t \to \mathbb{N}$ ($\mathcal{F}^t \subseteq \mathcal{F}$) also enters the picture, in line with greedy approaches [12]. At any iteration $t = 0, 1, \ldots$, the cost of including a still available feasible subset in the packing is the number of still available feasible subsets that the former intersects. Section 5 details near-Boolean functions, showing that there is a continuum of equivalent polynomials for MLE, with common degree and varying coefficients, and introduces the MLE of partition functions. Section 6 provides a novel modeling for coalition formation games, with $N$ as player set. Finally, Sect. 7 contains the conclusions.

## 2   Set Partitioning: Full-Dimensional Case

The $2^n$-set $\{0,1\}^n$ of vertices of the $n$-dimensional unit hypercube $[0,1]^n$ corresponds bijectively to power set $2^N$, as characteristic functions $\chi_A : N \to \{0,1\}$, $A \in 2^N$ are defined by $\chi_A(i) = 1$ if $i \in A$ and $\chi_A(i) = 0$ if $i \in N \backslash A = A^c$. Also, zeta function $\zeta : 2^N \times 2^N \to \mathbb{R}$ is the element of the incidence algebra [13,14] of Boolean lattice $(2^N, \cap, \cup)$ defined by $\zeta(A, B) = 1$ if $B \supseteq A$ and $\zeta(A, B) = 0$ if $B \not\supseteq A$. The collection $\{\zeta(A, \cdot) : A \in 2^N\}$ is a linear basis of the (free) vector space [13, p. 181] of real-valued functions $w$ on $2^N$, meaning that linear combination $w(B) = \sum_{A \in 2^N} \mu^w(A) \zeta(A, B) = \sum_{A \subseteq B} \mu^w(A)$ for all $B \in 2^N$ applies to any $w$, with Möbius inversion $\mu^w : 2^N \to \mathbb{R}$ given by

$$\mu^w(A) = \sum_{B \in 2^N} \mu(B, A) w(B) = \sum_{B \subseteq A} (-1)^{|A \backslash B|} w(B) = w(A) - \sum_{B \subset A} \mu^w(B),$$

where $\subset$ denotes strict inclusion while $\mu : 2^N \times 2^N \to \mathbb{R}$ is the Möbius function, i.e. the inverse of $\zeta$ in the incidence algebra of $(2^N, \cap, \cup)$, defined as follows:

$$\mu(B, A) = \begin{cases} (-1)^{|A \backslash B|} & \text{if } B \subseteq A, \\ 0 & \text{if } B \not\subseteq A, \end{cases} \text{ for all } A, B \in 2^N.$$

By means of this essential combinatorial *"analog of the fundamental theorem of the calculus"* [14], the MLE $f^w : [0,1]^n \to \mathbb{R}$ of $w$ takes values

$$f^w(\chi_B) = \sum_{A \in 2^N} \left( \prod_{i \in A} \chi_B(i) \right) \mu^w(A) = \sum_{A \subseteq B} \mu^w(A) = w(B) \text{ on vertices, and}$$

$$f^w(x) = \sum_{A \in 2^N} \left( \prod_{i \in A} x_i \right) \mu^w(A) \tag{1}$$

on any point $x = (x_1, \ldots, x_n) \in [0,1]^n$. Conventionally, $\prod_{i \in \emptyset} x_i := 1$ [1, p. 157].

Let $2_i^N = \{A : i \in A \in 2^N\} = \{A_1, \ldots, A_{2^{n-1}}\}$ be the $2^{n-1}$-set of subsets containing each $i \in N$. Unit simplex

$$\Delta_i = \left\{ \left( q_i^{A_1}, \ldots, q_i^{A_{2^{n-1}}} \right) : q_i^{A_k} \geq 0 \text{ for } 1 \leq k \leq 2^{n-1}, \sum_{1 \leq k \leq 2^{n-1}} q_i^{A_k} = 1 \right\}$$

has dimension $2^{n-1} - 1$ and generic point $q_i \in \Delta_i$.

**Definition 1.** *A fuzzy cover $q$ specifies, for each $i \in N$, a membership distribution over $2_i^N$, i.e.* $q = (q_1, \ldots, q_n) \in \Delta_N = \underset{1 \leq i \leq n}{\times} \Delta_i$.

Equivalently, $\mathbf{q} = \{q^A : \emptyset \neq A \in 2^N, q^A \in [0,1]^n\}$ is a $2^n - 1$-set whose elements $q^A = (q_1^A, \ldots, q_n^A)$ are $n$-vectors corresponding to non-empty subsets $A \in 2^N$ and specifying a membership $q_i^A$ for each $i \in N$, with $q_i^A \in [0,1]$ if $i \in A$ while $q_i^A = 0$ if $i \in A^c$. Since fuzzy covers are collections of points in $[0,1]^n$ and the MLE $f^w$ of $w$ is meant precisely to evaluate such points, the global worth $W(\mathbf{q})$ of $\mathbf{q} \in \Delta_N$ is the sum over all $q^A, A \in 2^N$ of $f^w(q^A)$ as defined by (1). That is,

$$W(\mathbf{q}) = \sum_{A \in 2^N} f^w(q^A) = \sum_{A \in 2^N} \left[ \sum_{B \subseteq A} \left( \prod_{i \in B} q_i^A \right) \mu^w(B) \right],$$

or equivalently $W(\mathbf{q}) = \sum_{A \in 2^N} \left[ \sum_{B \supseteq A} \left( \prod_{i \in A} q_i^B \right) \right] \mu^w(A). \tag{2}$

**Example:** for $N = \{1,2,3\}$, define $w$ by $w(\{1\}) = w(\{2\}) = w(\{3\}) = 0.2$, $w(\{1,2\}) = 0.8$, $w(\{1,3\}) = 0.3$, $w(\{2,3\}) = 0.6$, $w(N) = 0.7$, and $w(\emptyset) = 0$. Membership distributions over $2_i^N, i = 1,2,3$ are $q_1 \in \Delta_1, q_2 \in \Delta_2, q_3 \in \Delta_3$, with

$$q_1 = \begin{pmatrix} q_1^1 \\ q_1^{12} \\ q_1^{13} \\ q_1^N \end{pmatrix}, q_2 = \begin{pmatrix} q_2^2 \\ q_2^{12} \\ q_2^{23} \\ q_2^N \end{pmatrix}, q_3 = \begin{pmatrix} q_3^3 \\ q_3^{13} \\ q_3^{23} \\ q_3^N \end{pmatrix}.$$

If $\hat{q}_1^{12} = \hat{q}_2^{12} = 1$, then any membership $q_3 \in \Delta_3$ yields $W(\hat{q}_1, \hat{q}_2, q_3)$

$$= w(\{1,2\}) + \left(q_3^3 + q_3^{13} + q_3^{23} + q_3^N\right)\mu^w(\{3\}) = w(\{1,2\}) + w(\{3\}) = 1.$$

Therefore, there is a continuum of fuzzy covers achieving maximum worth, i.e. 1. In order to select the one $\hat{q} = (\hat{q}_1, \hat{q}_2, \hat{q}_3)$ where $\hat{q}_3^3 = 1$, attention must be placed only on *exact* ones, defined hereafter.

For any fuzzy covers $\mathbf{q} = \{q^A : \emptyset \neq A \in 2^N\}$ and $\hat{\mathbf{q}} = \{\hat{q}^A : \emptyset \neq A \in 2^N\}$, define $\hat{\mathbf{q}}$ to be a *shrinking* of $\mathbf{q}$ if there is a subset $A$ such that $\sum_{i \in A} q_i^A > 0$ and

$$\hat{q}_i^B = \begin{cases} q_i^B & \text{if } B \nsubseteq A \\ 0 & \text{if } B = A \end{cases} \quad \text{for all } B \in 2^N, i \in N,$$

$$\sum_{B \subset A} \hat{q}_i^B = q_i^A + \sum_{B \subset A} q_i^B \text{ for all } i \in A.$$

In words, a shrinking reallocates the whole membership mass $\sum_{i \in A} q_i^A > 0$ from $A \in 2^N$ to all proper subsets $B \subset A$, involving all and only those elements $i \in A$ with strictly positive membership $q_i^A > 0$.

**Definition 2.** *Fuzzy cover $\mathbf{q} \in \Delta_N$ is exact if there is no shrinking $\hat{\mathbf{q}}$ of $\mathbf{q}$ such that $W(\mathbf{q}) = W(\hat{\mathbf{q}})$ for all set functions $w$.*

**Proposition 1.** *If $\mathbf{q}$ is exact, then $\left|\{i : q_i^A > 0\}\right| \in \{0, |A|\}$ for all $A \in 2^N$.*

*Proof.* For $\emptyset \subset A^+(\mathbf{q}) = \{i : q_i^A > 0\} \subset A \in 2^N$, let $\alpha = |A^+(\mathbf{q})| > 1$ and note that $f^w(q^A) = \sum_{B \subseteq A^+(\mathbf{q})} \left(\prod_{i \in B} q_i^A\right)\mu^w(B)$. Let shrinking $\hat{\mathbf{q}}$, with $\hat{q}^{B'} = q^{B'}$ if $B' \notin 2^{A^+(\mathbf{q})}$, satisfy conditions

$$\sum_{B \in 2_i^N \cap 2^{A^+(\mathbf{q})}} \hat{q}_i^B = q_i^A + \sum_{B \in 2_i^N \cap 2^{A^+(\mathbf{q})}} q_i^B \text{ for all } i \in A^+(\mathbf{q}), B \neq \emptyset, \text{ and}$$

$$\prod_{i \in B} \hat{q}_i^B = \prod_{i \in B} q_i^B + \prod_{i \in B} q_i^A \text{ for all } B \in 2^{A^+(\mathbf{q})}, |B| > 1.$$

These $2^\alpha - 1$ equations with $\sum_{1 \leq k \leq \alpha} k\binom{\alpha}{k} > 2^\alpha$ variables $\hat{q}_i^B, \emptyset \neq B \subseteq A^+(\mathbf{q})$ admit a continuum of solutions, each providing precisely a shrinking $\hat{\mathbf{q}}$ where

$$\sum_{B \in 2^{A^+(\mathbf{q})}} f^w(\hat{q}^B) = f^w(q^A) + \sum_{B \in 2^{A^+(\mathbf{q})}} f^w(q^B) \Rightarrow W(\mathbf{q}) = W(\hat{\mathbf{q}}).$$

This entails that $\mathbf{q}$ is not exact. □

Partitions of $N$ are families $P = \{A_1, \ldots, A_{|P|}\} \subset 2^N$ of (non-empty) pairwise disjoint subsets called blocks [13], i.e. $A_k \cap A_l = \emptyset, 1 \leq k < l \leq |P|$, whose union satisfies $A_1 \cup \cdots \cup A_{|P|} = N$. Any $P$ corresponds to the collection $\{\chi_A : A \in P\}$ of those $|P|$ hypercube vertices identified by the characteristic functions of its blocks. Partitions $P$ can thus be regarded as $\mathbf{p} \in \Delta_N$ where $p_i^A = 1$ for all $A \in P, i \in A$, i.e. exact fuzzy covers where each $i \in N$ concentrates its whole membershisp on the unique $A \in 2_i^N$ such that $A \in P$.

**Definition 3.** *Fuzzy partitions are exact fuzzy covers.*

Objective function $W : \Delta_N \to \mathbb{R}$ given by expression (2) above includes among its extremizers (non-fuzzy) partitions. This expands a result from pseudo-Boolean optimization [1]. For all $\mathbf{q} \in \Delta_N, i \in N$, let $\mathbf{q} = q_i | \mathbf{q}_{-i}$, with $q_i \in \Delta_i$ and $\mathbf{q}_{-i} \in \Delta_{N\setminus i} = \times_{j \in N\setminus i} \Delta_j$. Then, for any $w$,

$$
W(\mathbf{q}) = \sum_{A \in 2_i^N} f^w(q^A) + \sum_{A' \in 2^N \setminus 2_i^N} f^w(q^{A'})
$$

$$
= \sum_{A \in 2_i^N} \sum_{B \subseteq A \setminus i} \left( \prod_{j \in B} q_j^A \right) \left( q_i^A \mu^w(B \cup i) + \mu^w(B) \right)
$$

$$
+ \sum_{A' \in 2^N \setminus 2_i^N} \sum_{B' \subseteq A'} \left( \prod_{j' \in B'} q_{j'}^{A'} \right) \mu^w(B').
$$

Now define

$$
W_i(q_i | \mathbf{q}_{-i}) = \sum_{A \in 2_i^N} q_i^A \left[ \sum_{B \subseteq A \setminus i} \left( \prod_{j \in B} q_j^A \right) \mu^w(B \cup i) \right] \text{ and}
$$

$$
W_{-i}(\mathbf{q}_{-i}) = \sum_{A \in 2_i^N} \left[ \sum_{B \subseteq A \setminus i} \left( \prod_{j \in B} q_j^A \right) \mu^w(B) \right]
$$

$$
+ \sum_{A' \in 2^N \setminus 2_i^N} \left[ \sum_{B' \subseteq A'} \left( \prod_{j' \in B'} q_{j'}^{A'} \right) \mu^w(B') \right], \text{ yielding}
$$

$$
W(\mathbf{q}) = W_i(q_i | \mathbf{q}_{-i}) + W_{-i}(\mathbf{q}_{-i}). \tag{3}
$$

**Proposition 2.** *For all $\mathbf{q} \in \Delta_N$, there are $\underline{\mathbf{q}}, \overline{\mathbf{q}} \in \Delta_N$ such that:*

*(i) $W(\underline{\mathbf{q}}) \leq W(\mathbf{q}) \leq W(\overline{\mathbf{q}})$, as well as*
*(ii) $\underline{q}^i, \overline{q}^i \in ex(\Delta_i)$ for all $i \in N$.*

*Proof.* For $i \in N, \mathbf{q}_{-i} \in \Delta_{N\setminus i}$, define $w_{\mathbf{q}_{-i}} : 2_i^N \to \mathbb{R}$ by

$$
w_{\mathbf{q}_{-i}}(A) = \sum_{B \subseteq A \setminus i} \left( \prod_{j \in B} q_j^A \right) \mu^w(B \cup i). \tag{4}
$$

Let $\mathbb{A}_{\mathbf{q}_{-i}}^+ = \arg\max w_{\mathbf{q}_{-i}}$ and $\mathbb{A}_{\mathbf{q}_{-i}}^- = \arg\min w_{\mathbf{q}_{-i}}$, with $\mathbb{A}_{\mathbf{q}_{-i}}^+ \neq \emptyset \neq \mathbb{A}_{\mathbf{q}_{-i}}^-$ at all $\mathbf{q}_{-i}$. Most importantly,

$$
W_i(q_i | \mathbf{q}_i) = \sum_{A \in 2_i^N} \left( q_i^A \cdot w_{\mathbf{q}_{-i}}(A) \right) = \langle q_i, w_{\mathbf{q}_{-i}} \rangle, \tag{5}
$$

where $\langle \cdot, \cdot \rangle$ denotes scalar product. Thus for given membership distributions of all $j \in N \setminus i$, global worth is affected by $i$'s membership distribution through a scalar product. In order to maximize (or minimize) $W$ by suitably choosing $q_i$ for given $\mathbf{q}_{-i}$, the whole of $i$'s membership mass must be placed over $\mathbb{A}^+_{\mathbf{q}_{-i}}$ (or $\mathbb{A}^-_{\mathbf{q}_{-i}}$), anyhow. Hence there are precisely $|\mathbb{A}^+_{\mathbf{q}_{-i}}| > 0$ (or $|\mathbb{A}^-_{\mathbf{q}_{-i}}| > 0$) available extreme points of $\Delta_i$. The following procedure selects (arbitrarily) one of them.

ROUNDUP$(w, \mathbf{q})$

*Initialize:* Set $t = 0$ and $\mathbf{q}(0) = \mathbf{q}$.

*Loop:* While there is some $i \in N$ with $q_i(t) \notin ex(\Delta_i)$, set $t = t + 1$ and:

(a) select some $A^* \in \mathbb{A}^+_{\mathbf{q}_{-i}(t)}$,

(b) define, for all $j \in N, A \in 2^N$,

$$q_j^A(t) = \begin{cases} q_j^A(t-1) & \text{if } j \neq i \\ 1 & \text{if } j = i \text{ and } A = A^* \\ 0 & \text{otherwise} \end{cases}.$$

*Output:* Set $\overline{\mathbf{q}} = \mathbf{q}(t)$.

Every change $q_i^A(t-1) \neq q_i^A(t) = 1$ (for any $i \in N, A \in 2_i^N$) induces a non-decreasing variation $W(\mathbf{q}(t)) - W(\mathbf{q}(t-1)) \geq 0$. Hence, the sought $\overline{\mathbf{q}}$ is provided in at most $n$ iterations. Analogously, replacing $\mathbb{A}^+_{\mathbf{q}_{-i}}$ with $\mathbb{A}^-_{\mathbf{q}_{-i}}$ yields the sought minimizer $\underline{\mathbf{q}}$ (see also [1, p. 163]).     $\square$

*Remark 1.* For $i \in N, A \in 2_i^N$, if all $j \in A \setminus i \neq \emptyset$ satisfy $q_j^A = 1$, then (4) yields $w_{\mathbf{q}_{-i}}(A) = w(A) - w(A \setminus i)$, while $w_{\mathbf{q}_{-i}}(\{i\}) = w(\{i\})$ regardless of $\mathbf{q}_{-i}$.

**Corollary 1.** *Some partition $P$ satisfies $W(\mathbf{p}) \geq W(\mathbf{q})$ for all $\mathbf{q} \in \Delta_N$, with $W(\mathbf{p}) = \sum_{A \in P} w(A)$.*

*Proof.* Follows from Propositions 1 and 2, with the above notation associating $\mathbf{p} \in \Delta_N$ to partition $P$.     $\square$

Defining global maximizers is clearly immediate.

**Definition 4.** *Fuzzy partition $\hat{\mathbf{q}} \in \Delta_N$ is a global maximizer if $W(\hat{\mathbf{q}}) \geq W(\mathbf{q})$ for all $\mathbf{q} \in \Delta_N$.*

As for local maximizers, consider a vector $\omega = (\omega_1, \ldots, \omega_n) \in \mathbb{R}^n_{++}$ of strictly positive weights, with $\omega_N = \sum_{j \in N} \omega_j$, and focus on the (Nash) equilibrium [15] of the game with elements $i \in N$ as players, each strategically choosing its membership distribution $q_i \in \Delta_i$ while being rewarded with fraction $\frac{\omega_i}{\omega_N} W(q_1, \ldots, q_n)$ of the global worth attained at any strategy profile $(q_1, \ldots, q_n) = \mathbf{q} \in \Delta_N$.

**Definition 5.** *Fuzzy partition $\hat{\mathbf{q}} \in \Delta_N$ is a local maximizer if for all $q_i \in \Delta_i$ and all $i \in N$ inequality $W_i(\hat{q}_i | \hat{\mathbf{q}}_{-i}) \geq W_i(q_i | \hat{\mathbf{q}}_{-i})$ holds (see expression (3)).*

This definition of local maximizer entails that the *neighborhood* $\mathcal{N}(\mathbf{q}) \subset \Delta_N$ of any $\mathbf{q} \in \Delta_N$ is

$$\mathcal{N}(\mathbf{q}) = \bigcup_{i \in N} \left\{ \tilde{\mathbf{q}} : \tilde{\mathbf{q}} = \tilde{q}_i | \mathbf{q}_{-i}, \tilde{q}_i \in \Delta_i \right\}.$$

**Definition 6.** *The $(i, A)$-derivative of $W$ at $\mathbf{q} \in \Delta_N$ is*

$$\partial W(\mathbf{q})/\partial q_i^A = W(\overline{\mathbf{q}}(i, A)) - W(\underline{\mathbf{q}}(i, A))$$
$$= W_i \left( \overline{q}_i(i, A) | \overline{\mathbf{q}}_{-i}(i, A) \right) - W_i \left( \underline{q}_i(i, A) | \underline{\mathbf{q}}_{-i}(i, A) \right),$$

*with $\overline{\mathbf{q}}(i, A) = \left( \overline{q}_1(i, A), \ldots, \overline{q}_n(i, A) \right)$ given by*

$$\overline{q}_j^B(i, A) = \begin{cases} q_j^B \text{ for all } j \in N \backslash i, B \in 2_j^N \\ 1 \text{ for } j = i, B = A \\ 0 \text{ for } j = i, B \neq A \end{cases},$$

*and $\underline{\mathbf{q}}(i, A) = \left( \underline{q}_1(i, A), \ldots, \underline{q}_n(i, A) \right)$ given by*

$$\underline{q}_j^B(i, A) = \begin{cases} q_j^B \text{ for all } j \in N \backslash i, B \in 2_j^N \\ 0 \text{ for } j = i \text{ and all } B \in 2_i^N \end{cases},$$

*thus $\nabla W(\mathbf{q}) = \{\partial W(\mathbf{q})/\partial q_i^A : i \in N, A \in 2_i^N\} \in \mathbb{R}^{n2^{n-1}}$ is the (full) gradient of $W$ at $\mathbf{q}$. The $i$-gradient $\nabla_i W(\mathbf{q}) \in \mathbb{R}^{2^{n-1}}$ of $W$ at $\mathbf{q} = q_i | \mathbf{q}_{-i}$ is set function $\nabla_i W(\mathbf{q}) : 2_i^N \to \mathbb{R}$ defined by $\nabla_i W(\mathbf{q})(A) = w_{\mathbf{q}_{-i}}(A)$ for all $A \in 2_i^N$, where $w_{\mathbf{q}_{-i}}$ is given by expression (4).*

*Remark 2.* Membership distribution $\underline{q}_i(i, A)$ is the null one: its $2^{n-1}$ entries are all 0, hence $\underline{q}_i(i, A) \notin \Delta_i$.

It is now possible to search for a local maximizer partition $\mathbf{p}^*$ from given fuzzy partition $\mathbf{q}$ as initial candidate solution, and while maintaining the whole search within the continuum of fuzzy partitions. This idea may be specified in alternative ways yielding different local search methods. One possibility is the following.

LOCALSEARCH$(w, \mathbf{q})$

*Initialize:* Set $t = 0$ and $\mathbf{q}(0) = \mathbf{q}$, with requirement $|\{i : q_i^A > 0\}| \in \{0, |A|\}$ for all $A \in 2^N$ (i.e. $\mathbf{q}$ is exact).

*Loop 1:* While $0 < \sum_{i \in A} q_i^A(t) < |A|$ for some $A \in 2^N$, set $t = t + 1$ and

(a) select a $A^*(t) \in 2^N$ satisfying

$$\frac{1}{|A^*(t)|} \sum_{i \in A^*(t)} w_{\mathbf{q}_{-i}(t-1)}(A^*(t)) \geq \frac{1}{|B|} \sum_{j \in B} w_{\mathbf{q}_{-j}(t-1)}(B)$$

for all $B \in 2^N$ such that $0 < \sum_{i \in B} q_i^B(t) < |B|$,

(b) for $i \in A^*(t)$ and $A \in 2_i^N$, define $q_i^A(t) = \begin{cases} 1 \text{ if } A = A^*(t), \\ 0 \text{ if } A \neq A^*(t), \end{cases}$

(c) for $j \in N \backslash A^*(t)$ and $A \in 2_j^N$ with $A \cap A^*(t) = \emptyset$, define

$$q_j^A(t) = q_j^A(t-1) + \left( w(A) \sum_{\substack{B \in 2_j^N \\ B \cap A^*(t) \neq \emptyset}} q_j^B(t-1) \right) \left( \sum_{\substack{B' \in 2_j^N \\ B' \cap A^*(t) = \emptyset}} w(B') \right)^{-1}$$

(d) for $j \in N \backslash A^*(t)$ and $A \in 2_j^N$ with $A \cap A^*(t) \neq \emptyset$, define $q_j^A(t) = 0$.

*Loop 2:* While $q_i^A(t) = 1, |A| > 1$ for some $i \in N$ and $w(A) < w(\{i\}) + w(A \backslash i)$, set $t = t + 1$ and define:

$$q_i^{\hat{A}}(t) = \begin{cases} 1 \text{ if } |\hat{A}| = 1 \\ 0 \text{ otherwise} \end{cases} \text{ for all } \hat{A} \in 2_i^N,$$

$$q_j^B(t) = \begin{cases} 1 \text{ if } B = A \backslash i \\ 0 \text{ otherwise} \end{cases} \text{ for all } j \in A \backslash i, B \in 2_j^N,$$

$$q_{j'}^{\hat{B}}(t) = q_{j'}^{\hat{B}}(t-1) \text{ for all } j' \in A^c, \hat{B} \in 2_{j'}^N.$$

*Output:* Set $\mathbf{q}^* = \mathbf{q}(t)$.

Both ROUNDUP and LOCALSEARCH yield a sequence $\mathbf{q}(0), \ldots, \mathbf{q}(t^*) = \mathbf{q}^*$ where $q_i^* \in ex(\Delta_i)$ for all $i \in N$. In the former at the end of each iteration $t$ the novel $\mathbf{q}(t) \in \mathcal{N}(\mathbf{q}(t-1))$ is in the neighborhood of its predecessor. In the latter $\mathbf{q}(t) \notin \mathcal{N}(\mathbf{q}(t-1))$ in general, as in $|P| \leq n$ iterations of *Loop 1* a partition $\{A^*(1), \ldots, A^*(|P|)\} = P$ is generated. For $t = 1, \ldots, |P|$, selected subsets $A^*(t) \in 2^N$ are any of those where the average over members $i \in A^*(t)$ of $(i, A^*(t))$-derivatives $\partial W(\mathbf{q}(t-1)) / \partial q_i^{A^*(t)}(t-1)$ is maximal. Once a block $A^*(t)$ is selected, then lines (c) and (d) make all elements $j \in N \backslash A^*(t)$ redistribute the entire membership mass currently placed on subsets $A' \in 2_j^N$ with non-empty intersection $A' \cap A^*(t) \neq \emptyset$ over those remaining $A \in 2_j^N$ such that, conversely, $A \cap A^*(t) = \emptyset$. The redistribution is such that each of these latter gets a fraction $w(A) / \sum_{B \in 2_j^N : B \cap A^*(t) = \emptyset} w(B)$ of the newly freed membership mass $\sum_{A' \in 2_j^N : A' \cap A^*(t) \neq \emptyset} q_j^{A'}(t-1)$. The subsequent *Loop 2* checks whether the partition generated by *Loop 1* may be improved by extracting some elements from existing blocks and putting them in singleton blocks of the final output. In the limit, set function $w$ may be such that for some element $i \in N$ global worth decreases when the element joins any subset $A \in 2_i^N, |A| > 1$, that is to say $w(A) - w(A \backslash i) - w(\{i\}) = \sum_{B \in 2^A \backslash 2^{A \backslash i} : |B| > 1} \mu^w(B) < 0$.

**Proposition 3.** LOCALSEARCH$(W, \mathbf{q})$ *outputs a local maximizer* $\mathbf{q}^*$.

*Proof.* It is plain that the output is a partition $P$ or, with the notation of Corollary 1 above, $\mathbf{q}^* = \mathbf{p}$. Accordingly, any element $i \in N$ is either in a singleton block $\{i\} \in P$ or else in a block $A \in P, i \in A$ such that $|A| > 1$. In the former case, any membership reallocation deviating from $p_i^{\{i\}} = 1$, given memberships $p_j, j \in N \backslash i$, yields a cover (fuzzy or not) where global worth is the same as at $\mathbf{p}$, because $\prod_{j \in B \backslash i} p_j^B = 0$ for all $B \in 2_i^N \backslash \{i\}$ (see Example 1 above). In the latter case, any membership reallocation $q_i$ deviating from $p_i^A = 1$ (given memberhips $p_j, j \in N \backslash i$) yields a cover which is best seen by distinguishing between $2_i^N \backslash A$ and $A$. Also recall that $w(A) - w(A \backslash i) = \sum_{B \in 2^A \backslash 2^{A \backslash i}} \mu^w(B)$. Again, all membership mass $\sum_{B \in 2_i^N \backslash A} q_i^B > 0$ simply collapses on singleton $\{i\}$ because $\prod_{j \in B \backslash i} p_j^B = 0$ for all $B \in 2_i^N \backslash A$. Hence $W(\mathbf{p}) - W(q_i | \mathbf{p}_{-i}) = w(A) - w(\{i\}) +$

$$- \left( q_i^A \sum_{B \in 2^A \backslash 2^{A \backslash i} : |B| > 1} \mu^w(B) + \sum_{B' \in 2^{A \backslash i}} \mu^w(B') \right)$$

$$= (p_i^A - q_i^A) \sum_{B \in 2^A \backslash 2^{A \backslash i} : |B| > 1} \mu^w(B).$$

Now assume that $\mathbf{p}$ is *not* a local maximizer, i.e. $W(\mathbf{p}) - W(q_i | \mathbf{p}_{-i}) < 0$. Since $p_i^A - q_i^A > 0$ (because $p_i^A = 1$ and $q_i \in \Delta_i$ is a deviation from $p_i$), then

$$\sum_{B \in 2^A \backslash 2^{A \backslash i} : |B| > 1} \mu^w(B) = w(A) - w(A \backslash i) - w(\{i\}) < 0.$$

Hence $\mathbf{p}$ cannot be the output of *Second Loop*.    □

In local search methods, the chosen initial candidate solution determines what neighborhoods shall be visited. The range of the objective function in a neighborhood is a set of real values. In a neighborhood $\mathcal{N}(\mathbf{p})$ of a $\mathbf{p} \in \Delta_N$ or partition $P$ only those $\sum_{A \in P : |A| > 1} |A|$ elements $i \in A$ in non-singleton blocks $A \in P, |A| > 1$ can modify global worth by reallocating their membership. In view of (the proof of) Proposition 3, the only admissible variations obtain by deviating from $p_i^A = 1$ with an alternative membership distribution $q_i^A \in [0, 1)$, with $W(q_i | \mathbf{p}_{-i}) - W(\mathbf{p})$ equal to $(q_i^A - 1) \sum_{B \in 2^A \backslash 2^{A \backslash i}, |B| > 1} \mu^w(B) + (1 - q_i^A) w(\{i\})$. Hence, choosing partitions as initial candidate solutions of LOCALSEARCH is evidently poor. A sensible choice should conversely allow the search to explore different neighborhoods where the objective function may range widely. A simplest example of such an initial candidate solution is $q_i^A = 2^{1-n}$ for all $A \in 2_i^N$ and all $i \in N$, i.e. the uniform distribution. On the other hand, the input of local search algorithms is commonly desired to be close to a global optimum, i.e. a maximizer in the present setting. This translates here into the idea of defining the input by means of set function $w$. In this view, consider $q_i^A = w(A) / \sum_{B \in 2_i^N} w(B)$, yielding $\frac{q_i^A}{q_i^B} = \frac{w(A)}{w(B)} = \frac{q_j^A}{q_j^B}$ for all $A, B \in 2_i^N \cap 2_j^N$ and $i, j \in N$. Finally note that with a suitable initial candidate solution, the search may be restricted to explore only a maximum number of fuzzy partitions, thereby containing the computational burden. In fact, if $\mathbf{q}(0)$ is the finest partition $\{\{1\}, \ldots, \{n\}\}$ or $q_i^{\{i\}}(0) = 1$ for all $i \in N$, then the search does not explore any neighborhood at all, and such an input coincides with the output. More reasonably, let $\mathbb{A}_{\mathbf{q}}^{\max} = \{A_1, \ldots, A_k\}$ denote the collection of $\supseteq$-maximal subsets where input memberships are strictly positive. That is, $q_i^{A_{k'}} > 0$ for all $i \in A_{k'}, 1 \le k' \le k$ as well as $q_i^B = 0$ for all $B \in 2^N \setminus (2^{A_1} \cup \cdots \cup 2^{A_k})$ and all $i \in B$. Then, the output shall be a partition $P$ each of whose blocks $A \in P$ satisfies $A \subseteq A_{k'}$ for some $1 \le k' \le k$. Hence, by suitably choosing the input, LOCALSEARCH outputs a partition with no less than some minimum desired number of blocks.

**Example:** for $N = \{1, 2, 3\}$, let $w(\emptyset) = 0$, $w(\{1\}) = 1 = w(\{3\})$, $w(\{2\}) = 2$, $w(\{1,3\}) = 3$, $w(\{1,2\}) = 4 = w(\{2,3\})$, $w(N) = 2$, with Möbius inversion $\mu^w(\{i\}) = w(\{i\}), i \in N$, $\mu^w(\{1,3\}) = 1 = \mu^w(\{1,2\}) = \mu^w(\{2,3\})$ and $\mu^w(N) = -5$. With input $\mathbf{q}(0)$ given by the uniform distribution $q_i^A(0) = \frac{1}{4}$ for all $i \in N$ and all $A \in 2_i^N$, in the first iteration $t = 1$ the situation concerning $w_{\mathbf{q}_{-i}}(A), i \in N, A \in 2_i^N$ is

$$\hat{w}_{\mathbf{q}_{-1}}(\{1\}) = 1, \ w_{\mathbf{q}_{-1}}(\{1,2\}) = \frac{5}{4}, \ w_{\mathbf{q}_{-1}}(\{1,3\}) = \frac{5}{4} \text{ and } w_{\mathbf{q}_{-1}}(N) = \frac{19}{16},$$

$$\hat{w}_{\mathbf{q}_{-2}}(\{2\}) = 2, \ w_{\mathbf{q}_{-2}}(\{1,2\}) = \frac{9}{4}, \ w_{\mathbf{q}_{-2}}(\{2,3\}) = \frac{9}{4} \text{ and } w_{\mathbf{q}_{-2}}(N) = \frac{35}{16},$$

$$\hat{w}_{\mathbf{q}_{-3}}(\{3\}) = 1, \ w_{\mathbf{q}_{-3}}(\{1,3\}) = \frac{5}{4}, \ w_{\mathbf{q}_{-3}}(\{2,3\}) = \frac{5}{4} \text{ and } w_{\mathbf{q}_{-3}}(N) = \frac{19}{16}.$$

In view of line (a), the selected block at the first iteration is $A^*(1) = \{2\}$. According to lines (b)–(d), $q_2^2(1) = 1$, $q_1^1(1) = \frac{3}{8}$ and $q_1^{13}(1) = \frac{5}{8}$ as well as $q_3^3(1) = \frac{3}{8}$ and $q_3^{13}(1) = \frac{5}{8}$. In the second iteration $t = 2$,

$$w_{\mathbf{q}_{-1}}(\{1\}) = 1 \text{ and } w_{\mathbf{q}_{-1}}(\{1,3\}) = \frac{13}{8},$$

$$w_{\mathbf{q}_{-3}}(\{3\}) = 1 \text{ and } w_{\mathbf{q}_{-3}}(\{1,3\}) = \frac{13}{8},$$

and thus the selected block at the second iteration is $A^*(2) = \{1,3\}$. In fact, $\mathcal{F}^* = \{\{1,3\}, \{2\}\}$ is an optimal solution, and thus the second loop does not produce any change.

## 3    Set Packing: Lower-Dimensional Case

If $\mathcal{F} \subset 2^N$, then $2_i^N \cap \mathcal{F} \ne \emptyset$ for every $i \in N$, otherwise the problem reduces to packing the proper subset $N \setminus \{i : 2_i^N \cap \mathcal{F} = \emptyset\}$. As outlined in Sect. 1, without additional notation simply let $\{\emptyset\} \in \mathcal{F} \ni \{i\}$ for all $i \in N$ with null weights $w(\emptyset) = 0 = w(\{i\})$ if $\{i\}$ is not an element of the original family of feasible subsets. Thus $(\mathcal{F}, \supseteq)$ is a poset (partially ordered set) with bottom $\emptyset$, and weight function $w : \mathcal{F} \to \mathbb{R}_+$ has well-defined Möbius inversion $\mu^w : \mathcal{F} \to \mathbb{R}$ [14]. Now memberships $q_i, i \in N$ are distributions over $\mathcal{F}_i = 2_i^N \cap \mathcal{F} = \{A_1, \ldots, A_{|\mathcal{F}_i|}\}$, with lower-dimensional (i.e. $|\mathcal{F}_i|$-dimensional) unit simplices

$$\bar{\Delta}_i = \left\{ \left( q_i^{A_1}, \ldots, q_i^{A_{|\mathcal{F}_i|}} \right) : q_i^{A_k} \geq 0 \text{ for } 1 \leq k \leq |\mathcal{F}_i|, \sum_{1 \leq k \leq |\mathcal{F}_i|} q_i^{A_k} = 1 \right\}$$

and corresponding fuzzy covers $\mathbf{q} \in \bar{\Delta}_N = \times_{1 \leq i \leq n} \bar{\Delta}_i$. Hence a fuzzy cover maximally consists of $|\mathcal{F}| - 1$ points in the unit $n$-dimensional hypercube. Accordingly, $[0, 1]^n$ may be replaced with $\mathcal{C}(\mathcal{F}) = co(\{\chi_A : A \in \mathcal{F}\}) \subset [0, 1]^n$, i.e. the convex hull [16] of characteristic functions corresponding to feasible subsets. Recursively (with $w(\emptyset) = 0$), Möbius inversion $\mu^w : \mathcal{F} \to \mathbb{R}$ is

$$\mu^w(A) = w(A) - \sum_{B \in \mathcal{F} : B \subset A} \mu^w(B),$$

while the MLE $f^w : \mathcal{C}(\mathcal{F}) \to \mathbb{R}$ of $w$ is

$$f^w(q^A) = \sum_{B \in \mathcal{F} \cap 2^A} \left( \prod_{i \in B} q_i^A \right) \mu^w(B).$$

Therefore, every fuzzy cover $\mathbf{q} \in \bar{\Delta}_N$ has global worth

$$W(\mathbf{q}) = \sum_{A \in \mathcal{F}} \sum_{B \in \mathcal{F} \cap 2^A} \left( \prod_{i \in B} q_i^A \right) \mu^w(B).$$

For all $i \in N, q_i \in \bar{\Delta}_i$, and $\mathbf{q}_{-i} \in \bar{\Delta}_{N \setminus i} = \times_{j \in N \setminus i} \bar{\Delta}_j$, let

$$W_i(q_i | \mathbf{q}_{-i}) = \sum_{A \in \mathcal{F}_i} q_i^A \left[ \sum_{B \in \mathcal{F}_i \cap 2^A} \left( \prod_{j \in B \setminus i} q_j^A \right) \mu^w(B) \right],$$

$$W_{-i}(\mathbf{q}_{-i}) = \sum_{A \in \mathcal{F}_i} \left[ \sum_{B \in \mathcal{F} \cap 2^{A \setminus i}} \left( \prod_{j \in B} q_j^A \right) \mu^w(B) \right]$$

$$+ \sum_{A' \in \mathcal{F} \setminus \mathcal{F}_i} \left[ \sum_{B' \in \mathcal{F} \cap 2^{A'}} \left( \prod_{j' \in B'} q_{j'}^{A'} \right) \mu^w(B') \right],$$

yielding again

$$W(\mathbf{q}) = W_i(q_i | \mathbf{q}_{-i}) + W_{-i}(\mathbf{q}_{-i}). \tag{6}$$

From (4) above, $w_{\mathbf{q}_{-i}} : \mathcal{F}_i \to \mathbb{R}$ now is

$$w_{\mathbf{q}_{-i}}(A) = \sum_{B \in \mathcal{F}_i \cap 2^A} \left( \prod_{j \in B \setminus i} q_j^A \right) \mu^w(B). \tag{7}$$

For each $i \in N$, denote by $ex(\bar{\Delta}_i)$ the set of $|\mathcal{F}_i|$ extreme points of simplex $\bar{\Delta}_i$. Like in the full-dimensional case, at any fuzzy cover $\hat{\mathbf{q}} \in \bar{\Delta}_N$ every $i \in N$ such that $\hat{q}_i \notin ex(\bar{\Delta}_i)$ may deviate by concentrating its whole membership on some $A \in \mathcal{F}_i$ such that $w_{\hat{\mathbf{q}}_{-i}}(A) \geq w_{\hat{\mathbf{q}}_{-i}}(B)$ for all $B \in \mathcal{F}_i$. This yields a non-decreasing variation $W(q_i|\hat{\mathbf{q}}_{-i}) \geq W(\hat{\mathbf{q}})$ in global worth, with $q_i \in ex(\bar{\Delta}_i)$. When all $n$ elements do so, one after the other while updating $w_{\mathbf{q}_{-i}(t)}$ as in ROUNDUP above, i.e. $t = 0, 1, \ldots$, then the final $\mathbf{q} = (q_1, \ldots, q_n)$ satisfies $\mathbf{q} \in \underset{i \in N}{\times} ex(\bar{\Delta}_i)$. Yet cases $\mathcal{F} \subset 2^N$ and $\mathcal{F} = 2^N$ are different in terms of exactness. Specifically, consider any $\emptyset \neq A \in \mathcal{F}$ such that $|\{i : q_i^A = 1\}| \notin \{0, |A|\}$ or $A_{\mathbf{q}}^+ = \{i : q_i^A = 1\} \subset A$, with $f^w(q^A) = \sum_{B \in \mathcal{F} \cap 2^{A_{\mathbf{q}}^+}} \mu^w(B)$. Then, $\mathcal{F} \cap 2^{A_{\mathbf{q}}^+}$ may admit no shrinking yielding an exact fuzzy cover with same global worth as (non-exact) $\mathbf{q}$.

**Proposition 4.** *The values taken on exact fuzzy covers do not saturate the whole range of $W : \bar{\Delta}_N \to \mathbb{R}_+$.*

*Proof.* For $N = \{1, 2, 3, 4\}$ and $\mathcal{F} = \{N, \{4\}, \{1, 2\}, \{1, 3\}, \{2, 3\}\}$, let $w(N) = 3$, $w(\{4\}) = 2$, $w(\{i, j\}) = 1$ for $1 \leq i < j \leq 3$. Define $\mathbf{q} = (q_1, \ldots, q_4)$ by $q_4^{\{4\}} = 1 = q_i^N, i = 1, 2, 3$, with non-exactness $|\{i : q_i^N > 0\}| = 3 < 4 = |N|$. As

$$W(\mathbf{q}) = w(\{4\}) + \sum_{1 \leq i < j \leq 3} w(\{i, j\}) = 2 + 1 + 1 + 1$$

and $A_{\mathbf{q}}^+ = \{1, 2, 3\}$, if distributions $\hat{q}_1, \hat{q}_2, \hat{q}_3$ place membership only over feasible subsets $B \in \mathcal{F} \cap 2^{A_{\mathbf{q}}^+}$, then global worth is $W(\hat{q}_1, \hat{q}_2, \hat{q}_3, q_4) < W(\mathbf{q})$. $\square$

Therefore, an arbitrary search for optimal fuzzy covers may provide a maximizer (global or local) which is not reducible to any feasible solution of the original set packing problem. Such feasible solutions are partitions $P$ all of whose blocks are feasible, and where singleton blocks with worth 0 are not included in the packing. In fact, similarly to the full-dimensional case, fairly simple conditions may be shown to be sufficient for a partition to be a local maximizer.

**Definition 7.** *Any $\hat{q}_i|\hat{\mathbf{q}}_{-i} = \hat{\mathbf{q}} \in \bar{\Delta}_N$ is a local maximizer of $W : \bar{\Delta}_N \to \mathbb{R}_+$ if $W_i(\hat{q}_i|\hat{\mathbf{q}}_{-i}) \geq W_i(q_i|\hat{\mathbf{q}}_{-i})$ for all $i \in N$ and all $q_i \in \bar{\Delta}_i$ (see expression (6) above).*

The neighborhood $\mathcal{N}(\mathbf{q}) \subset \bar{\Delta}_N$ of $\mathbf{q} \in \bar{\Delta}_N$ thus is

$$\mathcal{N}(\hat{\mathbf{q}}) = \bigcup_{i \in N} \left\{ \mathbf{q} : \mathbf{q} = q_i|\hat{\mathbf{q}}_{-i}, q_i \in \bar{\Delta}_i \right\}.$$

Any partition $P$ with $A \in \mathcal{F}$ for each block $A \in P$ clearly has associated $\mathbf{p}$ such that $\mathbf{p} \in \underset{i \in N}{\times} ex(\bar{\Delta}_i) \subset \bar{\Delta}_N$.

**Proposition 5.** *Any partition $P$ with associated $\mathbf{p}$ such that $\mathbf{p} \in \bar{\Delta}_N$ is a local maximizer if for all $A \in P$*

$$w(A) \geq w(\{i\}) + \sum_{\hat{B} \in \mathcal{F} \cap 2^{A \setminus i}} \mu^w(\hat{B}).$$

*Proof.* Firstly note that for all blocks $A \in P$, if any, such that $|A| = 1$ there is nothing to prove, as the summation reduces to $w(\emptyset) = 0$, and thus there only remains $w(\{i\}) \geq w(\{i\})$. Accordingly, let $A \in P$ and $|A| > 1$. For every $i \in A$, any membership reallocation $q_i \in \bar{\Delta}_i$ deviating from $p_i$ (i.e. $p_i^A = 1$), given memberships $\mathbf{p}_{-i}$ of other elements $j \in N \backslash i$ (i.e. $\bar{\Delta}_j \ni p_j^{A'} = 1$ for all $A' \in P$ and all $j \in A'$), yields $\mathbf{q} = (q_i | \mathbf{p}_{-i}) \in \bar{\Delta}_N$ which is best analyzed by distinguishing between $\mathcal{F}_i \backslash A$ and $A$. In particular,

$$w(A) = w(\{i\}) + \sum_{\substack{B \in \mathcal{F}_i \cap 2^A \\ |B| > 1}} \mu^w(B) + \sum_{\hat{B} \in \mathcal{F} \cap 2^{A \backslash i}} \mu^w(\hat{B}).$$

Again, all membership mass $\sum_{B \in \mathcal{F}_i \backslash A} q_i^B > 0$ collapses on singleton $\{i\}$, because $\prod_{i' \in B \backslash i} p_{i'}^B = 0$ for all $B \in \mathcal{F}_i \backslash A$ by the definition of $\mathbf{p}_{-i}$. Thus,

$$W(\mathbf{p}) - W(q_i | \mathbf{p}_{-i}) = w(A) - w(\{i\}) - \left( q_i^A \sum_{\substack{B \in \mathcal{F}_i \cap 2^A \\ |B| > 1}} \mu^w(B) + \sum_{\hat{B} \in \mathcal{F} \cap 2^{A \backslash i}} \mu^w(\hat{B}) \right)$$

$$= (p_i^A - q_i^A) \sum_{\substack{B \in \mathcal{F}_i \cap 2^A \\ |B| > 1}} \mu^w(B).$$

Now assume that $\mathbf{p}$ is *not* a local maximizer, i.e. $W(\mathbf{p}) - W(q_i | \mathbf{p}_{-i}) < 0$. Since $p_i^A - q_i^A > 0$ because $p_i^A = 1$ and $q_i \in \bar{\Delta}_i$ is a deviation from $p_i$, then

$$\sum_{\substack{B \in \mathcal{F}_i \cap 2^A \\ |B| > 1}} \mu^w(B) = w(A) - w(\{i\}) - \sum_{\hat{B} \in \mathcal{F} \cap 2^{A \backslash i}} \mu^w(\hat{B}) < 0,$$

and this contradicts precisely the premise $w(A) \geq w(\{i\}) + \sum_{\hat{B} \in \mathcal{F} \cap 2^{A \backslash i}} \mu^w(\hat{B})$ for all $A \in P$ and $i \in A$, thereby completing the proof.                              $\square$

## 4   Local Search

As outlined in Sect. 1, when $\mathcal{F} \subset 2^N$ every feasible subset $A \in \mathcal{F}$ is to be considered not only in terms of its weight $w(A)$, but also in terms of the number of feasible subsets $B \neq A$ such that $B \cap A \neq \emptyset$, as these latter are automatically excluded from the packing if $A$ is included. Formally, for any $\mathcal{F}' \subseteq \mathcal{F}$ a cost function $c : \mathcal{F}' \to \mathbb{N}$ counts the number of still available feasible subsets that each $A \in \mathcal{F}'$ intersects, itself included. That is to say, the cost of including $A$ in the packing is $c(A) = |\{B : B \in \mathcal{F}', B \cap A \neq \emptyset\}| \in \{1, \ldots, |\mathcal{F}'|\}$. Also, like for LOCALSEARCH in Sect. 2, those feasible subsets $A^*(t)$ to be included are selected iteratively, i.e. one after the other for $t = 0, 1, \ldots$, entailing that $\mathcal{F}^{t-1} = \{B : B \in \mathcal{F}, B \cap A^*(t') = \emptyset, 0 \leq t' < t\}$ is the family of feasible subsets still available for each iteration $t$. Accordingly, the underlying poset function $\hat{w}^t : \mathcal{F}^{t-1} \to \mathbb{R}_+$ used at $t$ takes into account both weights and costs

$c : \mathcal{F}^{t-1} \to \mathbb{N}$ by means of ratio $\hat{w}^t(A) = \frac{w(A)}{c(A)}$. Of course, $\emptyset \in \mathcal{F}^t$ for all $t$, thus Möbius inversion $\mu^{\hat{w}^t} : \mathcal{F}^{t-1} \to \mathbb{R}$ is well-defined. In particular, for any $\mathbf{q} \in \bar{\Delta}_N$, the analog of expression (7) above now is $\hat{w}^t_{\mathbf{q}_{-i}} : \mathcal{F}^{t-1}_i \to \mathbb{R}$ given by

$$
\hat{w}^t_{\mathbf{q}_{-i}}(A) = \sum_{B \in \mathcal{F}^{t-1}_i \cap 2^A} \left( \prod_{j \in B \setminus i} q^A_j \right) \mu^{\hat{w}^t}(B).
$$

Given this additional notation, LOCALSEARCH from Sect. 2 can be modified as follows.

LS-WITHCOST$(\hat{w}, \mathbf{q})$

*Initialize:* Set $t = 0$ and $\mathbf{q}(0) = \mathbf{q}$, with requirement $|\{i : q^A_i > 0\}| \in \{0, |A|\}$ for all $A \in \mathcal{F}$. Also set $\mathcal{F}^0 = \mathcal{F}$.

*Loop 1:* While $0 < \sum_{i \in A} q^A_i(t) < |A|$ for a $A \in \mathcal{F}^t$, set $t = t + 1$ and:

(a) select a $A^*(t) \in \mathcal{F}^{t-1}$ satisfying

$$
\min_{i \in A^*(t)} \hat{w}^t_{\mathbf{q}_{-i}(t-1)}(A^*(t)) \geq \min_{j \in B} \hat{w}^t_{\mathbf{q}_{-j}(t-1)}(B)
$$

for all $B \in \mathcal{F}^{t-1}$ such that $0 < \sum_{i \in B} q^B_i(t) < |B|$,

(b) for $i \in A^*(t)$ and $A \in \mathcal{F}^{t-1}_i$, define $q^A_i(t) = \begin{cases} 1 \text{ if } A = A^*(t), \\ 0 \text{ if } A \neq A^*(t), \end{cases}$

(c) for $j \in N \setminus A^*(t)$ and $A \in \mathcal{F}^{t-1}_j$ with $A \cap A^*(t) = \emptyset$, define

$$
q^A_j(t) = q^A_j(t-1) + \left( \hat{w}^t(A) \sum_{\substack{B \in \mathcal{F}^{t-1}_j \\ B \cap A^*(t) \neq \emptyset}} q^B_j(t-1) \right) \left( \sum_{\substack{B' \in \mathcal{F}^{t-1}_j \\ B' \cap A^*(t) = \emptyset}} \hat{w}^t(B') \right)^{-1}
$$

(d) for $j \in N \setminus A^*(t)$ and $A \in \mathcal{F}^{t-1}_j$ with $A \cap A^*(t) \neq \emptyset$, define $q^A_j(t) = 0$.

(e) define $\mathcal{F}^t = \{B : B \in \mathcal{F}^{t-1}, B \cap A^*(t) = \emptyset\}$.

*Loop 2:* While $q^A_i(t) = 1, |A| > 1$ for a $i \in N$ and

$$
w(A) < w(\{i\}) + \sum_{\hat{B} \in \mathcal{F} \cap 2^{A \setminus i}} \mu^w(\hat{B}),
$$

set $t = t + 1$ and define:

$$
q^{\hat{A}}_i(t) = \begin{cases} 1 \text{ if } |\hat{A}| = 1 \\ 0 \text{ otherwise} \end{cases} \text{ for all } \hat{A} \in \mathcal{F}_i,
$$

$$
q^B_j(t) = \begin{cases} 1 \text{ if } B = A \setminus i \\ 0 \text{ otherwise} \end{cases} \text{ for all } j \in A \setminus i, B \in \mathcal{F}_j,
$$

$$
q^{\hat{B}}_{j'}(t) = q^{\hat{B}}_{j'}(t-1) \text{ for all } j' \in A^c, \hat{B} \in \mathcal{F}_{j'}.
$$

*Output:* Set $\mathbf{q}^* = \mathbf{q}(t)$.

Both LOCALSEARCH and LS-WITHCOST generate in $|P| \leq n$ iterations of *Loop 1* a partition $\{A^*(1), \ldots, A^*(|P|)\} = P$. Selected blocks $A^*(t) \in \mathcal{F}^{t-1}$, $1 \leq t \leq |P|$ now are any of those feasible subsets where the minimum over elements $i \in A^*(t)$ of $(i, A^*(t))$-derivatives $\hat{w}^t_{\mathbf{q}_{-i}}(A^*(t))$ is maximal. The following *Loop 2* again checks whether the partition generated by *Loop 1* may be improved by extracting some elements from existing blocks and putting them in singleton blocks of the final output.

**Proposition 6.** LS-WITHCOST$(W, \mathbf{q})$ *outputs a local maximizer* $\mathbf{q}^*$.

*Proof.* Follows from Proposition 5 since *Loop 2* deals with $w$, not with $\hat{w}$.

Concerning input $\mathbf{q} = \mathbf{q}(0)$, consider setting

$$q_i^A = \frac{\frac{w(A)}{|\{A' : A' \in \mathcal{F}, A' \cap A \neq \emptyset\}|}}{\sum_{B \in \mathcal{F}_i} \frac{w(B)}{|\{B' : B' \in \mathcal{F}, B' \cap B \neq \emptyset\}|}} \quad \text{for all } A \in \mathcal{F}_i, i \in N, \text{ entailing}$$

$$\frac{q_i^A}{q_i^B} = \frac{w(A)|\{B' : B' \in \mathcal{F}, B' \cap B \neq \emptyset\}|}{w(B)|\{A' : A' \in \mathcal{F}, A' \cap A \neq \emptyset\}|} = \frac{q_j^A}{q_j^B}$$

for all $A, B \in \mathcal{F}_i \cap \mathcal{F}_j$ and $i, j \in N$.

Evidently, *Loop 1* may take exactly the same form as in LOCALSEARCH, that is with selected blocks $A^*(t) \in \mathcal{F}, t = 1, \ldots, |P|$ of the generated partition $P$ being any of those feasible subsets where the average, rather than the minimum, over elements $i \in A^*(t)$ of $(i, A^*(t))$-derivatives $\hat{w}^t_{\mathbf{q}_{-i}}(A^*(t))$ is maximal. This possibility may be useful in those settings where set packing appears in its weighted version, while using the minimum in place of the sum seems interesting for $k$-uniform (and $d$-regular) set packing problems (see Sect. 1). In fact, for the $k$-uniform case Möbius inversion is $\mu^{\hat{w}^t}(A) = \frac{1}{|\{B : B \in \mathcal{F}^{t-1}, B \cap A \neq \emptyset\}|}$ if $|A| = k$ and $\mu^{\hat{w}^t}(A) = 0$ if $|A| \in \{0, 1\}$ for all $A \in \mathcal{F}^{t-1}$ (recall the convention $\{\emptyset\} \in \mathcal{F} \ni \{i\}$ for all $i \in N$). It is also plain that in $k$-uniform set packing *Loop 2* is ineffective.

**Example:** let $N = \{1, 2, 3\}$ and $\mathcal{F} = \{\{1\}, \{3\}, \{1, 2\}, \{2, 3\}, \{1, 2, 3\}\}$, with $w(\{1\}) = 1$, $w(\{3\}) = 2 = w(\{1, 2\})$, $w(\{2, 3\}) = 3$, $w(N) = 3.5$. Accordingly,

$$\sum_{B \in \mathcal{F}_1} \frac{w(B)}{|\{B' : B' \in \mathcal{F}, B' \cap B \neq \emptyset\}|} = \frac{1}{3} + \frac{2}{4} + \frac{3.5}{5},$$

$$\sum_{B \in \mathcal{F}_2} \frac{w(B)}{|\{B' : B' \in \mathcal{F}, B' \cap B \neq \emptyset\}|} = \frac{2}{4} + \frac{3}{4} + \frac{3.5}{5},$$

$$\sum_{B \in \mathcal{F}_3} \frac{w(B)}{|\{B' : B' \in \mathcal{F}, B' \cap B \neq \emptyset\}|} = \frac{2}{3} + \frac{3}{4} + \frac{3.5}{5},$$

and thus input $\mathbf{q}(0)$ defined above, i.e.

$$q_i^A(0) = \frac{\frac{w(A)}{|\{A' : A' \in \mathcal{F}, A' \cap A \neq \emptyset\}|}}{\sum_{B \in \mathcal{F}_i} \frac{w(B)}{|\{B' : B' \in \mathcal{F}, B' \cap B \neq \emptyset\}|}} \quad \text{for all } A \in \mathcal{F}_i, i \in N, \text{ yields}$$

$$
\begin{pmatrix} q_1^1(0) \simeq 0.217 \\ q_1^{12}(0) \simeq 0.326 \\ q_1^N(0) \simeq 0.457 \end{pmatrix}, \quad \begin{pmatrix} q_2^{12}(0) \simeq 0.256 \\ q_2^{23}(0) \simeq 0.385 \\ q_2^N(0) \simeq 0.359 \end{pmatrix}, \quad \begin{pmatrix} q_3^3(0) \simeq 0.315 \\ q_3^{23}(0) \simeq 0.354 \\ q_3^N(0) \simeq 0.331 \end{pmatrix}.
$$

The enlarged $\mathcal{F}$ clearly is $\mathcal{F} = \{\emptyset, \{1\}, \{2\}, \{3\}, \{1,2\}, \{2,3\}, \{1,2,3\}\}$, and in the first iteration $t = 1$ the situation concerning $\hat{w}_{\mathbf{q}_{-i}}^t(A), i \in N, A \in \mathcal{F}_i$ is

$$
\hat{w}_{\mathbf{q}_{-1}}^1(\{1\}) = \frac{1}{3} \quad \text{and} \quad \hat{w}_{\mathbf{q}_{-3}}^1(\{3\}) = \frac{2}{3} \text{ (and of course } \hat{w}_{\mathbf{q}_{-2}}^1(\{2\}) = 0),
$$

$$
\hat{w}_{\mathbf{q}_{-1}}^1(\{1,2\}) = 0.256 \left( \frac{2}{4} - \frac{1}{3} \right) + \frac{1}{3} = 0.376,
$$

$$
\hat{w}_{\mathbf{q}_{-2}}^1(\{1,2\}) = 0.326 \left( \frac{2}{4} - \frac{1}{3} \right) = 0.054,
$$

$$
\hat{w}_{\mathbf{q}_{-2}}^1(\{2,3\}) = 0.354 \left( \frac{3}{4} - \frac{2}{3} \right) = 0.030,
$$

$$
\hat{w}_{\mathbf{q}_{-3}}^1(\{2,3\}) = 0.385 \left( \frac{3}{4} - \frac{2}{3} \right) + \frac{2}{3} = 0.699,
$$

$$
\hat{w}_{\mathbf{q}_{-1}}^1(N) = (0.359 \times 0.331) \left[ \frac{3.5}{5} - \left( \frac{2}{4} - \frac{1}{3} \right) - \left( \frac{3}{4} - \frac{2}{3} \right) - \frac{1}{3} - \frac{2}{3} \right]
$$
$$
+ 0.359 \left( \frac{2}{4} - \frac{1}{3} \right) + \frac{1}{3} = 0.328,
$$

$$
\hat{w}_{\mathbf{q}_{-2}}^1(N) = (0.457 \times 0.331) \left[ \frac{3.5}{5} - \left( \frac{2}{4} - \frac{1}{3} \right) - \left( \frac{3}{4} - \frac{2}{3} \right) - \frac{1}{3} - \frac{2}{3} \right]
$$
$$
+ 0.457 \left( \frac{2}{4} - \frac{1}{3} \right) + 0.331 \left( \frac{3}{4} - \frac{2}{3} \right) = 0.021,
$$

$$
\hat{w}_{\mathbf{q}_{-3}}^1(N) = (0.457 \times 0.359) \left[ \frac{3.5}{5} - \left( \frac{2}{4} - \frac{1}{3} \right) - \left( \frac{3}{4} - \frac{2}{3} \right) - \frac{1}{3} - \frac{2}{3} \right]
$$
$$
+ 0.359 \left( \frac{3}{4} - \frac{2}{3} \right) + \frac{2}{3} = 0.606.
$$

In view of line (a), the selected block at the first iteration is $A^*(1) = \{3\}$. According to lines (b)–(d), $q_3^3(1) = 1$, $q_1^1(1) = q_1^1(0) + \left( \frac{0.457}{3} \right) / \left( \frac{1}{3} + \frac{2}{4} \right) = 0.4$ and $q_1^{12}(1) = 0.6$ as well as $q_2^{12}(1) = 1$. In the second iteration $t = 2$,

$$
\hat{w}_{\mathbf{q}_{-1}}^2(\{1\}) = \frac{1}{2} \quad \text{and} \quad \hat{w}_{\mathbf{q}_{-2}}^1(\{2\}) = 0,
$$

$$
\hat{w}_{\mathbf{q}_{-1}}^2(\{1,2\}) = 1 + \frac{1}{2},
$$

$$
\hat{w}_{\mathbf{q}_{-2}}^2(\{1,2\}) = 0.6,
$$

and thus the selected block at the second iteration is $A^*(2) = \{1,2\}$. In fact, $\mathcal{F}^* = \{\{1,2\}, \{3\}\}$ and $\mathcal{F}^{**} = \{\{1\}, \{2,3\}\}$ are the two optimal solutions, with $W(\mathcal{F}^*) = W(\mathcal{F}^{**}) = 4$, and thus the second loop does not produce any change.

## 5    Near-Boolean Functions

Boolean functions $f : \{0,1\}^n \rightarrow \{0,1\}$ of $n$ variables constitute a subclass of pseudo-Boolean functions $f : \{0,1\}^n \rightarrow \mathbb{R}$. Following [17, p. 4], denote by $N = \{1,\ldots,n\}$ the set of indices of variables. As already explained, any pseudo-Boolean function has a unique expression as a multilinear polynomial in $n$ variables $x_1,\ldots,x_n \in [0,1]$ of the form $f(x_1,\ldots,x_n) = \sum_{A \subseteq N} \left( \alpha_A \prod_{i \in A} x_i \right)$, since $\alpha_A, A \in 2^N$ is in fact the Möbius inversion $\mu^w(A), A \in 2^N$ of a unique [14] set function $w : 2^N \rightarrow \mathbb{R}$ such that $w(A) = f(\chi_A)$, where $\chi_A$ is the characteristic function of $A$ (see above and [1, p. 162]). The $n$ variables thus range each in the unit interval $[0,1]$. Such a setting is here expanded by letting each $i$-th variable, $i = 1,\ldots,n$ range in a $2^{n-1} - 1$-dimensional unit simplex $\Delta_i$ whose extreme points are indexed by subsets $A \in 2_i^N$. The goal is to evaluate peculiar collections of fuzzy subsets of $N$ through the MLE given by expressions (1–2).

Let $ex(\Delta_N) = \times_{i \in N} ex(\Delta_i)$. For every $n$-collection $(q_1,\ldots,q_n) = \mathbf{q} \in ex(\Delta_N)$ of extreme points of simplices $\Delta_1,\ldots,\Delta_n$ as defined in Sect. 2, denote by $\hat{A}_1,\ldots,$ $\hat{A}_n \in 2^N$ the corresponding subsets, i.e. $q_i^{\hat{A}_i} = 1$ for all $i \in N$. Let $P(\mathbf{q})$ be the partition obtained by putting in a same block any two $i, j \in N$ such that $\hat{A}_i = \hat{A}_j$, i.e. $P(\mathbf{q}) = \{A : \hat{A}_i = \hat{A}_j \text{ for all } i, j \in A, \hat{A}_{j'} \neq \hat{A}_i \text{ for all } j' \in A^c\}$.

**Definition 8.** *Near-Boolean functions of $n$ variables $q_i \in ex(\Delta_i), i \in N$ are defined for a given set function $w : 2^N \rightarrow \mathbb{R}, w(\emptyset) = 0$, and have form*

$$F : ex(\Delta_N) \rightarrow \mathbb{R}, \text{ with } F(\mathbf{q}) = \sum_{A \in P(\mathbf{q})} w(A) \text{ for all } \mathbf{q} \in ex(\Delta_N).$$

**Definition 9.** *The MLE $\hat{F} : \Delta_N \rightarrow \mathbb{R}$ of near-Boolean functions $F$ is the polynomial given by expression (2), i.e.*

$$\hat{F}(\mathbf{q}) = \sum_{A \in 2^N} \left[ \sum_{B \supseteq A} \left( \prod_{i \in A} q_i^B \right) \right] \mu^w(A),$$

*with $q_i = (q_i^{A_1},\ldots,q_i^{A_{2^{n-1}}}) \in \Delta_i, i \in N$.*

### 5.1    Approximations

In line with [17], the issue of approximating a near-Boolean function $F$ for given set function $w$ by means of the least squares LS criterion concerns how to determine a near-Boolean function $F_k$ such that

$$\sum_{\mathbf{q} \in ex(\Delta_N)} [F(\mathbf{q}) - F_k(\mathbf{q})]^2 \tag{8}$$

attains its minimum over all near-Boolean functions $F_k$ with polynomial MLE $\hat{F}_k$ of degree $k$, that is

$$\hat{F}_k(\mathbf{q}) = \sum_{\substack{A \in 2^N \\ |A| \leq k}} \left[ \sum_{B \supseteq A} \left( \prod_{i \in A} q_i^B \right) \right] \mu^{w'}(A),$$

or, equivalently stated in terms of the underlying set function $w'$ for $F_k$, such that $\mu^{w'}(A) = 0$ if $|A| > k$.

Near-Boolean functions $F$ take values on $ex(\Delta_N)$, and $|ex(\Delta_N)| = 2^{n(n-1)}$. Therefore, they might be regarded as points in a $2^{n(n-1)}$-dimensional vector space, i.e. $F \in \mathbb{R}^{2^{n(n-1)}}$. In view of Proposition 1 formalizing exactness, this seems conceptually incorrect and with useless enumerative demand. Specifically, for every partition $P$ of $N$, with associated exact $\mathbf{p} \in ex(\Delta_N)$, there clearly exist many distinct non-exact $\mathbf{q} \in ex(\Delta_N)$ such that $P = P(\mathbf{q})$, entailing $F(\mathbf{q}) = F(\mathbf{p})$ (see Corollary 1 above). Counting these redundant non-exact $n$-collections of extreme points of simplices is worthless. Hence $k$-degree approximation is to be dealt with by replacing expression (8) with the following:

$$\sum_{\substack{\mathbf{p} \in ex(\Delta_N) \\ \mathbf{p} \text{ exact}}} [F(\mathbf{p}) - F_k(\mathbf{p})]^2. \tag{9}$$

Let $(\mathcal{P}^N, \wedge, \vee)$ denote the (geometric) lattice [13] of partitions of $N$. The number $|\mathcal{P}^N|$ of partitions of a $n$-set is given by *Bell number* $\mathcal{B}_n$ [13,14] (see below). Accordingly, near-Boolean functions might be regarded as points $F \in \mathbb{R}^{\mathcal{B}_n}$ in a $\mathcal{B}_n$-dimensional vector space, but this is again too large, as points in such a vector space correspond in fact to generic partition functions, whose Möbius inversion may take non-zero values on any partition $P \in \mathcal{P}^N$. Conversely, near-Boolean functions only involve those $h : \mathcal{P}^N \to \mathbb{R}$ such that $h(P) = h_w(P) = \sum_{A \in P} w(A)$ for some set function $w : 2^N \to \mathbb{R}$. The Möbius inversion of these partition functions may take non-zero values only on the $2^n - n$ modular elements [13,18] of lattice $(\mathcal{P}^N, \wedge, \vee)$, namely on those partitions with a number of non-singleton blocks $\leq 1$. This is shown below via recursion through the Möbius inversion of *additively separable* partition functions [19,20]. When regarded as points in a vector space (i.e. expressed as a linear combination of a basis, see above) these functions must be regarded as $h_w \in \mathbb{R}^{2^n - n}$.

It seems crucial emphasizing that while pseudo-Boolean functions admit a unique set function providing their best $k$-degree approximation, $0 \leq k \leq n$ [17], every near-Boolean function admits a continuum of set functions $w$ determining their unique best $k$-degree approximation. In particular, consider first the linear case, i.e. $k = 1$. The issue is to find a best LS approximation $F_1$ of any given $F$. That is, the set function $w'$ determining $F_1$ has to satisfy $w'(A) = \sum_{i \in A} w'(\{i\})$ for all $A \in 2^N$. Then,

$$h_{w'}(P) = \sum_{A \in P} w'(A) = \sum_{A \in P} \sum_{i \in A} w'(\{i\}) = w'(N)$$

for all $P \in \mathcal{P}^N$. Thus $h_{w'}$ is a constant partition function, or a valuation [13] of partition lattice $(\mathcal{P}^N, \wedge, \vee)$. Also, any further linear $v : 2^N \rightarrow \mathbb{R}$ such that $v(N) = w'(N)$ also satisfies $h_v(P) = h_{w'}(P)$ for all $P \in \mathcal{P}^N$. In other terms, there is a continuum of equivalent linear $v \neq w'$ such that $h_{w'} = h_v$, obtained each by distributing arbitrarily the whole of $w'(N)$ over singletons $\{i\} \in 2^N$. Cases $k > 1$ maintain this same feature: consider a set function $w'$ such that $\mu^{w'}(A) \neq 0$ for one or more (possibly all $\binom{n}{k}$) subsets $A \in 2^N$ such that $|A| = k$. Now fix arbitrarily $n$ values $v(\{i\}), i \in N$ with $\sum_{i \in N} w'(\{i\}) = \sum_{i \in N} v(\{i\})$. For all $A \in 2^N, |A| > 1$ Möbius inversion $\mu^v : 2^N \rightarrow \mathbb{R}$ can always be determined uniquely via recursion by

$$v(A) + \sum_{i \in A^c} v(\{i\}) = \sum_{B \subseteq A} \mu^v(B) + \sum_{i \in A^c} v(\{i\})$$

$$= w'(A) + \sum_{i \in A^c} w'(\{i\}) = \sum_{B \subseteq A} \mu^{w'}(B) + \sum_{i \in A^c} w'(\{i\}).$$

In other terms [19,20], if set function $w$ additively separates partition function $h$, i.e. $h = h_w$, and $v' = w - v$ is a linear set function, then $v + v'$ also additively separates $h$, i.e. $h_w = h_{v+v'}$. Hence, there is a continuum of equivalent set functions $w$ and $v + v'$ available for the sought $k$-degree approximation $F_k$, but still the $\mathcal{B}_n$ values taken by $F_k$ (more precisely, the $2^n - n$ values taken by $F_k$ on the modular elements of partition lattice $(\mathcal{P}^N, \wedge, \vee)$) are unique and independent from the chosen set function in the continuum of set functions $w, v + v'$ available, each determining an equivalent MLE $\hat{F}^k$ of $F^k$. This is detailed hereafter.

## 5.2   Equivalent Polynomials

Möbius inversion applies to any (locally finite) poset, provided a bottom element exists [14]. For the Boolean lattice $(2^N, \cap, \cup)$ of subsets of $N$ ordered by inclusion $\supseteq$ and the geometric lattice $(\mathcal{P}^N, \wedge, \vee)$ of partitions of $N$ ordered by coarsening $\geqslant$ [13,21], the bottom elements are, respectively, the empty set $\emptyset$ and the finest partition $P_\perp = \{\{1\}, \ldots, \{n\}\}$. For a lattice $(L, \wedge, \vee)$ ordered by $\geqslant$, with generic elements $x, y, z \in L$ and bottom element $x_\perp$, any lattice function $f : L \rightarrow \mathbb{R}$ has Möbius inversion $\mu^f : L \rightarrow \mathbb{R}$ given by $\mu^f(x) = \sum_{x_\perp \leqslant y \leqslant x} \mu_L(y, x) f(y)$, where $\mu_L$ is the Möbius function, defined recursively on ordered pairs $(y, x) \in L \times L$ by $\mu_L(y, x) = -\sum_{y \leqslant z < x} \mu_L(y, z)$ if $y < x$ (i.e. $y \leqslant x$ and $y \neq x$) as well as $\mu_L(y, x) = 1$ if $y = x$, while $\mu_L(y, x) = 0$ if $y \not\leqslant x$. The Möbius function of the subset lattice is $\mu_{2^N}(B, A) = (-1)^{|A \setminus B|}$, with $B \subseteq A$. Concerning the Möbius function $\mu_{\mathcal{P}^N}$ of $\mathcal{P}^N$, given any $P, Q \in \mathcal{P}^N$, if $Q < P = \{A_1, \ldots, A_{|P|}\}$, then for every $A \in P$ there are $B_1, \ldots, B_{k_A} \in Q$ such that $A = B_1 \cup \cdots \cup B_{k_A}$, with $k_A > 1$ for at least one $A \in P$. Segment $[Q, P] = \{P' : Q \leqslant P' \leqslant P\}$ is isomorphic to product $\times_{A \in P} \mathcal{P}(k_A)$, where $\mathcal{P}(k)$ is the lattice of partitions of a $k$-set. Let $m_k = |\{A : A \in P, k_A = k\}|$ for $k = 1, \ldots, n$. Then [14, pp. 359–360],

$$\mu_{\mathcal{P}^N}(Q, P) = (-1)^{-n + \sum_{1 \leqslant k \leqslant n} m_k} \prod_{1 < k < n} (k!)^{m_{k+1}}. \tag{10}$$

A partition function $h : \mathcal{P}^N \to \mathbb{R}$ may be said to be additively separable [19,20] if there is a set function $v : 2^N \to \mathbb{R}$ such that $h(P) = \sum_{A \in P} v(A)$ for all $P \in \mathcal{P}^N$, with the notation $h = h_v$. Möbius inversion $\mu^{h_v}$ may take non-zero values only on the $2^n - n$ modular elements of the partition lattice, namely the bottom $P_\perp$ and top $P^\top = \{N\}$, together with all those partitions of the form $\{A\} \cup P_\perp^{A^c}$ for $A \in 2^N$ such that $1 < |A| < n$, where $P_\perp^{A^c}$ is the finest partition of $A^c$ [13, Ex. 13, p. 71]. The Möbius inversion of an additively separable partition function $h_v$ is now detailed [19, Prop. 4.4, p. 138 and Appendix, p. 144], [20, Prop. 3.3, p. 452].

**Proposition 7.** *If $h = h_v$, then $h = h_w$ for a continuum of set functions $w : 2^N \to \mathbb{R}, w \neq v$.*

*Proof.* By direct substitution,

$$\mu^{h_v}(P) = \sum_{A \in P} \sum_{B \subseteq A} v(B) \sum_{Q \leqslant P : B \in Q} \mu_{\mathcal{P}^N}(Q, P) \quad \text{for all } P \in \mathcal{P}^N.$$

Now if $P \neq \{A\} \cup P_\perp^{A^c}$, then the recursive definition of $\mu_{\mathcal{P}^N}$ yields

$$\sum_{P' \leqslant P : A \in P'} \mu_{\mathcal{P}^N}(P', P) = \sum_{\{A\} \cup P_\perp^{A^c} \leqslant P' \leqslant P} \mu_{\mathcal{P}^N}(P', P) = 0,$$

and the same for $B \subset A$. Thus, $\mu^{h_v}$ may take non-zero values only on modular partitions, where it obtains recursively by $\mu^{h_v}(P_\perp) = \sum_{i \in N} v(\{i\})$ and $\mu^{h_v}(\{A\} \cup P_\perp^{A^c}) = \mu^v(A)$ for $1 < |A| < n$ as well as $\mu^{h_v}(P^\top) = \mu^v(N)$. Accordingly, any $w \neq v$ satisfying $\sum_{i \in N} v(\{i\}) = \sum_{i \in N} w(\{i\})$ and $\mu^v(A) = \mu^w(A)$ for all $A \in 2^N$ such that $|A| > 1$ also additively separates $h$, i.e. $h_v = h_w$. $\square$

The degree of a polynomial is the highest degree of its terms. In expression (2), for any given set function $w$, the degree is $\max\{|A| : \mu^w(A) \neq 0\}$, while every non-zero value of Möbius inversion $\mu^w$ is a coefficient of the polynomial. For any degree $k, 0 < k \leq n$, there exists a continuum of set functions such that $\max\{|A| : \mu^w(A) \neq 0\} = k$, each defining alternative but equivalent coefficients of the polynomial.

**Example:** let $N = \{1, 2, 3, 4\}$ and consider the (symmetric) set function $v : 2^N \to \mathbb{R}_+$ defined by $v(A) = |A|^2$, with Möbius inversion $\mu^v(A) = 1$ if $|A| = 1$ and $\mu^v(A) = 4 - 1 - 1 = 2$ if $|A| = 2$, while $\mu^v(A) = 0$ if $|A| \in \{0, 3, 4\}$. On partitions $P$ of $N$, the associated additively separable partition function $h_v$ thus takes values $h_v(P) = \sum_{A \in P} |A|^2$. For instance, on partitions $P^1 = 12|3|4$, $P^2 = 13|24$ and $P^3 = 123|4$ (with vertical bar | separating blocks), these values are $h_v(P^1) = 4 + 1 + 1 = 6$, $h_v(P^2) = 4 + 4 = 8$ and $h_v(P^3) = 9 + 1 = 10$. Now let set function $w : 2^N \to \mathbb{R}_+, w(\emptyset) = 0$ take values $w(\{1\}) = 0 = w(\{3\})$ and $w(\{2\}) = 2 = w(\{4\})$ on singletons, hence $\sum_{i \in N} v(\{i\}) = 4 = \sum_{i \in N} w(\{i\})$. In order to have $h_w(P) = h_v(P)$ for all partitions $P$, it must be $\mu^v(A) = \mu^w(A)$ for all $A \in 2^N$ such that $|A| > 1$. Hence

$$\mu^w(\{1,2\}) = 2 = \mu^v(\{1,2\}) \Rightarrow w(\{1,2\}) = 2 + w(\{1\}) + w(\{2\}) = 4,$$
$$\mu^w(\{1,3\}) = 2 = \mu^v(\{1,3\}) \Rightarrow w(\{1,3\}) = 2 + w(\{1\}) + w(\{3\}) = 2,$$
$$\mu^w(\{1,4\}) = 2 = \mu^v(\{1,4\}) \Rightarrow w(\{1,4\}) = 2 + w(\{1\}) + w(\{4\}) = 4,$$
$$\mu^w(\{2,3\}) = 2 = \mu^v(\{2,3\}) \Rightarrow w(\{2,3\}) = 2 + w(\{2\}) + w(\{3\}) = 4,$$
$$\mu^w(\{2,4\}) = 2 = \mu^v(\{2,4\}) \Rightarrow w(\{2,4\}) = 2 + w(\{2\}) + w(\{4\}) = 6,$$
$$\mu^w(\{3,4\}) = 2 = \mu^v(\{3,4\}) \Rightarrow w(\{3,4\}) = 2 + w(\{3\}) + w(\{4\}) = 4,$$

as well as

$$\mu^w(\{1,2,3\}) = 0 = \mu^v(\{1,2,3\}) \Rightarrow w(\{1,2,3\}) = 2 + 2 + 2 + 2 = 8,$$
$$\mu^w(\{1,2,4\}) = 0 = \mu^v(\{1,2,4\}) \Rightarrow w(\{1,2,4\}) = 2 + 2 + 2 + 2 + 2 = 10,$$
$$\mu^w(\{2,3,4\}) = 0 = \mu^v(\{2,3,4\}) \Rightarrow w(\{2,3,4\}) = 2 + 2 + 2 + 2 + 2 = 10,$$

and finally

$$\mu^w(N) = 0 = \mu^v(N) \Rightarrow w(N) = 2 + 2 + 2 + 2 + 2 + 2 + 4 = 16.$$

It is thus readily checked that the desired condition $h_w(P) = h_v(P)$ for all partitions $P$ holds. In particular, $h_w(P^1) = 4 + 2 = 6 = h_v(P^1)$ as well as $h_w(P^2) = 2 + 6 = 8 = h_v(P^2)$ and $h_w(P^3) = 8 + 2 = 10 = h_v(P^3)$.

## 5.3    MLE of Partition Functions

Although the lattice $(\mathcal{P}^N, \wedge, \vee)$ of partitions of a finite set $N$ is fundamental in combinatorial theory [13], and despite *"partitions are of central importance in the study of symmetric functions, a class of functions that pervades mathematics in general"* [22, p. 39], still the polynomial multilinear extension has thus far been investigated only for set functions [1,4]. (On symmetric function theory see [23, Chapt. 5], [24], [25, Chapt. 7].) Therefore, the purpose of this subsection is to briefly present the MLE of partition functions (and thus more generally of functions taking real values on a geometric lattice [13, p. 52]) by focusing on atoms of the partition lattice and on additive separability.

The rank function $r : \mathcal{P}^N \to \mathbb{Z}_+$ of the partition lattice is $r(P) = n - |P|$, with $r(P_\perp) = 0$ for the bottom element. Atoms are immediately above, with rank 1, in the associated *Hasse diagram*. This latter is ordered by coarsening $\geqslant$, with coarser partitions in upper levels [13,21], where $P \geqslant Q$ means that every block of $Q$ is included in some block of $P$. Hence atoms are those partitions consisting of $n - 1$ blocks, namely $n - 2$ singletons and one pair. These $\binom{n}{2}$ unordered pairs $\{i,j\} \in N_2$ are the same atoms as in subset lattice $(2^{N_2}, \cap, \cup)$, where $N_2 = \{\{i,j\} : 1 \leq i < j \leq n\}$ is the $\binom{n}{2}$-set of 2-cardinal subsets of $N$. For notational convenience, let $[ij] \in \mathcal{P}^N$ denote the atom where the unique 2-cardinal block is (unordered) pair $\{i,j\} \in [ij]$.

In order to replace subsets with partitions, let $\mathcal{P}_{(1)}^N = \{[ij] : 1 \leq i < j \leq n\}$ be the $\binom{n}{2}$-set of atoms of the partition lattice, i.e. $\mathcal{P}_{(1)}^N \cong N_2$. The analog of characteristic function $\chi_A, A \in 2^N$ is *indicator function* $I_P : \mathcal{P}_{(1)}^N \to \{0, 1\}$, with

$$I_P([ij]) = \begin{cases} 1 \text{ if } P \geqslant [ij], \\ 0 \text{ if } P \not\geqslant [ij], \end{cases} \text{ for all } P \in \mathcal{P}^N, [ij] \in \mathcal{P}_{(1)}^N.$$

In words, if pair $\{i, j\}$ is included in some block $A$ of $P$ (i.e. $\{i, j\} \subseteq A \in P$), then partition $P$ is coarser than atom $[ij]$, and the corresponding position $I_P([ij])$ of indicator array $I_P$ has entry 1. Otherwise, that position is 0. Hence $I_P$ is a Boolean $\binom{n}{2}$-vector just like $\chi_A$ is a Boolean $n$-vector. As detailed in the previous sections, the MLE of set functions extends them from the $2^n$-set $\{0, 1\}^n$ of vertices of the unit $n$-dimensional hypercube $[0, 1]^n$ to the whole of this latter. In order to do the same with partition functions, firstly it must be clear that those Boolean $\binom{n}{2}$-vectors $I_P, P \in \mathcal{P}^N$ corresponding to the indicator functions of partitions do not span the whole $2^{\binom{n}{2}}$-set $\{0, 1\}^{\binom{n}{2}}$ of vertices of the $\binom{n}{2}$-dimensional unit hypercube $[0, 1]^{\binom{n}{2}}$. Minimally, this is already observable when $N = \{1, 2, 3\}$ as in the above examples, since there are 5 partitions: the finest $\{\{1\}, \{2\}, \{3\}\}$ and coarsest $\{1, 2, 3\}$ ones, together with the $\binom{3}{2} = 3$ atoms $[12] = \{\{1, 2\}, \{3\}\}$, $[13] = \{\{1, 3\}, \{2\}\}$ and $[23] = \{\{2, 3\}, \{1\}\}$. This means that those three vertices of $[0, 1]^3$ corresponding to $[12] \vee [23]$, $[12] \vee [13]$ and $[13] \vee [23]$ all collapse on vertex $I_{P^\top}$ corresponding to the coarsest partition $[12] \vee [13] \vee [23] = \{1, 2, 3\}$ (here the top element among partitions is $P^\top = \{N\}$, and $I_{P^\top}$ denotes the $\binom{n}{2}$-vector with all entries equal to 1; also, $\vee$ is the coarsest-finer-than operator or join, while $\wedge$ is the finest-coarser-than operator or meet).

The number $|\mathcal{P}^N|$ of partitions of a $n$-set is the $n$-th *Bell number* $\mathcal{B}_n < 2^{\binom{n}{2}}$ $(n > 1)$ [13, pp. 70, 92] given by recursion $\mathcal{B}_0 = 1$ and $\mathcal{B}_n = \sum_{0 \leq k < n} \binom{n-1}{k} \mathcal{B}_k$. In view of the above argument, the MLE of partition functions extends them from the $\mathcal{B}_n$-set $\{I_P : P \in \mathcal{P}^N\} \subset \{0, 1\}^{\binom{n}{2}}$ of those vertices of the unit $\binom{n}{2}$-dimensional hypercube $[0, 1]^{\binom{n}{2}}$ corresponding to the indicator functions of partitions to the whole convex hull $co(\{I_P : P \in \mathcal{P}^N\})$ whose extreme points are precisely all $\mathcal{B}_n$ Boolean $\binom{n}{2}$-vectors given by indicator functions $I_P, P \in \mathcal{P}^N$. For notational convenience, let $\mathbb{P}_N = co(\{I_P : P \in \mathcal{P}^N\})$ be such a convex hull. When $N = \{1, 2, 3\}$ as above, convex polytope $\mathbb{P}_N$ is strictly included in $[0, 1]^3$ and its five extreme points [16,26] are $(0, 0, 0)$, $(1, 0, 0)$, $(0, 1, 0)$, $(0, 0, 1)$ and $(1, 1, 1)$. Also, $ex(\mathbb{P}_N) = \{I_P : P \in \mathcal{P}^N\}$ is the set of extreme points of $\mathbb{P}_N$.

Once this novel geometric perspective is clear, the MLE $f^h : \mathbb{P}_N \to \mathbb{R}$ of partition functions $h : \mathcal{P}^N \to \mathbb{R}$ readily obtains by means of the Möbius inversion $\mu^h : \mathcal{P}^N \to \mathbb{R}$ detailed in the previous subsection, just in the same way as for the MLE of set functions. In fact, consider $\binom{n}{2}$ variables $y_{[ij]_1}, \ldots, y_{[ij]_{\binom{n}{2}}} \in [0, 1]$ indexed by atoms $[ij]_1, \ldots, [ij]_{\binom{n}{2}} \in \mathcal{P}_{(1)}^N$. These are the analog of the $n$ variables $x_1, \ldots, x_n \in [0, 1]$ indexed by atoms $\{i\}, i \in N$ of Boolean lattice $(2^N, \cap, \cup)$ used for the MLE $f^w : [0, 1]^n \to \mathbb{R}$ of set functions $w : 2^N \to \mathbb{R}$ detailed in Sect. 2. For any partition function $h$, define its MLE $f^h$ by

$$f^h\left(y_{[ij]_1}, \ldots, y_{[ij]_{\binom{n}{2}}}\right) = \sum_{P \in \mathcal{P}^N}\left(\prod_{[ij] \leqslant P} y_{[ij]}\right)\mu^h(P), \qquad (11)$$

with $\mu^h$ given by

$$\mu^h(P) = \sum_{Q \in \mathcal{P}^N}\mu_{\mathcal{P}^N}(Q,P)h(Q) = \sum_{Q \leqslant P}\mu_{\mathcal{P}^N}(Q,P)h(Q) = h(P) - \sum_{Q < P}\mu^h(Q).$$

That is, $\mu^h$ is the Möbius inversion of $h$ in view of expression (10) defining the Möbius function of $\mathcal{P}^N$ from the previous subsection. Evidently, $f^h$ is multilinear. To see that (11) is indeed the sought extension, simply note that it coincides with $h$ on $ex(\mathbb{P}_N)$, i.e.

$$f^h(I_P) = \sum_{Q \in \mathcal{P}^N}\left(\prod_{[ij] \leqslant Q} I_P([ij])\right)\mu^h(Q) = \sum_{Q \leqslant P}\mu^h(Q) = h(P)$$

for all $P \in \mathcal{P}^N$, since

$$\prod_{[ij] \leqslant Q} I_P([ij]) = \begin{cases} 1 \text{ if } Q \leqslant P, \\ 0 \text{ if } Q \nleqslant P. \end{cases}$$

This is because for any partition $Q \in \mathcal{P}^N, Q \nleqslant P$ there is some atom $[ij]^*$ such that $Q \geqslant [ij]^* \nleqslant P$, entailing $I_P([ij]^*) = 0$ and thus $\prod_{[ij] \leqslant Q} I_P([ij]) = 0$. Note that the analog of convention $\prod_{i \in \emptyset} x_i := 1$ applying to set functions [1, p. 157] now is $\prod_{[ij] \leqslant P_\perp} y_{[ij]} := 1$, or more generally $\prod_{[ij] \in \emptyset} y_{[ij]} := 1$, hence for any $y = \left(y_{[ij]_1}, \ldots, y_{[ij]_{\binom{n}{2}}}\right) \in \mathbb{P}_N$ it holds

$$f^h(y) = h(P_\perp) + \sum_{P > P_\perp}\left(\prod_{[ij] \leqslant P} y_{[ij]}\right)\mu^h(P).$$

This enables to briefly check the functioning for additively separable partition functions $h_w$, in that the results obtained in the previous subsection yield

$$f^{h_w}(y) = h_w(P_\perp) + \sum_{\substack{P > P_\perp \\ P \text{ modular}}}\left(\prod_{[ij] \leqslant P} y_{[ij]}\right)\mu^{h_w}(P)$$

$$= \sum_{i \in N} w(\{i\}) + \sum_{\substack{A \in 2^N \\ |A| > 1}}\left(\prod_{\{i,j\} \subseteq A} y_{[ij]}\right)\mu^w(A).$$

Finally, for the sake of completeness, it can be mentioned that symmetric set functions $w$ satisfy $|A| = |B| \Rightarrow w(A) = w(B)$ (or $|A| = |B| \Rightarrow \mu^w(A) = \mu^w(B)$) for all $A, B \in 2^N$. On the other hand, symmetric partition functions $h$ satisfy $c^P = c^Q \Rightarrow h(P) = h(Q)$ (or $c^P = c^Q \Rightarrow \mu^h(P) = \mu^h(Q)$) for all $P, Q \in \mathcal{P}^N$, where $c^P = (c_1^P, \ldots c_n^P) \in \mathbb{Z}_+^n$ is the class of $P$, i.e. $c_k^P = |\{A : A \in P, |A| = k\}|$ is the number of $k$-cardinal blocks of $P$, for $1 \leq k \leq n$ [14]. Therefore, if $h_w$ is additively separated by a symmetric set function $w$, then $h_w$ is symmetric as well. Nevertheless, within the continuum of set functions additively separating $h_w$, only $w$ is symmetric, while all others addively separating set functions $v$ (i.e. such that $h_w = h_v$) are *not* symmetric.

# 6   Near-Boolean Games

In view of the above definition of local maximizers relying on equilibrium conditions for strategic $n$-player games, and having mentioned additive separablity of partition functions or global games [19], it seems now useful to regard variables as players in near-Boolean coalition formation games (for pseudo-Boolean functions and coalitional games see [17, Sect. 3]).

**Definition 10.** *A near-Boolean $n$-player game is a triple $(N, F, \pi)$ such that $N = \{1, \ldots, n\}$ is the player set and $F$ is a near-Boolean function taking real values on profiles $\boldsymbol{q} = (q_1, \ldots, q_n) \in ex(\Delta_N)$ of strategies, while payoffs are efficiently assigned by $\pi : ex(\Delta_N) \to \mathbb{R}^n$ to players, i.e. $\pi(\boldsymbol{q}) = (\pi_1(\boldsymbol{q}), \ldots, \pi_n(\boldsymbol{q}))$ satisfies $\sum_{i \in N} \pi_i(\boldsymbol{q}) = F(\boldsymbol{q})$ at all $\boldsymbol{q} \in ex(\Delta_N)$.*

**Definition 11.** *A fuzzy near-Boolean $n$-player game is a triple $(N, \hat{F}, \pi)$ such that $N$ is the player set and $\hat{F}$ is the MLE of a near-Boolean function taking values on profiles $\boldsymbol{q} = (q_1, \ldots, q_n) \in \Delta_N$ of strategies, while $\pi : \Delta_N \to \mathbb{R}^n$ efficiently assigns payoffs to players, i.e. $\sum_{i \in N} \pi_i(\boldsymbol{q}) = \hat{F}(\boldsymbol{q})$ at all $\boldsymbol{q} \in \Delta_N$.*

In both near-Boolean games and fuzzy ones the player set is finite. Given this, a main distinction is between games where players have either finite or else infinite sets of strategies, with near-Boolean games in the former class and fuzzy ones in the latter. In addition, players may play either deterministic (i.e. pure) or else random (i.e. mixed) strategies. In the latter case equilibrium conditions are stated in terms of expected payoffs, and by means of fixed point arguments for upper hemicontinuous correspondences such conditions are commonly fulfilled [15, p. 260]. The sets of deterministic strategies in fuzzy near-Boolean games are precisely the sets of random strategies in near-Boolean games. Nevertheless, the payoffs for the fuzzy setting are not, in general, expectations.

The framework where these games seem useful is coalition formation, which combines both strategic and cooperative games [27]. A generic strategy profile $\boldsymbol{q} \in ex(\Delta_N)$ of near-Boolean (non-fuzzy) games may well fail to be exact, but it is plain from Sect. 5 that the partition $P(\boldsymbol{q})$ of $N$ with associated exact $\mathbf{p} \in ex(\Delta_N)$ satisfies $F(\mathbf{p}) = F(\mathbf{q})$. In this view, near-Boolean games model strategic coalition formation in a very handy manner, in that they totally by-pass the need to define a mechanism mapping strategy profiles into partitions of players or coalition structures. More precisely, a mechanism is a mapping $M : ex(\Delta_N) \to \mathcal{P}^N$ such that when each player $i \in N$ specifies a coalition $A_i \in 2_i^N$, then $M(A_1, \ldots, A_n) = P$ is a resulting coalition structure. If the $n$ specified coalitions $A_i, i \in N$ are such that for some partition $P$ it holds $A_i = A$ for all $i \in A$ and all $A \in P$, then $M(A_1, \ldots, A_n) = P$. Otherwise, the generated partition $P' = M(A_1, \ldots, A_n)$ depends on what mechanism is chosen, and generally may be a rather fine one, i.e. possibly consisting of many small blocks. Conversely, near-Boolean games do not need any mechanism, in that even if players' strategies $(q_1, \ldots, q_n) = \mathbf{q}$ are such that $\mathbf{q}$ is not exact and thus does not yield a partition, still the global worth $F(\mathbf{q}) = F(\mathbf{p})$ is that attained at the partition $P(\mathbf{q})$ (with associated exact $\mathbf{p}$) whose blocks $A \in P$ each include maximal subsets of players choosing the same superset $A' \supseteq A$.

For a (non-fuzzy) near-Boolean game $(N, F, \pi)$, with underlying set function or coalitional game $w : 2^N \to \mathbb{R}_+, w(\emptyset) = 0$, consider the case where payoffs $\pi$ are defined as follows: for every block $A \in P(\mathbf{q})$ of the partition $P(\mathbf{q})$ associated with any strategy profile $\mathbf{q} \in ex(\Delta_N)$ and for every player $i \in A$, let

$$\pi_i(\mathbf{q}) = \sum_{B \in 2^A \setminus 2^{A \setminus i}} \frac{\mu^w(B)}{|B|}.$$

This is in fact the near-Boolean version of a well-known coalition formation game, with payoffs given by the *Shapley value* [28].

**Definition 12.** *For a near-Boolean function $F$, any $\mathbf{q} \in ex(\Delta_N)$ is a local max-imizer if for all $i \in N$ and all $q_i' \in ex(\Delta_i)$ inequality $F(\mathbf{q}) \geq F(q_i'|\mathbf{q}_{-i})$ holds.*

If payoffs are given by $\pi_i(\mathbf{q}) = \frac{\omega_i F(\mathbf{q})}{\sum_{j \in N} \omega_j}$ with $\omega_j > 0$ for all $j \in N$, then near-Boolean games are (pure) common interest potential games [29,30], meaning that the set of equilibria of $(N, F, \pi)$ coincides with the set of local maximizers of $F$, and players' preferences all agree on the set of strategy profiles.

## 7   Conclusions

Near-Boolean optimization is conceived to tackle problems where the instance includes a set function $w$ taking real values on a family of subsets of a finite set, and with the objective function assigning to any subfamily of pair-wise disjoint subsets the sum of their values quantified by $w$. Typically, such problems are set partitioning and set packing. Like in pseudo-Boolean optimization [1–3], the proposed method employs the polynomial multilinear extension of set functions to turn the discrete setting where the given problems are originally formulated into a continuous one, while extremizers are found where each variable is at an extreme point of its domain.

A main difference between pseudo-Boolean functions and near-Boolean ones is that the former coincide, in fact, with set functions, while for the latter the given $w$ may be replaced with a continuum of $w' \neq w$, and the associated polynomials all have common degree but different coefficients. This feature only characterizes additively separable partition functions [19,20], while for generic ones a novel MLE has been developed, with the corresponding derivatives and approximation issues (approached as in [17]) left for future work.

The near-Boolean setting also provides a novel type of coalition formation games, where the need to map strategy profiles into partitions of players by means of a mechanism is totally by-passed. In artificial intelligence, these games may model coalition structure generation in multiagent systems [31,32].

Finally, near-Boolean optimization seems generally interesting for objective function-based clustering [33], and for graph clustering in particular [34]. Specifically, for a given weighted graph, edge weights and vertices' weighted degrees may be used (in alternative ways) to obtain a quadratic MLE, i.e. a polynomial with degree 2. Then, optimization with respect to the resulting objective function provides a method for partitioning (i.e. clustering) the vertex set.

# References

1. Boros, E., Hammer, P.L.: Pseudo-Boolean optimization. Discrete Appl. Math. **123**, 155–225 (2002)
2. Crama, Y., Hammer, P.L.: Boolean Models and Methods in Mathematics, Computer Science, and Engineering. Encyclopedia of Mathematics and its Applications, vol. 134. Cambridge University Press, Cambridge (2010)
3. Crama, Y., Hammer, P.L.: Boolean Functions: Theory, Algorithms, and Applications. Cambridge University Press, Cambridge (2011)
4. Rossi, G.: Continuous set packing and near-Boolean functions. In: Proceedings of 5th International Conference on Pattern Recognition Applications and Methods, pp. 84–96 (2016)
5. Korte, B., Vygen, J.: Combinatorial Optimization: Theory and Algorithms. Springer, Heidelberg (2002)
6. Trevisan, L.: Non-approximability results for optimization problems on bounded degree instances. In: Proceedings 33rd ACM Symposium on Theory of Computing, pp. 453–461 (2001)
7. Papadimitriou, C.: Computational Complexity. Addison Wesley, Boston (1994)
8. Hazan, E., Safra, S., Schwartz, O.: On the complexity of approximating $k$-set packing. Comput. Complex. **15**(1), 20–39 (2006)
9. Sandholm, T.: Algorithm for optimal winner determination in combinatorial auctions. Artif. Intell. **135**(1–2), 1–54 (2002)
10. Milgrom, P.: Putting Auction Theory to Work. Cambridge University Press, Cambridge (2004)
11. Alidaee, B., Kochenberger, G., Lewis, K., Lewis, M., Wang, H.: A new approach for modeling and solving set packing problems. Eur. J. Oper. Res. **186**, 504–512 (2008)
12. Chandra, B., Halldórsson, M.M.: Greedy local improvement and weighted set packing. J. Algorithms **39**(2), 223–240 (2001)
13. Aigner, M.: Combinatorial Theory. Springer, Heidelberg (1997)
14. Rota, G.C.: On the foundations of combinatorial theory I: theory of Möbius functions. Z. Wahrscheinlichkeitsrechnung u. verw. Geb. **2**, 340–368 (1964)

15. Mas-Colell, A., Whinston, M.D., Green, J.R.: Microeconomic Theory. Oxford University Press, Oxford (1995)
16. Grünbaum, B.: Convex Polytopes, 2nd edn. Springer, Heidelberg (2001)
17. Hammer, P., Holzman, R.: Approximations of Pseudo-Boolean functions; applications to game theory. Math. Methods Oper. Res. - ZOR **36**, 3–21 (1992)
18. Stanley, R.: Modular elements of geometric lattices. Algebra Universalis **1**, 214–217 (1971)
19. Gilboa, I., Lehrer, E.: Global games. Int. J. Game Theory **20**(2), 120–147 (1990)
20. Gilboa, I., Lehrer, E.: The value of information - an axiomatic approach. J. Math. Econ. **20**, 443–459 (1991)
21. Stern, M.: Semimodular Lattices. Theory and Applications. Cambridge University Press, Cambridge (1999)
22. Knuth, D.E.: The Art of Computer Programming, Volume 4, Fascicle 3, Generating all Combinations and Partitions. Addison-Wesley, Boston (2005)
23. Kung, J.P.S., Rota, G.C., Yan, C.H.: Combinatorics: The Rota Way. Cambridge University Press, Cambridge (2009)
24. Rosas, M.H., Sagan, B.E.: Symmetric functions in noncommuting variables. Trans. Am. Math. Soc. **358**, 215–232 (2006)
25. Stanley, R.: Enumerative Combinatorics, vol. 2, 2nd edn. Cambridge University Press, Cambridge (2012)
26. Schrijver, A.: A Course in Combinatorial Optimization. Alexander Schrijver (2013). http://homepages.cwi.nl/~lex/files/dict.pdf
27. Slikker, M.: Coalition formation and potential games. Games Econ. Beh. **37**, 436–448 (2001)
28. Roth, A.: The Shapley Value. Cambridge University Press, Cambridge (1988)
29. Bowles, S.: Microeconomics: Behavior, Institutions, and Evolution. Princeton University Press, Princeton (2004)
30. Monderer, D., Shapley, L.S.: Potential games. Games Econ. Beh. **14**, 124–143 (1996)
31. Conitzer, V., Sandholm, T.: Complexity of constructing solutions in the core based on synergies among coalitions. Artif. Intel. **170**, 607–619 (2006)
32. Rahwan, T., Jenning, N.: An algorithm for distributing coalitional value calculations among cooperating agents. Artif. Intell. **171**, 535–567 (2007)
33. Rossi, G.: Multilinear objective function-based clustering. In: Proceedings of 7th International Joint Conference on Computational Intelligence, vol. 2(FCTA), pp. 141–149 (2015)
34. Schaeffer, S.E.: Graph clustering. Comput. Sci. Rev. **1**, 27–64 (2007)

# Approximate Inference in Related Multi-output Gaussian Process Regression

Ankit Chiplunkar[1,3(✉)], Emmanuel Rachelson[2],
Michele Colombo[1], and Joseph Morlier[2,3]

[1] Airbus Operations S.A.S., 31060 Toulouse, France
{ankit.chiplunkar,michele.colombo}@airbus.com
[2] ISAE-Supaero, Département d'Ingénierie des Systèmes Complexes (DISC),
31055 Toulouse, France
{emmanuel.rachelson,joseph.morlier}@isae-supaero.fr
[3] Université de Toulouse, CNRS, ISAE-SUPAERO,
Institut Clément Ader (ICA), 31077 Toulouse Cedex 4, France

**Abstract.** In Gaussian Processes a multi-output kernel is a covariance function over correlated outputs. Using a prior known relation between outputs, joint auto- and cross-covariance functions can be constructed. Realizations from these joint-covariance functions give outputs that are consistent with the prior relation. One issue with gaussian process regression is efficient inference when scaling upto large datasets. In this paper we use approximate inference techniques upon multi-output kernels enforcing relationships between outputs. Results of the proposed methodology for theoretical data and real world applications are presented. The main contribution of this paper is the application and validation of our methodology on a dataset of real aircraft flight tests, while imposing knowledge of aircraft physics into the model.

**Keywords:** Gaussian process · Kernel methods · Approximate inference · Multi-output regression · Flight-test data

## 1 Introduction

The main difference between the physical sciences and machine learning can be explained by the difference between deduction and induction. Physical sciences is deduction: where a very general formula is applied to a particular case. The

A. Chiplunkar—Ph.D. Candidate, Flight Physics Airbus Operations, 316 Route de Bayonne, 31060 Toulouse, France.

E. Rachelson—Associate Professor, Université de Toulouse, ISAE-SUPAERO, Département d'Ingénierie des Systèmes Complexes (DISC), 10 Avenue Edouard Belin, 31055 Toulouse Cedex 4, France.

M. Colombo—Loads and Aeroelastics Engineer, Flight Physics, 316 Route de Bayonne, 31060 Toulouse, France.

J. Morlier—Professor, Université de Toulouse, CNRS, ISAE-SUPAERO, Institut Clément Ader (ICA), 10 Avenue Edouard Belin, 31077 Toulouse Cedex 4, France.

© Springer International Publishing AG 2017
A. Fred et al. (Eds.): ICPRAM 2016, LNCS 10163, pp. 88–103, 2017.
DOI: 10.1007/978-3-319-53375-9_5

basics of newtonian physics when applied to a particular aircraft geometry give us inertial loads. The basics of aerodynamics when applied to particular set of aircraft geometry and aircraft states give out aerodynamic pressures. Physical sciences take global rules and apply them to local configurations, whereas machine learning is induction [9]. It looks at local features and data, tries to find similarity measures between them and gives a global formula for the environment. For this reason machine learning algorithms perform better in presence of more and more data.

Unfortunately, gathering highly accurate data for physical systems is a costly exercise. Highly accurate CFD simulations may run for weeks and flight test campaigns cost millions of dollars. In this regard there is a dichotomy between these two fields of science, where machine learning needs more data for good performance but procuring data from physical systems is a costly exercise. In this work we consider using prior information of relationships between several outputs thereby effectively increasing the number of data-points.

We use multiple-output Gaussian Process (GP) regression [12] to encode the physical laws of the system and effectively increase the amount of training data points. Inference on multiple output data is also known as co-kriging [14], multi-kriging [3] or Gradient Enhanced Kriging. Using a general framework [7] to calculate covariance functions between multiple-outputs, we extend the framework of gradient enhanced kriging to integral enhanced kriging, quadratic enhanced kriging or any functional relationship between inputs.

Let us start by defining a $P$ dimensional input space and a $D$ dimensional output space. Such that $\{(x_i^j, y_i^j)\}$ for $j \in [1; n_i]$ are the training datasets for the $i^{th}$ output. Here $n_i$ is the number of measurement points for the $i^{th}$ output, while $x_i^j \in \mathbb{R}^P$ and $y_i^j \in \mathbb{R}$. We next define $x_i = \{x_i^1; x_i^2; \ldots; x_i^{n_i}\}$ and $y_i = \{y_i^1; y_i^2; \ldots; y_i^{n_i}\}$ as the full matrices containing all the training points for the $i^{th}$ output such that $x_i \in \mathbb{R}^{n_i \times P}$ and $y_i \in \mathbb{R}^{n_i}$. Henceforth we define the joint output vector $Y = [y_1; y_2; y_3; \ldots; y_D]$ such that all the output values are stacked one after the other. Similarly, we define the joint input matrix as $X = [x_1, x_2, x_3, \ldots, x_D]$. If $\Sigma n_i = N$ for $i \in [1, D]$. Hence $N$ represents the total number of training points for all the outputs combined. Then $Y \in \mathbb{R}^N$ and $X \in \mathbb{R}^{N \times P}$.

For simplicity take the case of an explicit relationship between two outputs $y_1$ and $y_2$. Suppose we measure two outputs with some error, while the true physical process is defined by latent variables $f_1$ and $f_2$. Then the relation between the output function, measurement error and true physical process can be written as follows.

$$y_1 = f_1 + \epsilon_{n1}$$
$$y_2 = f_2 + \epsilon_{n2} \tag{1}$$

Where, $\epsilon_{n1}$ and $\epsilon_{n2}$ are measurement error sampled from a white noise gaussian $\mathcal{N}(0, \sigma_{n1})$ and $\mathcal{N}(0, \sigma_{n2})$ respectively. While the physics based relation can be expressed as follows.

$$f_1 = g(f_2, x_1) \tag{2}$$

Here $g$ is an operator defining the relation between $f_1$ and an independent latent output $f_2$.

While a joint model developed using correlated covariance functions gives better predictions, it incurs a huge cost on memory occupied and computational time. The main contribution of this paper is to apply approximate inference on these models of large datasets and reduce the heavy computational costs incurred. For a multi-output GP as defined earlier the covariance matrix is of size $N$, needing $\mathcal{O}\left(N^3\right)$ calculations for inference and $\mathcal{O}\left(N^2\right)$ for storage. In this work we compare performance of variational inference [15] and distributed GP [8] to approximate the inference in a joint-kernel.

The remaining paper proceeds as follows, Sect. 2 provides the theoretical framework for Gaussian Process regression. Section 3 extends the multi-output GP regression in presence of correlated covariances. In Sect. 4 various methods of approximating inference of a multi-output GP are derived. Finally, in Sect. 5 we demonstrate the approach on both theoretical and flight-test data.

## 2   Gaussian Process Regression

A gaussian process is an infinite dimensional multi-variate gaussian. Such that any subset of the process is a multi-variate gaussian distribution. A gaussian process can be fully parametrized by a mean and covariance function Eq. 3.

$$y(x) = GP(m(x,\theta), k(x,x',\theta)) \tag{3}$$

A random draw from a Gaussian Process gives us a random function around the mean function $m(x,\theta)$ and of the shape as defined by covariance function $k(x,x',\theta)$. Hence, Gaussian Process gives us a method to define a family of functions whose shape is defined by its covariance function. A popular choice of covariance function is a squared exponential function Eq. 4, because it defines a family of highly smooth (infinitely differentiable) non-linear functions as shown in Fig. 1(a).

$$k(x,x',\theta) = \theta_1^2 exp[-\frac{(x-x')^2}{2\theta_2^2}] \tag{4}$$

A covariance function is fully parametrized by its hyperparameters $\theta_i$'s. For the case of Squared exponential kernel the hyperparameters are its amplitude $\theta_1$ and length scale $\theta_2$.

Given a dataset $(x,y)$ regression deals with finding latent function $f$ between the inputs $x$ and outputs $y$. While performing polynomial regression we assume that our function $f$ comes from a family of polynomial functions. Since Gaussian Processes are such handy tools to define a family of non-linear functions. In a Gaussian Process Regression (GPR) we start with an initial family of functions defined by a GP called prior Fig. 1(a).

$$\mathbb{P}(f \mid x, \theta) = GP(y|0, K_{xx}) \tag{5}$$

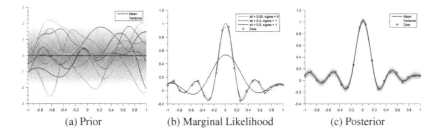

(a) Prior                    (b) Marginal Likelihood                    (c) Posterior

**Fig. 1.** Gaussian process regression.

Due to the bayesian setting of GP we can calculate the posterior mean and variance as shown in Eqs. 6 and 7. This means that we are effectively eliminating all the functions in the prior that do not pass through our data points Fig. 1(c).

$$m(y_*) = k_{x_* x}(K_{xx})^{-1} y \tag{6}$$

$$Cov(y_*) = k_{x_* x_*} - k_{x_* x} (K_{xx})^{-1} k_{xx_*} \tag{7}$$

We can also improve our predictions by choosing a better prior. This involves optimizing the Marginal Likelihood (ML) $\mathbb{P}(y \mid x, \theta)$ calculated as Eq. 8. The probability that our dataset $(x, y)$ comes from a family of functions defined by the prior is called the ML [12]. Hence, when we optimize the ML we are actually finding the optimal $\theta$ or family of functions that describe our data Fig. 1(b).

$$\mathbb{P}(y \mid x, \theta) = GP(y|0, K_{xx} + \sigma^2 I) \tag{8}$$

## 3    Multi-output Gaussian Process

Given a dataset for multiple outputs $\{(x_i, y_i)\}$ for $i \in [1; D]$ we define the joint output vector $Y = [y_1; y_2; y_3; \ldots; y_D]$ such that all the output values are stacked one after the other. Similarly, we define the joint input matrix as $X = [x_1; x_2; x_3; \ldots; x_D]$. For the sake of simplicity, suppose we measure two outputs $y_1$ and $y_2$ with some error, while the true physical process is defined by latent variables $f_1$ and $f_2$ Eq. 2. The operator $g(.)$ can be a known physical equation or a computer code between the outputs, it basically represents a transformation from one output to another.

### 3.1    Related Work

Earlier work developing such joint-covariance functions [2] have focused on building different outputs as a combination of a set of latent functions. GP priors are placed independently over all the latent functions thereby inducing a correlated covariance function. More recently it has been shown that convolution processes [1,3] can be used to develop joint-covariance functions for differential equations. In a convolution process framework output functions are generated by convolving several latent functions with a smoothing kernel function. In the current paper we assume one output function to be independent and evaluate the remaining auto- and cross-covariance functions exactly if the physical relation between them is linear [13] or use approximate joint-covariance for non-linear physics-based relationships between the outputs [7].

## 3.2 Multi-output Joint-Covariance Kernels

A GP prior in such a setting with 2 output variables is expressed in Eq. 9.

$$\begin{bmatrix} f_1 \\ f_2 \end{bmatrix} \sim GP \left[ \begin{pmatrix} 0 \\ 0 \end{pmatrix}, \begin{pmatrix} K_{11} & K_{12} \\ K_{21} & K_{22} \end{pmatrix} \right] \tag{9}$$

$K_{12}$ and $K_{21}$ are cross-covariances between the two inputs $x_1$ and $x_2$. $K_{22}$ is the auto-covariance function of independent output, while $K_{11}$ is the auto-covariance of the dependent output variable. The full covariance matrix $K_{XX}$ is also called the joint-covariance. While, the joint error matrix will be denoted by $\Sigma$;

$$\Sigma = \begin{bmatrix} \sigma_{n1}^2 & 0 \\ 0 & \sigma_{n2}^2 \end{bmatrix} \tag{10}$$

Where, $\epsilon_{n1}$ and $\epsilon_{n2}$ are measurement error sampled from a white noise gaussian $\mathcal{N}(0, \sigma_{n1})$ and $\mathcal{N}(0, \sigma_{n2})$.

For a linear operator $g(.)$ the joint-covariance matrix can be derived analytically [14], due to the affine property of Gaussian's Eq. 11.

$$\begin{bmatrix} f_1 \\ f_2 \end{bmatrix} \sim GP \left[ \begin{pmatrix} 0 \\ 0 \end{pmatrix}, \begin{pmatrix} g(g(K_{22}, x_2), x_1) & g(K_{22}, x_1) \\ g(K_{22}, x_2) & K_{22} \end{pmatrix} \right] \tag{11}$$

Using the known relation between outputs we have successfully correlated two GP priors from Eq. 2. This effectively means that when we randomly draw a function $f_2$ it will result in a correlated draw of $f_1$ such that the two draws satisfy the Eq. 2. We have effectively represented the covariance function $K_{11}$ in terms of the hyperparameters of covariance function $K_{22}$ using the known relation between outputs.

Without loss of generality we can assume that the independent output $f_2$ belongs to a family of functions defined by a Squared Exponential kernel. The joint-covariance between $f_1$ and $f_2$, means that a random draw of independent function $f_2$ will result in a correlated draw of the function $f_1$. The Fig. 2(a) shows random draws coming from a differential relationship between $f_1$ (red) and $f_2$ (blue) such that $f_1 = \frac{\partial f_2}{\partial x}$. We can see that the top figure is derivative of the bottom one since $f_{derivative}$ goes to zero where $f_{independent}$ goes to maxima or minima. Similarly, the Fig. 2(b) shows random draws coming from an integral relationship between $f_1$ (red) and $f_2$ (blue) such that $f_1 = \int f_2$. We can see that the top figure is integral of the bottom one since $f_{independent}$ goes to zero where $f_{integral}$ goes to maxima or minima.

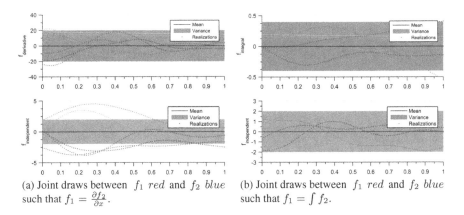

(a) Joint draws between $f_1$ *red* and $f_2$ *blue* such that $f_1 = \frac{\partial f_2}{\partial x}$.

(b) Joint draws between $f_1$ *red* and $f_2$ *blue* such that $f_1 = \int f_2$.

**Fig. 2.** Multi-output gaussian process random draws. (Color figure online)

### 3.3  GP Regression Using Joint-Covariance

We start with defining a zero-mean prior for our observations and make predictions for $y_1(x_*) = y_{*1}$ and $y_2(x_*) = y_{*2}$. The corresponding prior according to Eqs. 9 and 10 will be:

$$\begin{bmatrix} Y(X)) \\ Y(X_*)) \end{bmatrix} = GP\left[ \begin{bmatrix} 0 \\ 0 \end{bmatrix}, \begin{bmatrix} K_{XX} + \Sigma & K_{XX_*} \\ K_{X_*X} & K_{X_*X_*} + \Sigma \end{bmatrix} \right] \tag{12}$$

The posterior distribution is then given as a normal distribution with expectation and covariance matrix given by [12]

$$m(y_*) = K_{X_*X}(K_{XX})^{-1}Y \tag{13}$$

$$Cov(y_*) = K_{X_*X_*} - K_{X_*X}(K_{XX})^{-1}K_{XX_*} \tag{14}$$

Here, the elements $K_{XX}$, $K_{X_*X}$ and $K_{X_*X_*}$ are block covariances derived from Eq. 11. Due to the bayesian setting we have basically eliminated all the functions that do not pass through the points defined by the observed data.

The joint-covariance matrix depends on several hyperparameters $\theta$. They define a basic shape of the GP prior. To end up with good predictions it is important to start with a good GP prior. We minimize the negative log-marginal likelihood to find a set of good hyperparameters. This leads to an optimization problem where the objective function is given by Eq. 15

$$\log(\mathbb{P}(y \mid X, \theta)) = \log[GP(Y|0, K_{XX} + \Sigma)] \tag{15}$$

With its gradient given by Eq. 16

$$\frac{\partial}{\partial \theta} \log(\mathbb{P}(y \mid X, \theta)) = \frac{1}{2} Y^T K_{XX}^{-1} \frac{\partial K_{XX}}{\partial \theta} K_{XX}^{-1} Y$$

$$- \frac{1}{2} tr(K_{XX}^{-1} \frac{\partial K_{XX}}{\partial \theta}) \tag{16}$$

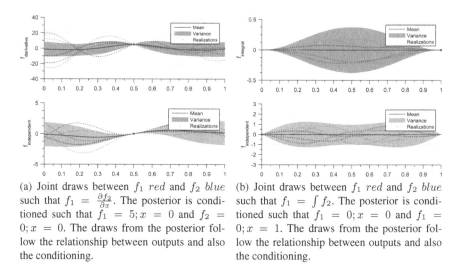

(a) Joint draws between $f_1$ *red* and $f_2$ *blue* such that $f_1 = \frac{\partial f_2}{\partial x}$. The posterior is conditioned such that $f_1 = 5; x = 0$ and $f_2 = 0; x = 0$. The draws from the posterior follow the relationship between outputs and also the conditioning.

(b) Joint draws between $f_1$ *red* and $f_2$ *blue* such that $f_1 = \int f_2$. The posterior is conditioned such that $f_1 = 0; x = 0$ and $f_1 = 0; x = 1$. The draws from the posterior follow the relationship between outputs and also the conditioning.

**Fig. 3.** Multi-output Gaussian process regression predictions. (Color figure online)

Here the hyperparameters of the prior are $\theta = \{l_2, \sigma_2^2, \sigma_{n1}^2, \sigma_{n2}^2\}$. These correspond to the hyperparameters of the independent covariance function $K_{22}$ and errors in the measurements $\sigma_{n1}^2$ and $\sigma_{n2}^2$. Calculating the negative log-marginal likelihood involves inverting the matrix $K_{XX} + \Sigma$. The size of the $K_{XX} + \Sigma$ matrix depends on total number of input points $N$, hence inverting the matrix becomes intractable for large number of input points.

The Fig. 3(a) shows mean and variance for a differential relationship between independent function $f_2$ (blue) and differential function $f_1$ (red) and such that $f_1 = \frac{\partial f_2}{\partial x}$. We have conditioned the two functions such that $f_1 = 5; x = 0$ and $f_2 = 0; x = 0$. This means that the function $f_2$ passes through 0 and has a derivative equal to 5 at $x = 0$. We have not maximized the marginal likelihood for this case and the hyperparameters of $K_{22}$ are $\theta_1 = 1; \theta_2 = 0.2$. All the corresponding draws from the posterior GP also follow the conditioning.

The Fig. 3(b) shows mean and variance for a integral relationship between independent function $f_2$ (blue) and differential function $f_1$ (red) and such that $f_1 = \int f_2.dx$. We have conditioned the two functions such that $f_1 = 0; x = 1$ and $f_1 = 0; x = 0$. This means that the function $f_2$ has an integral 0 at the two points $[0, 1]$. We have not maximized the marginal likelihood for this case and the hyperparameters of $K_{22}$ are $\theta_1 = 1; \theta_2 = 0.2$. All the corresponding draws from the posterior GP also follow the conditioning. We can observe that the draws will also have an integral 0 between the range $[0, 1]$.

In the next section we describe how to solve the problem of inverting huge $K_{XX} + \Sigma$ matrices using approximate inference techniques.

## 4    Approximating Inference

The above GP approach is intractable for large datasets. For a multi-output GP as defined in Sect. 3.2 the covariance matrix is of size $N$, where $\mathcal{O}\left(N^3\right)$ time is needed for inference and $\mathcal{O}\left(N^2\right)$ memory for storage. Thus, we need to consider approximate methods in order to deal with large datasets.

Inverting the covariance matrix takes considerable amount of time and memory during the process. Hence, almost all techniques to approximate inference try and approximate the inversion of covariance matrix $K_{XX}$. If a covariance matrix is diagonal or block-diagonal in nature then methods such as mixture of experts are used eg. distributed GP. Whereas if the covariance matrix is more spread out and has similar terms in its cross diagonals then low-rank approximations are used eg. variational approximation. The remaining section details the two methods for approximating covariance matrix which can be later used to resolve Eqs. 13, 14 and 16.

### 4.1    Variational Approximation on Multi-output GP

Sparse methods use a small set of $m$ function points as support. or inducing variables. Suppose we use $m$ inducing variables to construct our sparse GP. The inducing variables are the latent function values evaluated at inputs $x_M$. Learning $x_M$ and the hyperparameters $\theta$ is the problem we need to solve in order to obtain a sparse GP method. An approximation to the true log marginal likelihood in Eq. 15 can allow us to infer these quantities (Fig. 4).

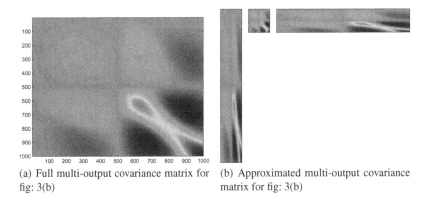

(a) Full multi-output covariance matrix for fig: 3(b)

(b) Approximated multi-output covariance matrix for fig: 3(b)

**Fig. 4.** Variational approximation of covariance matrix for Gaussian process regression.

We try to approximate the joint-posterior distribution $p(X|Y)$ by introducing a variational distribution $q(X)$. In the case of varying number of inputs for different outputs, we place the inducing points over the input space and extend the derivation of [15] to multi-output case.

$$q(X) = \mathcal{N}(X|\mu, A) \qquad (17)$$

Here $\mu$ and $A$ are parameters of the variational distribution. We follow the derivation provided in [15] and obtain the lower bound of true marginal likelihood.

$$F_V = log(\mathcal{N}[Y|0, \sigma^2 I + Q_{XX}]) - \frac{1}{2\sigma^2} Tr(\tilde{K}) \tag{18}$$

where $Q_{XX} = K_{XX_M} K_{X_M X_M}^{-1} K_{X_M X}$ and $\tilde{K} = K_{XX} - K_{XX_M} K_{X_M X_M}^{-1} K_{X_M X}$. $K_{XX}$ is the joint-covariance matrix derived using Eq. 11 using the input vector $X$ defined in Sect. 1. $K_{X_M X_M}$ is the joint-covariance function on the inducing points $X_M$, such that $X_M = [x_{M1}, x_{M2}, ..., x_{M2}]$. We assume that the inducing points $x_{Mi}$ will be same for all the outputs, hence $x_{M1} = x_{M2} = ... = x_{M2} = x_M$. While $K_{XX_M}$ is the cross-covariance matrix between $X$ and $X_M$.

Note that this bound consists of two parts. The first part is the log of a GP prior with the only difference that now the covariance matrix has a lower rank of $MD$. This form allows the inversion of the covariance matrix to take place in $\mathcal{O}\left(N(MD)^2\right)$ time. The second part as discussed above can be seen as a penalization term that regularizes the estimation of the parameters.

The bound can be maximized with respect to all parameters of the covariance function; both model hyperparameters and variational parameters. The optimization parameters are the inducing inputs $x_M$, the hyperparameters $\theta$ of the independent covariance matrix $K_{22}$ and the error while measuring the outputs $\sigma$. There is a trade-off between quality of the estimate and amount of time taken for the estimation process. On the one hand the number of inducing points determine the value of optimized negative log-marginal likelihood and hence the quality of the estimate. While, on the other hand there is a computational load of $\mathcal{O}\left(N(MD)^2\right)$ for inference. We increase the number of inducing points until the difference between two successive likelihoods is below a predefined quantity.

### 4.2    Distributed Inference on Multi-output GP

An alternative to sparse approximations is to learn local experts on subset of data. Traditionally, each subset of data learns a different model from another, this is done to increase the expressiveness in the model [11]. The final predictions are then made by combining the predictions of local experts [5].

An alternative way is to tie all the different experts using one single set of hyperparameters [8]. This is equivalent to assuming one single GP on the whole dataset such that there is no correlation across experts as seen in Fig. 5(b). This tying of experts acts as a regularization and inhibits overfitting. Although ignoring correlation among experts is a strong assumption, it can be justified if the experts are chosen randomly and with enough overlap.

If we partition the dataset into M subsets such as $\mathcal{D}^{(i)} = X^{(i)}, y^{(i)}, i = 1, ... M$.

$$\log p(y|X, \theta) \approx \sum_{k=1}^{M} \log p_k(y^{(i)}|X^{(i)}, \theta) \tag{19}$$

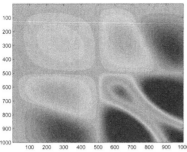
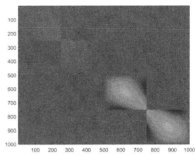

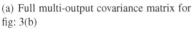

(a) Full multi-output covariance matrix for fig: 3(b)

(b) Distributed multi-output covariance matrix for fig: 3(b) the cross-diagonal terms (blue) are zero due to independence between experts

**Fig. 5.** Distributed approximation of covariance matrix for Gaussian process regression. (Color figure online)

The above Eq. 19 describes the formulation for marginal likelihood. Due to the independence assumption the marginal likelihood can be written as a sum of individual likelihoods and then can be optimized to find the best-fit hyperparameters. After learning the hyperparameters we can combine the predictions of local experts to give mean and variance predictions. The robust Bayesian Commitee Machine (rBCM) model combines the various experts using their confidence on the prediction point [8]. In such manner experts which have high confidence at the prediction points get more weight when compared to experts with low confidence.

$$m(Y_*) = (Cov(X_*))^{-2} \sum \beta_k \sigma_k^{-2} m_k(X_*) \qquad (20)$$

$$(Cov_((Y_*))^{-2} = \sum_k \beta_k \sigma_k^{-2} + (1 - \sum_k \beta_k)\sigma_{**}^{-2} \qquad (21)$$

In the above equations $m_k(X_*)$ and $\sigma_k$ are the mean and covariance predictions from expert $k$ at point $X_*$. $\sigma_{**}$ is the auto-covariance of the prior at prediction points $X_*$. $\beta_k$ determines the influence of experts on the final predictions [4] and is given as $\beta_k = \frac{1}{2}(\log \sigma_{**}^{-2} - \log \sigma_k^{-2})$.

## 5   Experiments

We empirically assess the performance of distributed Gaussian Process and Variational Inference with respect to the training time and accuracy. We start with a synthetic dataset where we try to learn the model over derivative relationship and compare the two inference techniques. We then evaluate the improvement on real world flight-test dataset.

The basic toolbox used for this paper is GPML provided with [12], we generate covariance functions to handle relationships as described in Eq. 11 using the

"Symbolic Math Toolbox" in MATLAB 2014b. Variational inference is wrapped from gpStuff toolbox [16] and distributed GP is inspired from [8]. All experiments were performed on an Intel quad-core processor with 4 Gb RAM.

## 5.1  Experiments on Theoretical Data

We consider a derivative relationship between two output functions as described in Eq. 2. Such that

$$g(f, x) = \frac{\partial f}{\partial x}$$

Since the differential relationship $g(.)$ is linear in nature we use the Eq. 11 to calculate the auto- and cross-covariance functions as shown in Table 1.

**Table 1.** Auto- and cross-covariance functions for a differential relationship.

| Initial covariance | $K_{22}$ | $\sigma^2 exp(\frac{-1}{2}\frac{d^2}{l^2})$ |
|---|---|---|
| Cross-covariance | $K_{12}$ | $\sigma^2 \frac{d}{l^2} exp(\frac{-1}{2}\frac{d^2}{l^2})$ |
| Auto-covariance | $K_{11}$ | $\sigma^2 \frac{d^2-l^2}{l^4} exp(\frac{-1}{2}\frac{d^2}{l^2})$ |

Data is generated from Eq. 22, a random function is drawn from GP to get $f_2$ whose derivative is then calculated to generate $f_1$. $y_1$ and $y_2$ are then calculated by adding noise according the Eq. 22. 10,000 points are generated for both the outputs $y_1$ and $y_2$. Values of $y_2$ are masked in the region $x \in [0, 0.3]$ the remaining points now constitute our training dataset.

$$f_2 \sim GP[0, K_{SE}(0.1, 1)]$$

$$\sigma_{n2} \sim \mathcal{N}[0, 0.2]$$

$$\sigma_{n1} \sim \mathcal{N}[0, 2] \tag{22}$$

$K_{SE}(0.1, 1)$ means squared exponential kernel with length scale 0.1 and variance as 1.

Figure 6 shows comparison between an independent fit GP and a joint multi-output GP whose outputs are related through a derivative relationship described in Sect. 5.1. For Fig. 6(a) using variational inference algorithm we optimize the lower bound of log-marginal likelihood, for independent GP's on $y_1$ and $y_2$. For Fig. 6(b) using variational inference we optimize the same lower bound but with a joint-covariance approach as described in Sect. 4.1 using $y_1$, $y_2$ and $g(.)$. We settled on using 100 equidistant inducing points for this exercise [6] and have only optimized the hyperparameters to learn the model.

Figure 6(a) shows the independent fit of two GP for the differential relationship, while Fig. 6(b) shows the joint GP fit. The GP model with joint-covariance gives better prediction even in absence of data of $y_2$ for $x \in [0, 0.3]$.

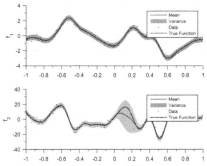 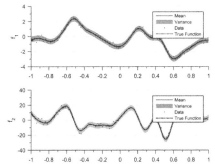

(a) Independent fit for two GP's mean is represented by solid black line. $2\Sigma$ confidence band is represented by light red for $f_1$ and light blue for $f_2$. The dashed black line represents the true latent function values; noisy data is denoted by blue dots. Experiment was run on 10,000 points but only 100 data points are plotted to increase readibility. Inference is performed using variational inference algorithm and equidistant 100 inducing points. We can observe the huge difference between the real data and the predicted mean values at zone with no data.

(b) Joint multi-output GP's for two outputs, mean is represented by solid black line. $2\Sigma$ confidence band is represented by light red for $f_1$ and light blue for $f_2$. The dashed black line represents the true latent function values; noisy data is denoted by blue dots. Experiment was run on 10,000 points but only 100 data points are plotted to increase readibility. Inference is performed using variational inference algorithm and equidistant 100 inducing points. We can observe the improved prediction between zone with no data because information is being shared between the two outputs.

**Fig. 6.** Experimental results for differential relationship while using variational approximation. (Color figure online)

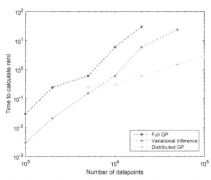 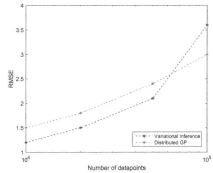

(a) Comparison of time to calculate negative log marginal likelihood for a Full GP, Variational inference and Distributed GP with increasing number of datapoints. We observe for datapoints greater that 10e5 the distributed GP algorithm starts outperforming variational inference

(b) Comparison of RMSE for variational inference and distributed GP algorithm. We observe for datapoints greater that 10e4 the distributed GP algorithm starts outperforming variational inference.

**Fig. 7.** Comparison of run time and RMSE between distributed GP and variational inference.

For the second experiment we compare the Root Mean Squared Error (RMSE) and run-times of distributed GP and Variational Inference algorithms while performing approximate inference. We progressively generate from 10e3 to 10e5 data-points according to the Eq. 22 and Sect. 5.1. We separated 75% of the data as training set and 25% of the data as the test set, the training and test sets were chosen randomly. The variational inference relationship kernel as described in Sect. 4.1 with 100 equidistant inducing points was used. We learn the optimal values of hyper-parameters for all the sets of training data. The distributed GP algorithm as described in Sect. 4.2 was used with randomly chosen 512 points per expert. We learn the optimal values of hyper-parameters for all the sets of training data. The accuracy is plotted as RMSE values with respect to the test set. The runtime is defined as time taken to calculate negative log marginal likelihood Eqs. 18 and 19. The RMSE values are calculated for only the dependent output $y_1$ and then plotted in the Fig. 7(a).

In Fig. 7(a) the time to calculate negative log-marginal likelihood with increasing number of training points is calculated. As expected the full GP takes more time when compared to variational inference or distributed GP algorithms. The Variational inference algorithm has better run-time till 10e4 data-points after that distributed GP takes lesser time. In Fig. 7(b) the RMSE error with test set is compared between the variational inference and distributed GP algorithm. Here too the variational inference algorithm performs better lesser number of datapoints but distributed GP starts performing better when we reach more than 10e4 datapoints.

One thing to note is that we have fixed the number and position of inducing points while performing this experiment. While a more optimized set of inducing points will have better results, for datasets of the order 10e5 distributed GP algorithm starts outperforming variational inference. Moreover, upon observing Fig. 5(b) we can say that the covariance matrix in a joint GP setting is not diagonal in nature and hence an approximation technique which can compensate between low-rank approximation and diagonal approximation should be investigated [10].

## 5.2   Experiments on Flight Test Data

We perform experiments on flight loads data produced during flight test phase at Airbus. Loads are measured across the wing span using strain gauges. Shear load $T_z$ and bending moment $M_x$ as described in Fig. 8 are used as two outputs for this exercise. $\eta$ or point of action of forces and angle of attack $\alpha$ are the two inputs. The aircraft is in quasi-equillibrium in all conditions and there are no dynamic effects observed throughout this dataset. All data is normalized according to airbus policy.

The relation between $T_z$ and $M_x$ can be written as:

$$M_x(\eta, \alpha) = \int_{\eta}^{\eta_{edge}} T_Z(x, \alpha)(x - \eta) dx. \tag{23}$$

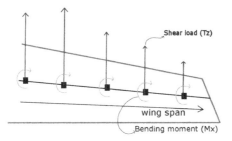

**Fig. 8.** Wing load diagram.

The Eq. 23 is applicable for the $\eta$ axis. Here, $\eta_{edge}$ denotes the edge of the wing span. The forces are measured at 5 points on the wing span and at 8800 points on the $\alpha$ dimension. We compare plots of relationship-imposed multioutput GP and independent GP. Then we compare the measures of negative-log marginal likelihood and RMSE for varying number of inducing points.

Figure 9(a) shows the independent (blue shade) and joint fit (red shade) of two GP. The top figure shows $T_Z$ with the variance of dependent GP plotted in red and variance of independent GP plotted in blue. Bottom figure shows plots for $M_X$. Since the number of input points is less than 10e5 we use variational inference. 100 inducing points in the input space are used to learn and plot the

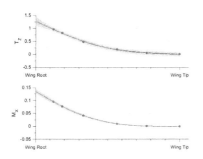

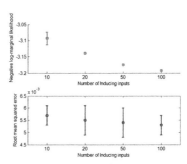

(a) $2\sigma$ confidence interval and mean of the dependent GP are represented in red shade and solid red line. $2\sigma$ confidence interval and mean of the independent GP are represented in blue shade and solid blue line. Experiment was run on 8800 data points Noisy data is denoted by circles only 1 $\alpha$ step is plotted. Confidence interval improves upon adding the relationship kernel.

(b) Progression of RMSE and log-likelihood upon increasing number of inducing points. Top plot shows the value of mean and variance of negative log-marginal likelihood. The bottom figure in blue shows the mean and variance of root mean squared error. 10 sets of experiments were run on 75% of the data as training set and 25% of the data as the test set, the training and test sets were chosen randomly.

**Fig. 9.** Experimental results for aircraft flight loads. (Color figure online)

figure. The variance of red is smaller than that of blue showing the improvement in confidence when imposing relationships in the GP architecture. The relationship between $T_Z$ and $M_X$ gives rise to better confidence during the loads prediction. This added information is very useful when identifying faulty sensor data since Eq. 23 will push data points which do not satisfy the relationship out of the tight confidence interval.

Figure 9(b) shows improvement in the negative log-marginal likelihood and RMSE plots upon increasing number of inducing points. 10 sets of experiments were run on 75% of the data as training set and 25% of the data as the test set, the training and test sets were chosen randomly. We learn the optimal values of hyper-parameters and inducing points for all the 10 sets of experiments of training data. Finally, RMSE values are evaluated with respect to the test set and negative log-marginal likelihood are evaluated for each learned model. The RMSE and log-likelihood improve upon increasing the number of inducing points.

## 6    Conclusions and Future Work

This paper presents approximate inference methods for physics-based multiple output GP's. We extend the variational inference and distributed GP inference techniques to be applied on physics-based multi-output GP's and reduce the computational load for inference.

Section 5.1 demonstrates the advantage of using multi-output GP in presence of prior known relationships on a theoretical dataset. Significant improvement in prediction can be observed in presence of missing data due to transfer of information occuring due to the prior relationship. Then, we compare the difference in accuracy and run times upon using distributed GP or variational inference to approximate inference. Although variational inference performs better for datasets of size less than 10e4, distributed GP becomes significantly advantageous for datasets greater than 10e5.

Section 5.2 shows real world application of joint kernel on flight loads data. We demonstrate that adding prior physics-based relationships allows us to create a robust, physically consistent and interpretable surrogate model for these loads. Aircraft flight domain consists of various maneuvers, this is further complicated by several relationships between outputs. In the future we wish to develop better strategies to exploit the clustered nature of flight-domain and more than two relationships.

## References

1. Alvarez, M.A., Luengo, D., Lawrence, N.D.: Latent force models. In: Dyk, D.A.V., Welling, M. (eds.) AISTATS, JMLR Proceedings, vol. 5, pp. 9–16. JMLR.org (2009)
2. Bonilla, E., Chai, K.M., Williams, C.: Multi-task Gaussian process prediction. In: Platt, J., Koller, D., Singer, Y., Roweis, S. (eds.) Advances in Neural Information Processing Systems 20, pp. 153–160. MIT Press, Cambridge (2008)

3. Boyle, P., Frean, M.: Dependent Gaussian processes. In: Advances in Neural Information Processing Systems, vol. 17, pp. 217–224. MIT Press (2005)
4. Cao, Y., Fleet, D.J.: Generalized product of experts for automatic and principled fusion of Gaussian process predictions. CoRR abs/1410.7827 (2014). http://arXiv.org/abs/1410.7827
5. Chen, T., Ren, J.: Bagging for Gaussian process regression. Neurocomputing **72**(7), 1605–1610 (2009)
6. Chiplunkar, A., Rachelson, E., Colombo, M., Morlier, J.: Sparse physics-based Gaussian process for multi-output regression using variational inference. In: Proceedings of the 5th International Conference on Pattern Recognition Applications and Methods, pp. 437–445 (2016)
7. Constantinescu, E.M., Anitescu, M.: Physics-based covariance models for Gaussian processes with multiple outputs. Int. J. Uncertain. Quantif. **3**, 47–71 (2013)
8. Deisenroth, M.P., Ng, J.W.: Distributed Gaussian processes. arXiv preprint arXiv:1502.02843 (2015)
9. Domingos, P.: A few useful things to know about machine learning. Commun. ACM **55**(10), 78–87 (2012)
10. March, W.B., Xiao, B., Biros, G.: Askit: approximate skeletonization kernel-independent treecode in high dimensions. SIAM J. Sci. Comput. **37**(2), A1089–A1110 (2015)
11. Rasmussen, C.E., Ghahramani, Z.: Infinite mixtures of Gaussian process experts. In: Advances in Neural Information Processing Systems, vol. 2, pp. 881–888 (2002)
12. Rasmussen, C.E., Williams, C.K.I.: Gaussian Processes for Machine Learning (Adaptive Computation and Machine Learning). MIT Press, Cambridge (2005)
13. Solak, E., Murray-smith, R., Leithead, W.E., Leith, D.J., Rasmussen, C.E.: Derivative observations in gaussian process models of dynamic systems. In: Becker, S., Thrun, S., Obermayer, K. (eds.) Advances in Neural Information Processing Systems 15, pp. 1057–1064. MIT Press, Cambridge (2003)
14. Stein, M.L.: Interpolation of Spatial Data: Some Theory for Kriging. Springer, New York (1999)
15. Titsias, M.K.: Variational learning of inducing variables in sparse gaussian processes. In: Artificial Intelligence and Statistics, vol. 12, pp. 567–574 (2009)
16. Vanhatalo, J., Riihimäki, J., Hartikainen, J., Jylänki, P., Tolvanen, V., Vehtari, A.: GPstuff: Bayesian modeling with Gaussian processes. J. Mach. Learn. Res. **14**(1), 1175–1179 (2013)

# An Online Data Validation Algorithm for Electronic Nose

Mina Mirshahi, Vahid Partovi Nia$^{(\boxtimes)}$, and Luc Adjengue$^{(\boxtimes)}$

Department of Mathematics and Industrial Engineering,
Polytechnique Montreal, Montreal, QC, Canada
{mina.mirshahi,vahid.partovinia,luc.adjengue}@polymtl.ca
http://www.polymtl.ca

**Abstract.** An electronic nose (e-nose) is a device that analyzes the chemical components of an odour. The e-nose consists of an array of gas sensors for chemical detection, and a mechanism for pattern recognition to return the odour concentration. Odour concentration defines the identifiability and perceivability of an odour. Given that accurate prediction of odour concentration requires valid measurements, automatic assessment of sampled measurements is of prime importance. The impairment of the e-nose, and environmental factors (including wind, humidity, temperature, etc.) may introduce significant amount of noise. Inevitably, the pattern recognition results are affected. We propose an online algorithm to evaluate the validity of sensor measurements during the sampling before using the data for pattern recognition phase. The proposed algorithm is computationally efficient and straightforward to implement.

**Keywords:** Artificial olfaction · Computational complexity · Electronic nose · Gas sensor · Outlier detection · Robust covariance estimation

## 1 Introduction

### 1.1 Background

The recognition of chemicals in the environment is an essential need for the living organisms. Odours are detected through millions of olfactory receptors that are located at the top of nasal cavities. The human olfactory system consists of three main components: (1) an array of olfactory receptors (2) the olfactory bulb that receives neural inputs about odours detected by the receptors and (3) the brain. The olfactory system collects a sample from its environment and transmit it to the brain, where it is recognized as a specific odour.

An olfactory system is able to detect a broad range of smells. However, the human olfacotry system fails to respond to many air pollutants; people can have different sensitivity to many air pollutants and even be accustomed to toxic smells.

The contamination of air by harmful chemicals is referred to as air pollution and is one of the biggest concerns worldwide. This is mainly because the air

A. Fred et al. (Eds.): ICPRAM 2016, LNCS 10163, pp. 104–120, 2017.
DOI: 10.1007/978-3-319-53375-9_6

pollution has direct influence on the environmental and human health. Auditing odourants is a crucial element in assessment of indoor and outdoor air quality. There are various odour measurement techniques such as dilution-to-threshold, olfactometers, and referencing techniques [25]. The dependence of these methods on human evaluation makes them less accurate and sometimes undesirable.

The concept of an artificial olfaction was introduced by [29]. The primary artificial olfaction rely on a gas multisensor array. The term electronic nose (e-nose) appeared for the first time in the early 1990s [12]. E-nose is designed for recognizing simple or complex odours in its environment and it comprises two main elements of hardware and software. The hardware usually include a set of gas sensors (such as metal oxide semi-conductors, conducting polymers, etc.) with partial specificity, air conditioner, flow controller, electronics, and many more components. The software consists of statistical methods for pre-processing the data and pattern recognition methods for predicting the odour concentration.

The gas sensors of e-nose should have certain features. Similar to human nose receptors, the gas sensors of e-nose need to be highly sensitive with respect to chemical compounds and less sensitive towards temperature and humidity. In addition, the sensors should be able to respond to various chemical compounds. Among the other features, one can name durability, selectivity, and easy calibration.

Gas sensor's performance is affected by various elements. One of the most serious deterioration in sensors is owing to a phenomenon called *drift*. Drift is the low frequency change in a sensor that causes offset measurements. Sensor drift, therefore, need to be detected and compensated to guarantee accurate sensor measurement. Several methods have been introduced to overcome the drift phenomenon including [2, 5, 28, 34].

The multivariate response of gas sensor arrays undergoes different pre-processing procedures before the prediction is performed using statistical tools such as regression, classification, or clustering. [3, 15, 23] have discussed methods for analyzing the gas sensor array data.

## 1.2   Motivation

The e-nose is capable of reproducing the human sense of smell using an array of gas sensors and pattern recognition methods. Pattern recognition methods use a set of labelled data to predict the odour concentration for each set of sensor measurements. The labelled data consist of a sub-sample of sensors' outputs considered for further analyses of its concentration in olfactometry.

One of the application of e-nose is in environmental activities; e-noses provide industries with odour management plan to minimize the effect of odour in the environment. To this end, e-noses are installed in outdoor fields such as compost sites, landfill sites, waste water plants, etc., where the environmental condition can greatly fluctuate. Consequently, the occurrence of unwanted variability is very typical.

During the sampling process, sensors in the e-nose device may report incorrect values or some of the sensors stop functioning for a short period of time. These anomalies are ought to be diagnosed and reported in real time using a computationally efficient algorithm, which is the focus of this research.

We propose an online data validation algorithm which compares e-nose measurements with a set of reference samples and allocate them accordingly to different zones. The zones are distinguished from each other using distinct colors like green, yellow, red, etc., to represent the extent of the validity of the measurement. The main focus of this work is summarized in the below flowchart, Fig. 1.

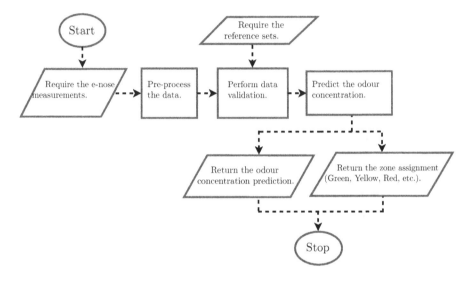

**Fig. 1.** A schematic flowchart of the proposed online task for an e-nose.

### 1.3    Data Preparation

The e-nose relies on a sensor array consists of several gas sensors. The number of the sensors depends on the purpose of analysis. Each sensor represents an attribute; the more the sensors are, the better the e-nose discriminate among analytes. Nonetheless the inclusion of too many sensors can lead to unnecessary data and a complex system.

The e-nose under the study includes 11 senors each designed to be responsive to a specific chemical compound in the air. However, senors react to almost all gases as they may not be highly selective. As a result, some of the sensors are highly positively correlated with each other, see Figs. 2 and 3 (left panel). Consider the data matrix $\mathbf{X}_{n \times p}$ with its rows being $n$ independent realization of 11 sensor values, $\mathbf{x}_{p \times 1}^{\top}$ in which $\mathbf{a}^{\top}$ indicates the transpose of the vector $\mathbf{a}$.

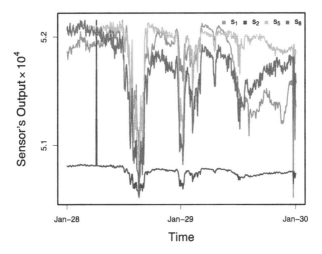

**Fig. 2.** Senor's output during three days of sampling for 4 randomly selected sensors.

The covariance matrix of $\mathbf{x}_{p \times 1}$, say $\boldsymbol{\Sigma} = [\sigma_{ij}]_{i,j=1,2,\ldots,p}$, is defined as

$$\boldsymbol{\Sigma}_{p \times p} = \text{Cov}(\mathbf{x}) = \text{E}\{(\mathbf{x} - \boldsymbol{\mu})(\mathbf{x} - \boldsymbol{\mu})^{\top}\},$$

where $\boldsymbol{\mu}$ represents the mean of $\mathbf{x}$, and E is the mathematical expectation. The covariance, $\sigma_{ij}$, measures the degree to which two attributes are linearly associated. However, in order to have a better idea about the relationship between two attributes, one needs to eliminate the effect of other attributes. The partial correlation is the correlation between two attributes, while controlling for the effects of other attributes. The inverse of covariance matrix is commonly known as precision or concentration matrix. The entries of $\boldsymbol{\Sigma}^{-1}$ have an interpretation in terms of partial correlation. Non-zero elements of $\boldsymbol{\Sigma}^{-1}$ implies conditional dependence. Therefore, the sparse estimation of $\boldsymbol{\Sigma}^{-1}$ pinpoints the block structure of attributes. Sparse estimation of $\boldsymbol{\Sigma}^{-1}$ set some of the $\boldsymbol{\Sigma}^{-1}$ entries to zero. Investigation of the inherent dependence between the sensor values is then performed by means of the partial correlation.

Here, the *graphical lasso* [11] is considered for a better understanding of the existing relationship between the sensor values. [11] proposed estimating the covariance matrix such that its inverse, $\boldsymbol{\Sigma}^{-1}$, is sparse by applying a *lasso penalty* [33]. In Fig. 3 (right panel), the undirected graph connects two attributes which are conditionally correlated given all other attributes. The sensors 5 to 8 are correlated with each other conditioning on the effect of the others. This is also reflected in the heatmap of the correlation matrix Fig. 3 (left panel). This dependence must be taken into account while modelling the data. Gaussianity of the data is another crucial assumption that should be verified. The validity of this assumption for the sensor values is tested using various methods such as analyzing the distribution of individual sensor values, scatter plot of the

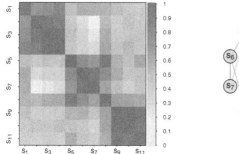

**Fig. 3.** Left panel, heatmap of the correlation matrix of the sensor values ($s_1$–$s_{11}$). Right panel, the undirected graph of partial correlation using the graphical lasso. The undirected graph of the right panel approves the block structure of the heatmap of the left panel.

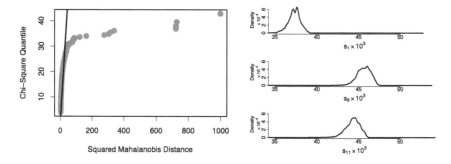

**Fig. 4.** Left panel, the Q-Q plot of squared Mahalanobis distance supposed to follow the chi-squared distribution for Gaussian data. Right panel, the marginal density for some randomly chosen sensor values. Both graphs confirm the non-Gaussianity of data.

linear projection of data using principal components, estimating the multivariate kurtosis and skewness, and also multivariate Mardia test, see Fig. 4.

## 2    Data Analysis

We aim to verify the validity of e-nose measurements by considering some reference samples for the purpose of comparison. These reference samples are collected when the e-nose functions normally, and the conditions are fully under control. The e-nose measurements are compared with reference samples and are allocated to various zones accordingly. These zones are distinguished by various colors, like green, yellow, red, etc., to indicate the status of e-nose measurements [26].

Two distinct reference sets, if applicable, are recommended for data validation. *Reference* 1 consists of data in a period of sampling defined by an expert after installation of the e-nose. The data in this period of sampling is called as *proposed set*. *Reference* 2, upon its availability, is manually gathered samples from the field that are brought to the laboratory for quantification of their odour concentration. The data in this period of sampling is called *calibration set*, to emphasize that it can be incorporated for data modelling using a supervised learning algorithm.

If a new datum does not follow the overall pattern of data previously observed, then it is marked as an outlier and is assigned to Red zone. This zone represents a dramatic change in the pattern of samples and is referred to as "risky" observations. If the new datum is not an outlier and it is also located within the data polytope of the Reference 1 or the Reference 2, it is allocated to Green or Blue zone respectively. These zones represent the "safe" observations. If the new datum is not an outlier, but outside of the area of Green and Blue zones, they are assigned to Yellow zone. This zone displays potentially "critical" observations.

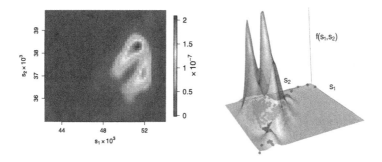

**Fig. 5.** Validity assessment for about 700 samples based on 2 sensor values. Left panel, the plot illustrates the contour map of estimated density function for the 2 sensors. Right panel, the density function of the samples demonstrated in $3D$ with zones identified for each of the samples in the sensor 1 ($s_1$) versus sensor 2 ($s_2$) plane. Higher density is assigned to the Green, Blue, and Orange zones compared to the Yellow and Red zones. (Color figure online)

If large proportion of samples belong to the Yellow and Red zones, the reliability of the system should be suspected. Undesirable measurements can be the outcome of physical complications, such as sensor loss in the e-nose, or sudden changes in the chemical pattern of the environment. Zone assignment, therefore, require some outlier detection algorithms. For the Green and the Blue zones, the new samples are projected onto a subspace with lower dimension. Dimension reduction methods such as principal component analysis (PCA) can serve for this purpose [20]. PCA attempts to explain the data covariance matrix, $\hat{\boldsymbol{\Sigma}}$, by a set of components; these components are the linear combination of the primary attributes. PCA, basically, converts a set of possibly correlated attributes into a set of linearly uncorrelated axes through orthogonal linear transformations.

The first $k$ $(k < p)$ principal components are the eigenvectors of the covariance matrix $\boldsymbol{\Sigma}$ associated with the $k$ largest eigenvalues. The classical estimation of covariance matrix, $\hat{\boldsymbol{\Sigma}}$, is strongly influenced by outliers [30]. As producing outlier is typical of sensor data, robust covariance estimation must be applied to avoid misleading results.

Robust principal component analysis [17] is employed for dimension reduction purpose throughout this article. This robust PCA computes the covariance matrix through projection pursuit [24] and minimum covariance determinant [7] methods. The robust PCA procedure can be summarized as follows:

1. The matrix of data is pre-processed such that the data spread in the subspace of at most $\min(n - 1, p)$.
2. In the spanned subspace, the most obvious outliers are diagnosed and removed from data. The covariance matrix is calculated for the remaining data, $\hat{\boldsymbol{\Sigma}}_0$.
3. $\hat{\boldsymbol{\Sigma}}_0$ is used to decide about the number of principal components to be retained in the analysis, say $k_0$ $(k_0 < p)$.
4. The data are projected onto the subspace spanned by the first $k_0$ eigenvectors of $\hat{\boldsymbol{\Sigma}}_0$.
5. The covariance matrix of the projected points is estimated robustly using minimum covariance determinant method and its $k$ leading eigenvalues are computed. The corresponding eigenvectors are the robust principal components.

The Red zone represents the outliers of the samples as being measured by the e-nose through time. One common approach for detecting outliers in multivariate data is to use the Mahalonobis ditstance

$$D_m(\mathbf{x}_i) = \sqrt{(\mathbf{x}_i - \hat{\mu})^\top \hat{\boldsymbol{\Sigma}}^{-1}(\mathbf{x}_i - \hat{\mu})}. \tag{1}$$

The large value of $D_m(\mathbf{x}_i)$ for $i = 1, 2, \ldots, n$, indicates that the observation $\mathbf{x}_i$ locates away from the centre of the data $\hat{\mu}$. As the estimation of $\mu$ and $\boldsymbol{\Sigma}$ is itself affected by outliers, the use of Eq. (1) is inadvisable for outlier detection. Even the robust plugin estimation of $\mu$ and $\boldsymbol{\Sigma}$ do not lead to any improvement as long as the associated outlier detection cut-offs are based on elliptical distributions. [4, 18] suggested an outlier detection method which does not assume any elliptical distribution for data. Their method is formed on a modified version of Stahel-Donoho outlyingness measure [9, 32] and is called *adjusted outlyingness* (AO) criterion. For the observation $\mathbf{x}_i$, the Stahel-Donoho measure is

$$\mathrm{SD}(\mathbf{x}_i) = \sup_{\mathbf{a} \in \mathbf{R}^p} \frac{|\mathbf{a}^\top \mathbf{x}_i - \mathrm{median}(\mathbf{X}_n \mathbf{a})|}{\mathrm{mad}(\mathbf{X}_n \mathbf{a})}, \tag{2}$$

where $\mathrm{mad}(\mathbf{X}_n \mathbf{a}) = 1.483 \, \mathrm{median}_i |\mathbf{a}^\top \mathbf{x}_i - \mathrm{median}(\mathbf{X}_n \mathbf{a})|$ is the median absolute deviation. The SD measure essentially looks for outliers by projecting each observation on many univariate directions. As it is not applicable to look for all possible directions, it is suggested that considering $250p$ directions, where $p$ is the number of attributes, suffices and produces efficient results. Taking into account the effect of skewness in the SD measure results in the following AO

$$
AO_i = \sup_{\mathbf{a}\in\mathbf{R}^p} \begin{cases} \frac{\mathbf{a}^\top \mathbf{x}_i - \text{median}(\mathbf{X}_n\mathbf{a})}{w_2 - \text{median}(\mathbf{X}_n\mathbf{a})} & \text{if } \mathbf{a}^\top \mathbf{x}_i > \text{median}(\mathbf{X}_n\mathbf{a}), \\ \frac{\text{median}(\mathbf{X}_n\mathbf{a}) - \mathbf{a}^\top \mathbf{x}_i}{\text{median}(\mathbf{X}_n\mathbf{a}) - w_2} & \text{if } \mathbf{a}^\top \mathbf{x}_i < \text{median}(\mathbf{X}_n\mathbf{a}), \end{cases} \tag{3}
$$

where $w_1$ and $w_2$ are the lower and upper whiskers of the adjusted boxplot [19]. If the $AO_i$ exceeds the upper whisker of the adjusted boxplot, it is then detected as an outlier.

The sample that is rendered as an outlier by AO measure, belongs to the Red zone. For the specification of the remaining zones, we need to define the polytopes of the samples in Reference 1 and Reference 2. These polytopes are built using the convex hull of the robust principal component *scores*. More specifically, the boundary of the Green zone is defined by computing the convex hull of the robust principal component scores of the Reference 1.

Before determining the color tag for each new data, the samples are checked for missing values and are imputed if needed by *multivariate imputation* methods such as [21]. The idea behind the validity assessment is visualized in Fig. 5. For simplicity, only 2 sensors are used for all computations in Fig. 5 and a $2D$ presentation of zones is plotted using the sensors' coordinates. Suppose that $\mathbf{X}_{n\times 11}$ represents the matrix of sensor values for $n$ samples, $\mathbf{y}_n$ the vector of corresponding odour concentration values and $\mathbf{x}_l^\top$ is the $l$th row of $\mathbf{X}_{n\times 11}$, $l = 1, 2, \ldots, n$. Furthermore, suppose that $n_1$ refers to the number of samples in the proposed set of the sampling and $n_2$ refers to the number of samples in the calibration set. The samples of the proposed set are always available, but not necessarily the calibration set. Two different scenarios occur based on the availability of the calibration set.

If the calibration set is accessible, then Scenario 1 happens. Otherwise, we only deal with Scenario 2. Scenario 1 is a general case which is explained more in detail. The data undergo a pre-processing stage, including imputation and outlier detection, before any further analyses. Having done the pre-processing stage, data are stored as Reference 1, $\mathbf{X}_{n_1\times 11}$, and Reference 2, $\mathbf{X}_{n_2\times 11}$. The first $k$, e.g. $k = 2, 3$, robust principal components of $\mathbf{X}_{n_1\times 11}$ are calculated and the corresponding *loading* matrix is denoted by $\mathbf{L}_1$. The pseudo code of two algorithms for Scenario 1 is provided below. Scenario 2 is a special case of Scenario 1 in which Sub-Algorithm (Scenario 1) is used with $\text{ConvexHull}^{(2)} = \varnothing$ that eliminates the Blue and Orange zones. Consequently, there is no model for odour concentration prediction in the Main Algorithm.

In Sect. 3, a set of simulated data is used to verify the relevancy of our proposed algorithm and the choice of statistical methods. The applicability of our algorithm is also tested based on 8 months sampling from the e-nose in Sect. 4.

**Algorithm.** Sub-Algorithm (Scenario 1).
___
1: **if** the point $\mathbf{x}_l^\top$, $l = 1, 2, \ldots, N$ is identified as an outlier by $AO$ measure **then**
2:     $\mathbf{x}_l^\top$ is in Red zone,
3: **else if** $\mathbf{x}_l^\top \mathbf{L}_1 \in \text{ConvexHull}^{(1)}$ AND $\mathbf{x}_l^\top \mathbf{L}_1 \notin \text{ConvexHull}^{(2)}$ **then**
4:     $\mathbf{x}_l^\top$ is in Green zone,
5: **else if** $\mathbf{x}_l^\top \mathbf{L}_1 \notin \text{ConvexHull}^{(1)}$ AND $\mathbf{x}_l^\top \mathbf{L}_1 \in \text{ConvexHull}^{(2)}$ **then**
6:     $\mathbf{x}_l^\top$ is in Blue zone,
7: **else if** $\mathbf{x}_l^\top \mathbf{L}_1 \in \text{ConvexHull}^{(1)}$ AND $\mathbf{x}_l^\top \mathbf{L}_1 \in \text{ConvexHull}^{(2)}$ **then**
8:     $\mathbf{x}_l^\top$ is in Orange zone,
9: **else**
10:     $\mathbf{x}_l^\top$ is in Yellow zone.
11: **end if**
___

**Algorithm.** Main Algorithm (Scenario 1).
___
**Require:** $\mathbf{X}_{n_1 \times 11}$, $\mathbf{X}_{n_2 \times 11}$, and the loading matrix $\mathbf{L}_1$ using robust PCA over Reference 1, $\mathbf{X}_{n_1 \times 11}$.
1: $\text{ConvexHull}^{(1)} \leftarrow$ the convex hull of the projected values of the Reference 1, $\mathbf{X}_{n_1 \times 11}\mathbf{L}_1$.
2: Train a supervised learning model on Reference 2, $\mathbf{X}_{n_2 \times 11}$, and its odour concentration vector, $\mathbf{y}_{n_2}$.
3: $\text{ConvexHull}^{(2)} \leftarrow$ the convex hull of the projected values of the Reference 2, $\mathbf{X}_{n_2 \times 11}\mathbf{L}_1$.
4: Do **Sub-Algorithm** for new data $\mathbf{x}^*$.
5: Predict the odour concentration for new data $\mathbf{x}^*$ using the trained supervised learning model.
___

## 3   Simulation

We examine the methodology on two sets of simulated data to highlight the importance of the assumptions such as non-elliptical contoured distribution and robust estimation considered in our methodology. In each example, we stored the simulated data in the matrix $\mathbf{X}_{n \times 2}$, where $\mathbf{x}_l^\top = (x_{l1}, x_{l2})$; $l = 1, 2, \ldots, n$.

In the first example, the data is simulated from a mixture distribution with 10% contamination. The elements of mixture distribution are chosen arbitrarily from Gaussian and the Student's t-distribution.

We simulated data from the bivariate skew t-distribution [14] in the second example in order to test the effect of skewness on our algorithm.

Using classical approaches for outlier detection without considering the actual data distribution, mistakenly renders many observations as outliers, Figs. 6 and 7 (top right panel). The parameters of interest, the mean vector and the covariance matrix, need to be estimated robustly, otherwise the confidence region misrepresents the underlying distribution. In Figs. 6 and 7 (bottom left panel), the classical confidence region is pulled toward the outlier observations. On the contrary, the robust confidence region perfectly unveil the distribution of the majority of observations because of the robust and efficient estimation of the mean and the covariance matrix. Consequently, the classical principal components are affected by the inefficient estimation of the covariance matrix. We proposed using methods that deal with asymmetric data appropriately. Adjusted outlyingness (AO)

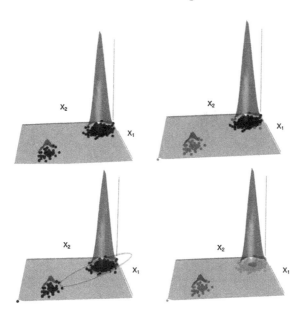

**Fig. 6.** Top left panel, the simulated data from the mixture distribution $f(x) = (1 - \varepsilon)f_1(x) + \varepsilon f_2(x)$ with contamination proportion of $\varepsilon = \frac{1}{10}$, and $f_1$ and $f_2$ being the Gaussian and Student's t-distribution respectively. Top right panel, the outliers of data are identified and highlighted with red using the classical Mahalonobis distance and 95th percentile of the chi-squared distribution with two degrees of freedom. Bottom left panel, the 95% confidence region for the data is computed using the classical estimates of parameters (cyan) and the robust estimates (gold). Bottom right panel, the Main Algorithm is implemented and the zones are plotted using their associated color tag. (Color figure online)

measure identifies the outliers of the data correctly. Considering a sub-sample of data as Reference 1 in each of the examples, the result of the Main Algorithm can be observed in the right bottom panel of Figs. 6 and 7.

## 4    Experiment

In order to evaluate the performance of our data validation method, we implement the Main Algorithm on a collection of e-nose measurements. We decide to keep the first 3 robust principle components of the data $PC1$, $PC2$, $PC3$ for simplification and the easy visualization. The 3 principal components correspond to the 3 largest eigenvalues of the robust covariance matrix. Prior to the implementation of the Main Algorithm, the data undergoes a pre-processing stage including the imputation of the missing values.

The validity of the e-nose measurements are identified using the Main Algorithm for the 8 months of sampling. In favor of more readable graphs, only a subset of 500 samples out of 200 thousands of observations are plotted. In Fig. 8,

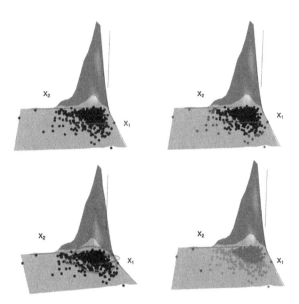

**Fig. 7.** Top left panel, the simulated data from the bivariate skew t-distribution. Top right panel, the outliers of data are identified and highlighted with red using the classical Mahalonobis distance and 95th percentile of the chi-squared distribution with two degrees of freedom. Bottom left panel, the 95% confidence region for the data is computed using the classical estimates of parameters (cyan) and the robust estimates (gold). Bottom right panel, the Main Algorithm is implemented and the zones are plotted using their associated color tag. (Color figure online)

the sample points are drawn in gray and each zone is highlighted using its corresponding color. The circles in Fig. 8 are also illustrated on $PC1$ and $PC2$ plane for a better demonstration of the zones.

The interpretation of a zone is heavily depends on its definition. For instance, the Green, Blue, and Orange zones, represent samples that are very close the samples that have already been observed in either Reference 1 or Reference 2. As the observations in reference sets were entirely under control, the Green, Blue, and Orange zones affirm the validity of the samples. In addition, the accuracy of the gas concentration predicted for these zones is certified. On the other hand, the gas concentration prediction for samples in the Red zone is less accurate compared with that of the Green, Blue, and Orange zones.

The data that are significantly dissimilar to the already observed data deserve further attention. These data are outliers and are reported in the Red zone. Similarly, the gas concentration predictions associated with samples in the Red zone can be very mis-leading. Generating a remarkable percentage of samples belonging to the Yellow and the Red zones refers to the possible failure of the e-nose equipment.

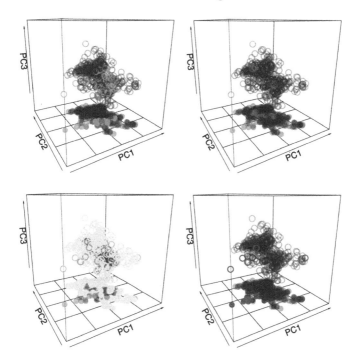

**Fig. 8.** A random sample of size $n = 500$ is plotted over the first three robust principal components coordinates. From top left panel to bottom right panel, the colored blobs represent green, blue, yellow, and red zones respectively. (Color figure online)

# 5   Computational Complexity

Here, we discuss the computational complexity of our proposed algorithm (Main Algorithm). First, a brief introduction to computational complexity is given to facilitate the understanding.

The computational complexity of an algorithm is studied asymptotically by the big O-notation [1]. The big O-notation explains how quickly the run-time of an algorithm grows relative to its input. For instance, sum of $n$ values require $(n-1)$ operations. Consequently, the mean requires $n$ operations reserving one for the division of the sum by $n$. As they are both bounded by a linear function, they have computational complexity of order $O(n)$. In other words, the performance of the sum and mean grow linearly and in a direct proportion to the size of the input. Note that not all algorithms are computationally linear. Computational complexity of covariance matrix, for instance, is $O(np^2)$ where $n$ is the sample size and $p$ is the number of attributes. Since each covariance calls for sum of the pairwise cross-products each of complexity $O(n)$. In total, there are $\frac{p(p-1)}{2}$ off-diagonal cross products and $p$ square sums for the diagonal entries of the covariance matrix. This yields $n\{p(p-1)+p\}$ operations. For a fixed number of attributes $p$, the computation is of order $O(n)$. Likewise, for a fixed number of observations the computation is of order $O(p^2)$. Another nontrivial example for non-linear algorithm is PCA or the robust PCA. Computation of robust principal components involves various operations that has been briefly discussed in Sect. 2. Computational complexity of robust PCA is discussed below. Computation of robust PCA comprises the following steps:

1. Reducing the data space to an affine subspace spanned by the $n$ observations using singular value decomposition of $(\mathbf{X} - \mathbf{1}_n\hat{\boldsymbol{\mu}})^\top (\mathbf{X} - \mathbf{1}_n\hat{\boldsymbol{\mu}})$, where $\mathbf{1}_n$ is the column vector of $n$ dimension with all entries equal to 1. This step is of order $O(p^3)$, see [13,16].
2. Finding the least outlying points using the Stahel-Donoho affine-invariant outlyingness [9,32]. Adjusting this outlyingness measure by the minimum covariance determinant location and scale estimators is of order $O(pn \log n)$, see [17,18]. Then the covariance matrix of the non-outliers data, $\hat{\boldsymbol{\Sigma}}_0$, is calculated which is computationally less expensive.
3. Performing the principal component analysis on $\hat{\boldsymbol{\Sigma}}_0$ and choosing the number of projection components (say $k_0 < p$) to be retained. Computing the $\hat{\boldsymbol{\Sigma}}_0$ needs $np^2$ operations. Thus its complexity is $O(np^2)$. The spectral decomposition of the covariance matrix is achieved by applying matrix-diagonalization method, such as singular value decomposition or Cholesky decomposition. This results in $O(p^3)$ computational complexity. Determining the $k_0$ largest eigenvalues and their corresponding eigenvectors has time complexity of $O(k_0 p^2)$ [10]. As a result, the time complexity of this step is $O(np^2)$.
4. Projecting the data onto the subspace spanned by the first $k_0$ eigenvectors, i.e. $(\mathbf{X} - \mathbf{1}_n\hat{\boldsymbol{\mu}})\mathbf{P}_{p \times k_0}$ where $\mathbf{P}_{p \times k_0}$ is the matrix of eigenvectors corresponding to the first $k_0$ eigenvalues. This step has $O(npk_0)$ time complexity.
5. Computing the covariance matrix of the projected points using the method of fast minimum covariance determinant has the computational complexity which is sub-linear in $n$, for fixed $p$. This is $O(n)$ [31]. The calculation of the spectral decomposition of the final covariance matrix is bounded by $O(nk_0)$ time complexity.

**Remark 1.** The computational complexity of robust PCA is $O(\max\{pn \log n, np^2\})$, or $O(p^2 n \log n)$ considering the worst case complexity.

To ascertain the complexity of the Main Algorithm, one needs to analyze each step separately. The measurement validation in e-nose broadly necessitates the calculation of certain steps of the Main Algorithm including Step Require, Step 1, Step 3, and Step 4. All these tasks excluding Step 4 of the Main Algorithm (Sub-Algorithm) must be run only once. Step 4 duplicates upon the arrival of the new observations.

First, we start by evaluating the complexity of Step Require, Step 1, and Step 3 that should be run once. Afterwards Step 4 is analyzed in a similar fashion. Note that for the e-nose data, the number of samples is generally much greater than the number of sensors $p$. In addition, as the number of sensors $p$ is fixed in an e-nose equipment, the computational complexity is reported as the function of number of samples only.

The Main Algorithm starts with the robust PCA over the Reference 1. As a result, Step Require has $O(\{n_1 \log n_1\})$ complexity assuming $p$ to be fixed. Step 1 requires $O(n_1 k_0)$ computing time for computing $\mathbf{X}_{n_1 \times 11} \mathbf{L}_1$ where $k_0$ stands for the number of eigenvectors retained in the loading matrix $\mathbf{L}_1$. Computing the convex hull of these projected values for $k_0 \leq 3$ is of order $O(n_1 \log n_1)$. For $k_0 > 3$, the computational complexity of hull increases exponentially with $k_0$, see [6,27]. Similarly, the same complexity is valid for Step 3. Performing some pre-processing steps on the Reference sets including outlier detection using AO measure has $O(n_1 \log n_1)$ complexity [18] assuming that $n_1 > n_2$, which is common in practice. As a result, Step Require, Step 1, and Step 3 which is performed only once take $O(n_1 \log n_1)$ run-time.

Now, we analyze Step 4 in terms of its computational complexity. Step 4 mainly does the following three tasks.

(i) Accumulating the new observations with the past history, $\mathbf{X}_{1:t \times p}^{\top} = [\mathbf{X}_{1:t-1 \times p}^{\top} : \mathbf{x}_{t \times p}]$ where $n_1 < t \leq n$, and identifying outliers using AO measure. This has computational complexity of $O(t \log t)$.

(ii) Projecting the observations onto the space of Reference 1, $\mathbf{x}_l^{\top} \mathbf{L}_1$. This is a simple matrix product and has the computational complexity of $O(k_0 p)$.

(iii) Verifying whether the projection of data, $\mathbf{x}_l^{\top} \mathbf{L}_1$, locates within the convex hull of either Reference 1 or Reference 2 which is equivalent to solving a linear optimization with linear constraints [8,22]. The algorithm used for this purpose has computational complexity which varies quadratically with respect to the number of vertices of the convex hull, and has $O(n_1^2 k_0)$ complexity in the worst case. The R code used for solving this linear program resembles the MATLAB code[1] and is available upon the request.

Thus, the computational complexity of Step 4 is $O(t \log t)$ as in practice the convex hull of Reference 1 is computed, in Step 1, and kept fixed prior to this step.

**Remark 2.** The computational complexity of Main Algorithm is $O(t \log t)$.

The mean CPU time in seconds for Step Require, Step 1, and Step 3 that need to be run once and Step 4 which duplicates for each new sample, are reported in Fig. 9. This figure confirms that the run-times for the ensemble of the steps Require, 1, 3 and 4 agree with the computational complexity evaluated theoretically earlier. This implies that measurement validation can be achieved with $O(t \log t)$ time complexity employing our proposed method.

## 6   Conclusion

An electronic nose device, which mainly consists of a multi-sensor array, attempts to mimic the human olfactory system. The sensor array is composed of various

---

[1] http://www.mathworks.com/matlabcentral/fileexchange/10226-inhull.

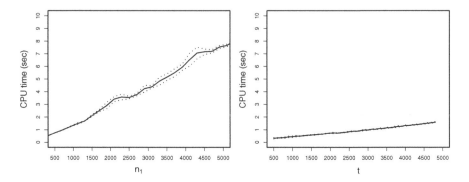

**Fig. 9.** The solid line shows the mean CPU time in seconds as a function of input being run on 1.3 GHz i5 processor. The dashed lines depict the lower and the upper bound of the 95% confidence interval for the mean CPU time. Left panel, the run-time corresponding to Step Require, Step 1, and Step 3 as the function of the number of samples in Reference 1, $n_1$. Right panel, the run-time associated with Step 4 as a function of the total number of samples upto the moment, $t$. In each iteration, 100 new observations are sampled.

sensors selected to react to a wide range of chemicals to distinguish between mixtures of analytes. Employing the pattern recognition methods, the sensor's output are compared with reference samples to predict odour concentration. Consequently, the accuracy of predicted odour concentration depends heavily on the validity of sensor's output. An automatic procedure that detects the samples' validity in an online fashion has been a technical shortage and is addressed in this work. A measurement validation process provides the possibility of attaching a margin of error to the predicted odour concentrations. Furthermore, it allows taking the subsequent actions such as re-sampling to re-calibrate the models or checking the e-nose device for possible sensor failures. The proposed measurement validation algorithm initiates a new development in automatic odour detection by minimizing the manpower intervention.

**Acknowledgement.** The project was funded by the natural sciences and engineering research council of Canada (NSERC) through the industrial partnership Engage program. Vahid Partovi Nia is partially supported by the Canada excellence research chair in data science for real-time decision making.

# References

1. Arora, S., Barak, B.: Computational Complexity: A Modern Approach. Cambridge University Press, Cambridge (2009)
2. Artursson, T., Eklov, T., Lundstrom, I., Martensson, P., Sjostrom, M., Holmberg, M.: Drift correction methods for gas sensors using multivariate methods. J. Chemometr. **14**, 711–723 (2000)
3. Bermak, A., Belhouari, S.B., Shi, M., Martinez, D.: Pattern recognition techniques for odor discrimination in gas sensor array. Encycl. Sens. **X**, 1–17 (2006)

4. Brys, G., Hubert, M., Rousseeuw, P.J.: A robustification of independent component analysis. Chemometrics **19**, 364–375 (2006)
5. Carlo, S.D., Falasconi, M.: Drift correction methods for gas chemical sensors in artificial olfaction systems: techniques and challenges. Adv. Chem. Sens. **14**, 305–326 (2012)
6. Chan, T.M.: Output-sensitive results on convex hulls, extreme points, and related problems. Discrete Comput. Geom. **16**(4), 369–387 (1996)
7. Croux, C., Haesbroeck, G.: Principal components analysis based on robust estimators of the covariance or correlation matrix: infulence functions and efficiencies. Biometrika **87**, 603–618 (2000)
8. Dobkin, D.P., Reiss, S.P.: The complexity of linear programing. Theor. Comput. Sci. **11**, 1–18 (1980)
9. Donoho, D.L.: Breakdown properties of multivariate location estimators. Ph.D. Qualifying Paper Harvard University (1982)
10. Du, Q., Fowler, J.E.: Low-complexity principal component analysis for hyperspectral image compression. Int. J. High Perform. Comput. Appl. **22**, 438–448 (2008)
11. Friedman, J., Hastie, T., Tibshirani, R.: Sparse inverse covariance estimation with the graphical lasso. Biostatistics **9**, 432–441 (2008)
12. Gardner, J., Bartlett, P.: A brief history of electronic noses. Sens. Actuators B: Chem. **18**, 211–220 (1994)
13. Golub, G.H., Loan, C.F.V.: Matrix Computations, 3rd edn. The John Hopkins University Press, Baltimore (1996)
14. Gupta, A.: Multivariate skew t-distribution. Statistics **37**, 359–363 (2003)
15. Gutierrez-Osuna, R.: Pattern analysis for machine olfaction: a review. IEEE Sens. J. **2**, 189–202 (2002)
16. Holmes, M.P., Gray, A.G., Isbell, C.L.: Fast SVD for large-scale matrices. In: Workshop on Efficient Machine Learning at NIPS, vol. 58 (2007)
17. Hubert, M., Rousseeuw, P.J., Branden, K.V.: ROBPCA: a new approach to robust principal component analysis. Thechnometrics **47**, 64–79 (2005)
18. Hubert, M., Van der Veeken, S.: Outlier detection for skewed data. J. Chemom. **22**, 235–246 (2008)
19. Hubert, M., Vandervieren, E.: An adjusted boxplot for skewed distributions. Comput. Stat. Data Anal. **52**, 5186–5201 (2008)
20. Jolliffe, I.: Principal Component Analysis. Springer, Heidelberg (2002)
21. Josse, J., Pagès, J., Husson, F.: Multiple imputation in principal component analysis. Adv. Data Anal. Classif. **5**, 231–246 (2011)
22. Kan, A.R., Telgen, J.: The complexity of linear programming. Stat. Neerl. **35**(2), 91–107 (1981)
23. Kermiti, M., Tomic, O.: Independent component analysis applied on gas sensor array measurement data. IEEE Sens. J. **3**, 218–228 (2003)
24. Li, G., Chen, Z.: Projection-pursuit approach to robust dispersion matrices and principal components: primary theory and Monte Carlo. J. Am. Stat. Assoc. **80**, 759–766 (1985)
25. McGinley, P.C., Inc, S.: Standardized odor measurement practices for air quality testing. In: Air and Waste Management Association Symposium on Air Quality Measurement Methods and Technology, San Francisco, CA (2002)
26. Mirshahi, M., Partovi Nia, V., Adjengue, L.: Statistical measurement validation with application to electronic nose technology. In: Proceedings of the 5th International Conference on Pattern Recognition Applications and Methods, pp. 407–414 (2016)

27. Ottmann, T., Schuierer, S., Soundaralakshmi, S.: Enumerating extreme points in higher dimensions. In: Mayr, E.W., Puech, C. (eds.) STACS 1995. LNCS, vol. 900, pp. 562–570. Springer, Heidelberg (1995). doi:10.1007/3-540-59042-0_105

28. Padilla, M., Perera, A., Montoliu, I., Chaudry, A., Persaud, K., Marco, S.: Drift compensation of gas sensor array data by orthogonal signal correction. J. Chemom. Intell. Lab. Syst. **100**, 28–35 (2010)

29. Persaud, K., Dodd, G.: Analysis of discrimination mechanisms in the mammalian olfactory system using a model nose. Nature **299**, 352–355 (1982)

30. Prendergast, L.: A note on sensitivity of principal component subspaces and the efficient detection of influential observations in high dimensions. Electron. J. Stat. **2**, 454–467 (2008)

31. Rousseeuw, P.J., Driessen, K.V.: A fast algorithm for the minumum covariance determinant estimator. Technometrics **41**, 212–223 (1999)

32. Stahel, W.A.: Robust estimation: infinitesimal optimality and covariance matrix estimators. Ph.D. thesis, ETH, Zurich (1981)

33. Tibshirani, R.: Regression shrinkage and selection via the lasso. J. Roy. Stat. Soc. Ser. B **58**, 267–288 (1996)

34. Zuppa, M., Distante, C., Persaud, K.C., Siciliano, P.: Recovery of drifting sensor responses by means of DWT analysis. J. Sens. Actuators **120**, 411–416 (2007)

# Near-Duplicate Retrieval: A Benchmark Study of Modified SIFT Descriptors

Afra'a Ahmad Alyosef[(⊠)] and Andreas Nürnberger

Department of Technical and Business Information Systems,
Faculty of Computer Science, Otto von Geruicke University Magdeburg,
Magdeburg, Germany
{afraa.ahmad-alyosef,andreas.nuernberger}@ovgu.de

**Abstract.** Local feature detectors and descriptors are widely used for image near-duplicate retrieval tasks. However, most studies and evaluations published so far focused on increasing retrieval accuracy by improving descriptor properties and similarity measures. There has been almost no comparisons considering the modification of the descriptors and the impact on accuracy *and* performance, which is especially of interest for interactive retrieval systems that require fast system responses. Therefore, we evaluate in this paper accuracy and performance of variations of SIFT descriptors (reduced SIFT versions, RC-SIFT$-64D$, the original SIFT$-128D$) and SURF$-64D$ in two cases: Firstly, using benchmarks of various sizes. Secondly, using one particular benchmark but extracting varying amounts of descriptors. Another aspect that has been almost neglected in previous benchmarks is the combination of different affine transformations in near-duplicate images. A problem that many real-world systems have to face. Therefore, we provide in addition results of a comparative performance analysis using benchmarks generated by combining several image affine transformations.

## 1 Introduction

Finding near-duplicate images is still a very challenging task, due to the various scenarios in which near-duplicate images could have been created: using different cameras or slightly different positions; different camera settings or lenses; different lighting conditions; post processing of images using image processing software, may be even to hide illegal use of copyrighted material. Therefore, the features and similarity models used to find near-duplicate images have to be quite robust.

The image near-duplicate retrieval process can be divided into several stages depending on the used techniques and the goal of the retrieval task that should be supported. However, the first step is to represent images by means of one or more kinds of expressive features. The goal is to reduce the amount of processed information. The scale invariant feature transform (SIFT) provides keypoints and descriptors that are used in many NDR approaches [6,8,9,11]. This is mainly due to its invariance to scale and rotation variation and its robust performance even

© Springer International Publishing AG 2017
A. Fred et al. (Eds.): ICPRAM 2016, LNCS 10163, pp. 121–138, 2017.
DOI: 10.1007/978-3-319-53375-9_7

if the images differ in perspective, noise, and illumination [1]. The huge amount of descriptors that are required to represent a large scale image dataset and the high dimensionality of these descriptors imposes strong demands on memory and computing power in order to support near-duplicate retrieval tasks. To reduce the amount of extracted data, we proposed a method in [22] to compress the region around the SIFT descriptor. This compression leads to a decrease in time and memory usage of feature indexing and matching. We showed in [22] that the region compressed SIFT (RC-SIFT) descriptors are invariant to affine transformation change and perform robust as the original SIFT features to viewpoint change, scale change and blurring change. In this work, we evaluate the performance of the RC-SIFT$-64D$ [22] descriptor in solving near-duplicate retrieval tasks in two cases: Firstly, for benchmarks of various sizes. Secondly, when a specific benchmark is used but descriptor databases of various sizes are extracted from images. After that, the robustness of the RC-SIFT$-64D$ descriptor is evaluated by various combinations of image affine transformations.

The remainder of this paper is organized as follows. Section 2 provides a short definition of near-duplicate images. Section 3 gives an overview of prior work related with the SIFT algorithm and image NDR algorithms. Section 4 details the proposed method to produce the region compressed SIFT descriptor. Section 5 presents the settings of our experiments and the measures used to describe the performance. Section 6 discusses the results of experiments. Finally, Sect. 7 draws conclusions of this work and discusses possible future work.

## 2    Near-Duplicate Images

To clarify the meaning of near duplicate images, we define first briefly the concept of exact duplicate images: Two images are considered as exact duplicate iff there is no difference between both of them [7], i.e. all corresponding pixels are identical. Two images are defined to be near-duplicates (ND) [7,10] if they show the same scene (the same object) but they differ (slightly) in some properties that can be represented by affine transformations (such as noise, blurring, compression, contrast etc.) or time conditions (lighting or illumination conditions) or the images are even taken from different perspective. Unfortunately, so far the range of transformations in which images are still considered near-duplicates is not yet clearly defined in the literature. Moreover, the evaluation of NDR algorithms is still challenging and focuses mostly on comparisons of rankings or performance.

## 3    Related Works

The SIFT detector and descriptor has been shown superior performance to several other low dimensional descriptors [25]. Therefore, it has been widely used in image near-duplicate retrieval field [8,9], image classification [2] and processing medical images [3] i.g., checking the existence of cancerous growth.

To accelerate the feature indexing process several methods have been proposed to reduce the length of the original SIFT descriptor vector. This is achieved either by ignoring some patches of the original descriptor [13] to get $96D$, $64D$ and $32D$ descriptors or by employing principle component analysis to obtain $64D$ SIFT descriptors [21]. This approach is in need of an off-line training stage to compute the suitable eigenvalue vector. The issue of extracting variable amounts of SIFT features is addressed in [26] by pruning the extracted features based on their contrast property. In [22], we proposed a method to compress the descriptor without the need for a training stage and without ignoring any part of the region around the keypoint. The details of this method are also described in Sect. 4.

To accelerate the features matching process, various methods have been proposed to structure, index or quantize SIFT features. In [1] the best-bin-first algorithm based on a $kd-$tree has been used to speed up the process of matching in 128 dimensional space. However, this method is not appropriate for large scale feature databases due to the required time for backtracking through the tree which leads to decreased $kd-$tree efficiency. To overcome this problem, direct clustering specifically, $k-$means clustering have been used in [14–17], to group the SIFT descriptors into $k$ groups. The obtained cluster centers construct a bag of words; each descriptor is assigned to its closest word in this bag. In this way, images are represented in form of vectors of bag of words. The concept of bag of words is extended in [4,5] and combined with further training steps to improve the retrieval of relevant scenes or objects. In [20] a bag of words is built to construct image vectors. The dimensionality of these vectors is jointly optimized and reduced by applying principle component analysis. A vocabulary tree and the inverted file concept are constructed based on hierarchical $k-$means clustering in [2,3] to refine the splitting of features into groups. In [27] retrieval performance is improved through re-ranking the retrieved images based on the scale and orientation properties of the extracted features.

The next subsection gives an overview of the SIFT detector and descriptor algorithm to simplify the description of the region compressed SIFT descriptor later on.

### 3.1   SIFT$-128D$ Descriptor

As described in [1] the original SIFT detector and descriptor algorithm consists of four major stages: scale invariant peak detection, feature localization, orientation assignment and descriptor construction. In the first stage the locations and scales of interest points (called keypoints) are identified. This is achieved by building a Gaussian pyramid and searching for the local maxima or minima in the difference of Gaussian (DoG) images. The second stage determines the location of the candidate keypoints and rejects the keypoint that have low contrast or are poorly localized on an edge. The third stage assigns the dominant orientation for each keypoint based on the properties of its local image patch. In the final stage keypoint descriptor is computed based on the local gradient and orientation data of a patch around a keypoint. This descriptor is built in form of $n \times n$ array of orientation histogram. For each bin in this histogram $r$ orientations are assigned, so that each descriptor has $n \times n \times r$ element. The size of descriptor is determined by the width of a histogram $n$ and the number of orientations $r$. The standard length of the SIFT descriptor [1] is 128 elements. Figure 1(a) shows the final form of the SIFT$-128D$ descriptor around a keypoint.

Since, the sparsity of descriptors may increase as the dimensionality of the SIFT descriptor increase [12] and this may affect the accuracy of descriptor indexing in image NDR, we proposed an approach [22] to compress the dimensionality of the SIFT descriptor. In the next section, we describe this approach in detail.

## 4    Region Compressed SIFT Descriptor

In [22], we proposed an approach to compress the dimensionality of the original SIFT descriptor from $128D$ to $64D$ without ignoring any part of the local patch around a keypoint and without the need for a training stage. This approach aims to reduce the usage of memory and the amounts of processed data. Moreover, it improves the retrieval task in the near-duplicate retrieval field. To achieve this, SIFT features are first extracted in the same way described in [1] (see Sect. 3.1). After that, the SIFT local descriptor is computed over a local image region around each keypoint. The original SIFT descriptor has the dimensionality of $4 \times 4 \times 8$ and it is computed in form of three dimensional histograms centered at the keypoint. This gradient orientation histogram explain that a keypoint may be located at any allowed position in the local patch around a keypoint in vertical and horizontal location (i.e. $4 \times 4$ locations). For each location eight directions are assigned. We proposed in [22] that for each two possible horizontal shifting in the same direction with respect to the keypoint, only one vertical shifting is available so that, for all possible horizontal shiftings (i.e., four horizontal shiftings) in all directions only two vertical shifting exists. For each of this ($4 \times 2$) locations eight directions are assigned. As a result we obtain $4 \times 2 \times 8$ histogram i.e., $64D$ SIFT descriptor. We called our method for extracting and compressing SIFT descriptor "Region Compressed SIFT" (RC-SIFT). The histogram at each keypoint can be presented by a triplet of elements $H_y$, $H_x$ and $H_\theta$ where:

$$H_y = y - \frac{N_y - 1}{2} \tag{1}$$

$$H_x = x - \frac{N_x - 1}{2} \tag{2}$$

$$H_\theta = \frac{2\pi}{N_\theta} \tag{3}$$

Where $N_y$ and $N_x$ are the number of bins in $H_y$ and $H_x$, respectively. The variables $y$ and $x$ are defined as $y = 0, ..., N_y - 1$, and $x = 0, ..., N_x - 1$ and $N_\theta$ defines the number of orientations in each bin of a histogram and its values: $\theta = 0, ..., N_\theta - 1$. The best performance [22] is found when $N_y = 2$, $N_x = 4$ and $N_\theta = 8$ and when $N_y = 4$, $N_x = 2$ and $N_\theta = 8$. These two forms of the RC-SIFT$-64D$ descriptor are presented in Fig. 1(b) and (c) respectively. Contrary

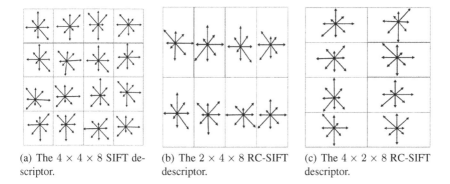

(a) The $4 \times 4 \times 8$ SIFT descriptor.

(b) The $2 \times 4 \times 8$ RC-SIFT descriptor.

(c) The $4 \times 2 \times 8$ RC-SIFT descriptor.

**Fig. 1.** The Different forms of the SIFT descriptor. (a) presents the original SIFT$-128D$ descriptor. Whereas (b) and (c) show the RC-SIFT$-64D$ descriptors of forms $4 \times 2 \times 8$ (referred as RC-SIFT$-64$(R)) and $2 \times 4 \times 8$ (referred as RC-SIFT$-64$(C)) respectively. The symbols RC-SIFT$-64$(R) and RC-SIFT$-64$(C) are used in the all presented tables.

to the methods proposed in [13, 21], which ignore some parts of SIFT descriptor or need for a training stage as described in Sect. 3, we compressed the SIFT descriptor without ignoring any region of the local patch around a keypoint and without the need for an off-line training stage.

In the following, we present an extensive benchmark study to verify the performance of the RC-SIFT$-64D$ in solving near-duplicate retrieval tasks in the following scenarios:

- Various benchmarks: Check the performance using various image databases. We apply our experiments on UKbench [2] and Caltech-Buildings benchmark [24] (see Sect. 6).
- Benchmarks of various sizes: Verify the performance using benchmarks of various sizes produced from UKbench benchmark (see Subsect. 6.1).
- Descriptor databases of various sizes: Evaluate the performance for a variable number of extracted features (see Subsect. 6.2).
- Combination of image affine transformations: The robustness of the RC-SIFT descriptor is verified when a combination of affine transformations applied to images. The following combination are applied in this work: illumination and rotation changes (see Subsect. 6.3), illumination change and adding noise (see Subsect. 6.3) and combination of adding noise and rotation (see Subsect. 6.3).
- Combination of Blurring and image affine transformations: The robustness of the RC-SIFT and all other proposed descriptors is verified against combinations of blurring and affine transformations (see Subsect. 6.4).

## 5   Evaluation

The performance of the RC-SIFT$-64D$ descriptor is compared to the original SIFT$-128D$, the SURF$-64D$ [18] and the SIFT$-64D$ [13] descriptors mentioned in Sect. 3 by solving different image near-duplicate retrieval tasks. To achieve this, large scale image benchmarks of different sizes and resolutions are used. In the following subsections the used image benchmarks and the evaluation measures are described.

### 5.1 Benchmark Datasets

In this work, the experiment is performed on two different benchmark datasets. The first benchmark is UKbench [2] (this dataset can be download from [28]). From this benchmark various image datasets of sizes (10200, 6000, 4000 and 2000) are formed as described in Subsect. 6.1. The resolution of these images is $640 \times 480$. This benchmark consist of indoor/outdoor images of different scenes in groups of four images for each scene. The images of each scene vary in one or more of the following conditions: view point, scale, lightness, appear new objects and occlusion of objects. The second benchmark is the Caltech-Buildings [24, 29] image dataset which contains 250 images for 50 different buildings around the Caltech campus.(i.e. in groups of five images for each building taken at different perspectives and scales). Moreover, this benchmark contains of high resolution image (i.e. the resolution of each image is $2048 \times 1536$).

### 5.2 Evaluation Measures

To evaluate the performance of the proposed descriptors, the descriptors of each kind are firstly indexed using the vocabulary tree concept as described in [22]. In our experiment the initial number of clusters is $k = 10$. The similarity between two images is computed by traversing each normalized vector of the query image $q\_img$ in the vocabulary tree of the database images $db\_img$ and it is given as [2]:

$$s(q\_img, db\_img) = \left\| \frac{q\_img}{\|q\_img\|} - \frac{db\_img}{\|db\_img\|} \right\| \tag{4}$$

All implementations are build using windows platform and Visual C++ programing language with "Opencv" functions. Matlab functions are used to apply combination of image affine transformations and blurring. The Matlab library VLFeat is employed to index the extracted descriptors.

The results of the experiments are evaluated by computing the *recall* value. Considering $N_q$ is the number of relevant images to a specific query image in the database, $N_{qr}$ the number of relevant images obtained in matching results, then the *recall* is defined as follows:

$$Recall = \frac{N_{qr}}{N_q} \tag{5}$$

The mean recall $MR$ for a set of query images is computed as

$$MR = \frac{1}{Q} \sum_{q=1}^{Q} Recall(q) \tag{6}$$

Where $Q$ is the total number of query images. To measure the amount of difference between the $MR$ and the recall value of each query image, the variance of the recall values $VR$ is computed as:

$$VR = \frac{1}{Q} \sum_{q=1}^{Q} (Recall(q) - MR)^2 \tag{7}$$

However, the computation of the recall ignores the ranking of the relevant images in the results. Therefore, we compute the mean average precision $MAP$ which characterizes the relation between the relevant images and their ranking in the results [23] and it is defined as:

$$MAP = \sum_{q=1}^{Q} \frac{Ap(q)}{Q} \tag{8}$$

where $Ap(q)$ is the average precision for image $q$ and is given as:

$$AP(q) = \frac{1}{n} \sum_{i=1}^{n} p(i) \times r(i) \tag{9}$$

where $r(i) = 1$ if the $i^{th}$ retrieved image based on the query image $q$ is one of the relevant images and $r(i) = 0$ otherwise, $p(i)$ is the precision at the $i^{th}$ element.

# 6   Result and Analysis

The results of the SIFT$-64D$ and the original SIFT$-128D$ are evaluated in different cases using various kinds of image benchmarks as described in the following.

## 6.1   UKbench Benchmark

From this benchmark [2] we construct four image datasets of different size to test the robustness of the RC-SIFT descriptors in solving the task of image near-duplicate retrieval. For the experiment, we select the first image of each scene as a query image while the remaining three images of each scene are used as a basic database for retrieval task. The constructed benchmarks have the sizes 10200, 6000, 4000 and 2000 images and they are referred as $UKBench10$, $UKBench6$, $UKBench4$ and $UKBench2$, respectively. The features and descriptors are extracted using the original SIFT$-128D$, SURF$-64D$, SIFT$-64D$ [13] and our RC-SIFT$-64D$(R) and RC-SIFT$-64D$(C) descriptors. After that, the descriptors of each kind are indexed separately using a vocabulary tree of depth $L = 4$ and initial clusters $k = 10$. To achieve the retrieval task, the distance between a query image and database images is computed as described in Eq. 4 using the $L1-$norm and $L2-$norm. However, in our experiment the $L1-$norm obtains better results than the $L2-$norm. Therefore, we present the results obtained when the $L1-$norm is used. A query image is retrieved if its corresponding images in a database appear in the top three, ten or fifty retrieved images.

Table 1 summarizes the results of all proposed descriptors using benchmarks of various sizes. In this table a query image is retrieved if its relevant images in the benchmark appear in the top three retrieved images. It shows that the RC-SIFT$-64D$ obtained slightly better results than SIFT$-128D$. The values of variance are small for all descriptors but the smallest values are found for SURF-64$D$ and SIFT-64$D$ [13]. The best mean average precision is found for the RC-SIFT$-64D$ and then for the SIFT$-128D$ descriptors. Tables 2 and 3 present the performance of various descriptors when the belonging images appear in the top ten or fifty results, respectively. In the both cases the best performance is shown by RC-SIFT$-64D$ and SIFT$-128D$.

The results presented in Tables 1, 2 and 3 show that, if the mean recall increase the variance values increase for both SURF-64$D$ and SIFT-64$D$ [13]. Whereas, for both of RC-SIFT$-64D$ and SIFT$-128D$ the variance of recall decrease as the mean recall value increases. Table 4 provides a qualitative comparison between all proposed descriptors. For this example it shows that the best results are found when the RC-SIFT$-64D$ is used. However, there are of course other examples where the SIFT$-128D$ preforms best. Moreover, we note in many cases that despite the equivalent recall results of SIFT$-128D$ and RC-SIFT$-64D$ descriptors, the RC-SIFT$-64D$ obtains better mean average precision values than the SIFT$-128D$ descriptor. Table 5 presents an example of the results where the performance of SIFT$-128D$ and RC-SIFT$-64D$ is equivalent but the ranking of the results found by RC-SIFT$-64D$ is better than SIFT$-128D$.

**Table 1.** The retrieval performance of SIFT$-128D$, SIFT$-64D$, SURF$-64D$ and our RC-SIFT-64$D$ using benchmarks of various sizes $UKBench10$, $UKBench6$, $UKBench4$ and $UKBench2$, each of them contains images of various scenes with groups of four images belong to the same scene. The first image of each scene is used as a query image. The mean recall $MR$, the variance of recall $VR$ and mean average precision $MAP$ are computed in percent based on the top three retrieved images. The symbols RC-SIFT$-64D$(R) and RC-SIFT$-64D$(C) are used to refer the compression of forms $4 \times 2 \times 8$ and $2 \times 4 \times 8$, respectively.

| Method | $UKBench10$ | | | $UKBench6$ | | | $UKBench4$ | | | $UKBench2$ | | |
|---|---|---|---|---|---|---|---|---|---|---|---|---|
| | MR | VR | MAP | MR | VR | MAP | MR | VR | MAP | MR | VR | MAP |
| SIFT-128$D$ | 49.3 | 15.1 | 47.5 | 55.3 | 14.4 | 53.5 | 53.1 | 14.3 | 51.3 | 51.6 | 13.4 | **49.7** |
| SURF-64$D$ | 24.3 | 13.2 | 22.9 | 26.3 | 12.3 | 24.6 | 25.0 | 11.1 | 23.4 | 26.1 | 11.2 | 25.5 |
| SIFT-64$D$ | 27.2 | 11.2 | 25.2 | 29.9 | 11.5 | 27.9 | 27.1 | 10.9 | 25.2 | 25.6 | 10.0 | 24.0 |
| RC-SIFT-64$D$(R) | **50.7** | 14.8 | **48.8** | **57.1** | 13.7 | **55.2** | **54.5** | 13.6 | **52.7** | 54.9 | 12.5 | **53.1** |
| RC-SIFT-64$D$(C) | 49.9 | 14.6 | **47.9** | 56.3 | 14.0 | **54.1** | 53.1 | 13.7 | **51.7** | 51.8 | 12.8 | 49.7 |

**Table 2.** The retrieval performance of SIFT$-128D$, SIFT$-64D$, SURF$-64D$ and our RC-SIFT-64$D$ using benchmarks of various sizes ($UKBench10$, $UKBench6$, $UKBench4$ and $UKBench2$), each of them containing images of scenes with groups of four images belong to the same scene. The first image of each scene is used as a query image. The $MR$, $VR$ and $MAP$ are computed based on the top ten retrieved images.

| Method | $UKBench10$ | | | $UKBench6$ | | | $UKBench4$ | | | $UKBench2$ | | |
|---|---|---|---|---|---|---|---|---|---|---|---|---|
| | MR | VR | MAP | MR | VR | MAP | MR | VR | MAP | MR | VR | MAP |
| SIFT-128$D$ | 58.7 | 15.2 | 50.1 | 64.8 | 13.7 | 57.3 | **62.3** | 14.1 | **54.9** | 61.0 | 13.3 | **53.3** |
| SURF-64$D$ | 30.2 | 14.7 | 23.3 | 34.2 | 14.6 | 28.3 | 31.9 | 13.7 | 24.2 | 33.7 | 12.9 | 28.2 |
| SIFT-64$D$ | 36.2 | 14.0 | 28.2 | 39.0 | 14.3 | 31.2 | 35.4 | 13.5 | 28.1 | 30.1 | 12.3 | 26.6 |
| RC-SIFT-64$D$(R) | **60.7** | 14.8 | **52.7** | **67.1** | 13.1 | **59.4** | **64.6** | 13.4 | **57.0** | **64.9** | 12.7 | **57.4** |
| RC-SIFT-64$D$(C) | **59.2** | 14.9 | **50.6** | **65.1** | 13.5 | **58.0** | 62.0 | 13.6 | 54.5 | **61.4** | 12.8 | **53.5** |

**Table 3.** The retrieval performance of SIFT$-128D$, SIFT$-64D$, SURF$-64D$ and our RC-SIFT-64$D$ using benchmarks of various sizes $UKBench10$, $UKBench6$, $UKBench4$ and $UKBench2$, each of them contains images of scenes with groups of four images belong to the same scene. The $MR$, $VR$ and $MAP$ are computed based on the top fifty retrieved images and the task is to retrieve the belonging to the same scene images in the top fifty results.

| Method | $UKBench10$ | | | $UKBench6$ | | | $UKBench4$ | | | $UKBench2$ | | |
|---|---|---|---|---|---|---|---|---|---|---|---|---|
| | MR | VR | MAP | MR | VR | MAP | MR | VR | MAP | MR | VR | MAP |
| SIFT-128$D$ | 69.4 | 13.0 | 51.2 | 75.0 | 11.1 | 58.4 | 73.0 | 11.5 | 56.0 | 72.4 | 11.5 | **54.5** |
| SURF-64$D$ | 45.1 | 15.6 | 25.7 | 50.8 | 14.9 | 30.0 | 47.0 | 15.0 | 26.8 | 47.2 | 14.0 | 26.9 |
| SIFT-64$D$ | 49.1 | 15.1 | 29.4 | 52.0 | 14.8 | 32.3 | 47.9 | 14.9 | 29.2 | 46.3 | 14.1 | 28.0 |
| RC-SIFT-64$D$(R) | **72.2** | 11.8 | **53.9** | **77.6** | 9.8 | **60.6** | **75.5** | 10.3 | **58.1** | **76.1** | 9.6 | **58.6** |
| RC-SIFT-64$D$(C) | **70.2** | 13.0 | **52.1** | **75.6** | 10.9 | **59.0** | **73.1** | 11.3 | **56.2** | 72.7 | 11.0 | 54.9 |

## 6.2    Caltech-Buildings Benchmark

In this case, because of the high resolution of images of this benchmark [24], we determine three different threshold to extract different numbers of descriptors from the images. The used number of features in this experiment are 2500, 1000 and 500 and they are referred as *Caltech*–2500, *Caltech*–1000 and *Caltech*–500, respectively. We compute the performance of all proposed descriptors in solving the task of image near-duplicate retrieval with the three descriptors databases of different sizes. For the experiment, we select the first image of each scene as a query image while the remaining four images of each scene are used as a basic database for retrieval task. In this experiment a vocabulary tree of depth $L = 3$ and initial clusters $k = 10$ is used. In addition, the $L1 - norm$ is used to normalize the vectors of images. Table 6 presents the results of all descriptors when the related images appear in the top four results. It shows a comparable performance of the RC-SIFT$-64$ and the SIFT$-128$ descriptors. However, Tables 7 and 8 present a little bit enhancement in the performance of the RC-SIFT$-64$ compared to the SIFT$-128$ descriptor. Moreover, the results show that the performance of the SIFT$-64$ and the SURF$-64$ descriptors for this benchmark

**Table 4.** Performance comparison between all proposed methods in solving the image near-duplicate retrieval task. The results presentthat RC-SIFT$-64D$ shows the best performance.

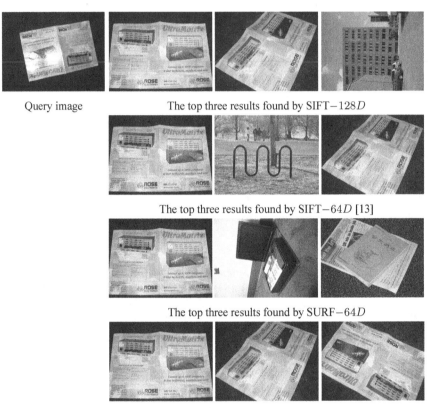

Query image                    The top three results found by SIFT$-128D$

The top three results found by SIFT$-64D$ [13]

The top three results found by SURF$-64D$

The top three results found by RC-SIFT$-64D$

is better than their performance for the benchmarks constructed based on the UKbench benchmark.

In the next step the robustness and invariant properties of the SIFT$-128$, SIFT$-64$, SURF$-64$ and RC-SIFT$-64$ are verified against a combination of different kinds of image transformations and blurring.

### 6.3   Combination of Image Affine Transformations

Various experiments are accomplished to verify the robustness of the original SIFT$-128D$, SIFT$-64D$, SURF$-64D$ and RC-SIFT$-64$ descriptors against combinations of image transformations in the field of image NDR. In this work we discuss the following kinds of combinations: a combination of illumination increase or decrease with rotation change, illumination increase or decrease with

**Table 5.** Equivalent performance of the SIFT$-128D$ and the RC-SIFT$-64D$ descriptor but different ranking of the retrieved results. In this example RC-SIFT$-64D$ presents better raking of the results than SIFT$-128D$.

Query image                    The top three results found by SIFT$-128D$

The top three results found by RC-SIFT$-64D$

**Table 6.** The retrieval performance of SIFT$-128D$, SIFT$-64D$, SURF$-64D$ and RC-SIFT-64$D$ when various number of features are extracted from the images (i.e. 500, 1000 and 2500 features for each image). This is done using the Caltech-Buildings. A query image is retrieved when one or more of its related images is obtained in the top four results. The results are presented in percent.

| Method | $Caltech - Buil500$ | | | $Caltech - Buil1000$ | | | $Caltech - Buil2500$ | | |
|--------|------|-----|------|------|-----|------|------|-----|------|
|        | MR   | VR  | MAP  | MR   | VR  | MAP  | MR   | VR  | MAP  |
| SIFT-128$D$ | **44.0** | 8.2 | 39.9 | **43.0** | 6.5 | **40.2** | **39.5** | 8.2 | **36.7** |
| SIFT-64$D$ | 39.5 | 8.3 | 35.4 | 38.2 | 6.5 | 34.1 | 36.7 | 7.7 | 31.9 |
| SURF-64$D$ | 33.2 | 7.5 | 29.3 | 31.7 | 6.7 | 28.1 | 29.8 | 7.2 | 26.1 |
| RC-SIFT64$D$(R) | **44.0** | 7.8 | **40.3** | 42.8 | 7.2 | 39.8 | **39.0** | 8.4 | **36.4** |
| RC-SIFT64$D$(C) | **44.0** | 7.8 | **40.1** | **43.3** | 7.1 | **40.4** | **39.0** | 8.5 | **36.4** |

**Table 7.** The retrieval performance of SIFT$-128D$, SIFT$-64D$, SURF$-64D$ and RC-SIFT-64$D$ when the $Caltech - 2500$, $Caltech - 1000$ and $Caltech - 500$ benchmarks are used. A query image is retrieved if one or more of its related images is obtained in the top ten results.

| Method | $Caltech - Buil500$ | | | $Caltech - Buil1000$ | | | $Caltech - Buil2500$ | | |
|--------|------|-----|------|------|-----|------|------|-----|------|
|        | MR   | VR  | MAP  | MR   | VR  | MAP  | MR   | VR  | MAP  |
| SIFT-128$D$ | 57.0 | 11.0 | 44.9 | **53.0** | 11.4 | **43.6** | 48.5 | 12.6 | **40.3** |
| SIFT-64$D$ | 51.0 | 13.0 | 39.2 | 47.3 | 14.1 | 39.1 | 41.7 | 13.2 | 36.2 |
| SURF-64$D$ | 39.7 | 12.8 | 33.0 | 34.2 | 14.5 | 26.7 | 33.8 | 12.6 | 23.8 |
| RC-SIFT64$D$(R) | **58.2** | 10.6 | **45.6** | **53.0** | 10.4 | **43.8** | 49.0 | 12.3 | **40.6** |
| RC-SIFT64$D$(C) | **57.7** | 10.8 | **45.2** | **52.8** | 10.6 | **43.6** | 49.0 | 12.6 | **40.3** |

**Table 8.** The retrieval performance of SIFT−128$D$, SIFT−64$D$, SURF−64$D$ and RC-SIFT-64$D$ when 500, 1000 and 2500 features are extracted from the Caltech-Buildings benchmark images. The performance is verified in the top fifty retrieved images.

| Method | $Caltech - Buil500$ | | | $Caltech - Buil1000$ | | | $Caltech - Buil2500$ | | |
|---|---|---|---|---|---|---|---|---|---|
| | MR | VR | MAP | MR | VR | MAP | MR | VR | MAP |
| SIFT-128$D$ | 74.0 | 8.2 | 47.4 | 71.5 | 9.3 | **45.3** | 66.5 | 10.4 | **43.0** |
| SIFT-64$D$ | 67.0 | 14.6 | 42.7 | 59.8 | 14.3 | 43.5 | 53.0 | 14.5 | 37.0 |
| SURF-64$D$ | 50.4 | 14.9 | 40.3 | 48.9 | 14.0 | 35.3 | 48.5 | 14.9 | 25.6 |
| RC-SIFT64$D$(R) | **75.0** | 7.0 | **47.6** | **73.5** | 8.9 | **46.0** | **67.0** | 10.6 | **43.1** |
| RC-SIFT64$D$(C) | **75.3** | 7.0 | **48.0** | **73.2** | 8.7 | 44.7 | **67.0** | 11.0 | **43.0** |

adding noise and finally, rotation change with adding noise. To achieve this, the first 500 images of each scene of the UKbench [2] benchmark (referred as $UKbench5$) are picked. Afterwards, we convolve the images of the $UKbench5$ benchmark with a combination of different kinds of the image affine transformation. The descriptors are indexed using a vocabulary tree of depth $L = 3$ and initial centers $k = 10$. The similarity is computed using the $L1-$norm. A query image is considered to be retrieved if its corresponding database image appears in the top of the retrieved images.

**Table 9.** The performance comparison of SIFT−128$D$, SIFT−64$D$, SURF−64$D$ and our RC-SIFT-64$D$(R)and RC-SIFT-64$D$(C) using a ground truth illuminated and rotated benchmarks (generated from $UKbench5$). For each query image we check if its corresponding image in the used benchmark appears as the first retrieved image in the result. The results are presented for two levels of illumination increase (i.e. 50, 120) for each five rotation values are applied: $40°, 135°, 215°, 250°, 300°$. The results are presented in percent.

| Method | Illumination increase 50 | | | | | Illumination increase 120 | | | | |
|---|---|---|---|---|---|---|---|---|---|---|
| | $40°$ | $135°$ | $215°$ | $250°$ | $300°$ | $40°$ | $135°$ | $215°$ | $250°$ | $300°$ |
| SIFT-128$D$ | **76.2** | **78.0** | **78.0** | **78.0** | **76.0** | **31.0** | 29.4 | **30.0** | 29.4 | 29.1 |
| SIFT-64$D$ | 75.6 | 76.0 | 76.2 | 76.2 | 75.3 | 29.6 | 29.2 | 29.2 | **29.6** | 29.0 |
| SURF-64$D$ | 75.5 | 76.9 | 76.9 | 76.9 | **75.8** | 28.9 | 29.2 | 29.0 | 29.1 | 28.7 |
| RC-SIFT64 (R) | **75.9** | 77.7 | **77.8** | **78.0** | **76.0** | **31.0** | 29.6 | **29.6** | **29.6** | **29.3** |
| RC-SIFT64 (C) | 75.8 | **77.8** | **77.8** | 75.7 | 75.7 | **30.0** | 29.2 | 29.5 | 29.5 | **29.2** |

**Combination of Illumination and Rotation.** To evaluate the robustness of the descriptors with respect to combinations of illumination and rotation changes, the illumination of the $UKbench5$ images is increased using the values $50, 70, 100, 120$ [13, 22]. After that, the illuminated images are rotated at different angles in a clockwise direction (i.e. $40°, 135°, 215°, 250°, 300°$) to generate 20 benchmarks each of them contains 500. To verify the robustness of the descriptors to illumination decrease and rotation, the values $30, 50, 70, 90$ are subtracted

**Table 10.** The performance evaluation of SIFT$-128D$, SIFT$-64D$, SURF$-64D$ and our RC-SIFT-64$D$(R)and RC-SIFT-64$D$(C) in the case of combination of illumination decrease and rotation. For each query image the retrieval task is achieved if its corresponding image in the illuminated and rotated benchmark appears in the top of the result. The results are presented for two levels of illumination decrease (i.e. 30, 90) for each five rotation values are applied: $40°, 135°, 215°, 250°, 300°$.

| Method | Illumination decrease 30 | | | | | Illumination decrease 90 | | | | |
|--------|------|------|------|------|------|------|------|------|------|------|
| | $40°$ | $135°$ | $215°$ | $250°$ | $300°$ | $40°$ | $135°$ | $215°$ | $250°$ | $300°$ |
| SIFT-128$D$ | **79.8** | **75.8** | **76.2** | **78.6** | 77.3 | **52.8** | 49.0 | **50.8** | 48.8 | 47.3 |
| SIFT-64$D$ | 78.6 | **75.6** | **76.0** | **78.6** | 77.5 | 52.6 | **49.6** | **51.2** | 50.3 | 47.6 |
| SURF-64$D$ | 78.2 | 74.8 | 75.8 | 78.0 | 77.4 | 50.7 | 48.7 | 50.2 | **50.8** | 47.7 |
| RC-SIFT64 (R) | **79.5** | **75.8** | **76.0** | **78.8** | **78.0** | **53.1** | **49.6** | **51.2** | **51.2** | **49.3** |
| RC-SIFT64 (C) | 79.2 | 75.5 | 75.8 | 78.0 | **77.6** | **52.8** | **49.2** | 50.3 | 50.6 | **48.7** |

from all channels of the pixels of each image after that the same previous rotation angles are applied to generate 20 benchmarks too. Tables 9 and 10 show robust performance for all used rotation angles when small amount of illumination change is applied to images. However, these tables present a decrease of performance for all rotation angles when the illumination change increase. From these results we deduce that the increasing values of combination affect negatively the stability of the extracted descriptors. Moreover, the comparison of these results with the results of applying the illumination or the rotation change separately [22] clarify that the performance of all proposed descriptors decrease when the rotation and illumination changes are combined (for rotation change the performance is more than 92% and up to 100% for illumination change [2, 22]).

**Table 11.** The performance of SIFT$-128D$, SIFT$-64D$, SURF$-64D$ and our RC-SIFT$-64D$ when a combination of salt and pepper noise and illumination increase is applied on $UKbench5$ images. The results are presented for two level of noise densities (i.e. 15% and 35%), for each the lightness increases using the values: $Li = 50$ and $Li = 120$). A query image is retrieved if its corresponding image in the used noised and illuminated benchmark appears in the top of the retrieved image. In this table $SP$ and $Li$ refer to the salt pepper noise and lightness change, respectively.

| Method | Li 50 | SP 15% | SP 15% | | Li 120 | SP 35% | SP 15% | |
|--------|-------|--------|--------|--------|--------|--------|--------|--------|
| | | | Li 50 | Li 120 | | | Li 50 | Li 120 |
| SIFT-128$D$ | 100 | 82.6 | **68.0** | 32.8 | **91.2** | 20.2 | 5.8 | 3.0 |
| SIFT-64$D$ | 99.8 | 82.2 | **67.6** | 32.6 | 90.2 | 15.2 | 5.0 | **3.1** |
| SURF-64$D$ | 99.5 | 81.2 | 67.2 | 32.3 | 88.1 | 14.5 | 4.7 | 2.8 |
| RC-SIFT-64$D$(R) | 99.5 | **83.4** | **67.6** | **34.7** | 90.2 | **20.8** | **6.2** | **3.4** |
| RC-SIFT-64$D$(C) | 97.0 | **83.1** | 67.1 | 34.2 | 90.2 | 20.5 | 5.8 | **3.1** |

**Table 12.** The performance of SIFT$-128D$, SIFT$-64D$, SURF$-64D$ and our RC-SIFT$-64D$ using when a combination of salt and pepper noise and increasing darkness is applied on $UKbench5$ images. The results are presented for two level of noise densities (i.e. 15% and 35%), for each the darkness increases using the values $Dr = 50$ and $Dr = 120$. A query image is retrieved if its corresponding image in the used noised and illuminated benchmark appears in the top of the retrieved image. In this table $SP$ and $Dr =$ refer to the salt pepper noise and lightness, respectively.

| Method | Dr 30 | SP 15% | SP 15% | | Dr 90 | SP 35% | SP 15% | |
|---|---|---|---|---|---|---|---|---|
| | | | Dr 30 | Dr 90 | | | Dr 30 | Dr 90 |
| SIFT-128$D$ | **100** | 82.6 | **73.6** | **36.7** | **91.2** | 20.2 | 8.0 | 6.2 |
| SIFT-64$D$ | **99.8** | 82.2 | 73.0 | **36.7** | 90.2 | 15.2 | 8.0 | 5.8 |
| SURF-64$D$ | 99.5 | 81.2 | 73.0 | 35.3 | 88.1 | 14.5 | 7.1 | 5.3 |
| RC-SIFT-64$D$(R) | 99.5 | **83.4** | **73.4** | **37.0** | 90.2 | **20.8** | **10.2** | **7.0** |
| RC-SIFT-64$D$(C) | 97.0 | **83.1** | 73.1 | 36.6 | 90.2 | 20.5 | 9.6 | 6.8 |

**Combination of Noise and Illumination Change.** To test the robustness of the proposed descriptors to the illumination change and added noise, salt and pepper noise with density of 15% and 35% is applied to $UKbench5$ images. After that, the brightness of noised images is increased using the values $50, 70, 100, 120$ [13,22] or decreased by subtracting the values $30, 50, 70, 90$ from all channels of the pixels of the image. As a result we obtain 16 benchmarks which contains various levels of additional noise and illumination change. Tables 11 and 12 present the performance of various descriptors in two cases: Firstly when adding noise and illumination change are applied separately and secondly in case of combination. These tables show a decrease in the performance of all presented descriptors in the case of combination. Moreover, the performance in the case of combination is always lower than the minimum performance obtained by applying the affine transformation separately. However, in case of using the salt and pepper noise with density of 35% all presented descriptors are not stable anymore. Figure 2 presents the difference between the locations of the extracted descriptors by the RC-SIFT$-64D$ before and after applying a combination of illumination increase and noise. The most extracted features in Fig. 2(a) and (b) locate at comparable positions in the both images. Whereas, the locations of features in Fig. 2(a) differ from the those in Fig. 2(b). Therefore, the image in Fig. 2(b) appears as the first retrieved results of the query image Fig. 2(a) whereas, the image in Fig. 2(c) does not appear in the top retrieved results of the query image Fig. 2(a).

**Addition of Noise and Rotation.** A combination of adding noise and rotation is achieved by firstly adding the salt and pepper noise with density of 15% or 35% to the $UKbench5$ benchmark (the detail of adding noise is described in [22]). Secondly, the noised images are rotated at different angles in a clockwise direction (i.e. $40^o, 135^o, 215^o, 250^o, 300^o$) to generate ten benchmarks of

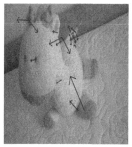  

(a) An image of the *UKbench5* benchmark (i.e. the query image).

(b) The image in 2(a) after applying a combination of illumination increase with value 50 and noise of density 15%.

(c) The image in 2(a) after applying a combination of illumination increase with value 50 and noise of density 35%.

**Fig. 2.** It is shown in (a) and (b), that many extracted descriptors (presented in blue) have the same locations in both images. Therefore, the image retrieval task is achieved successfully in this case. Whereas, the extracted descriptors in (a) and (c) are located in different positions thus, the benchmark image (c) does not appear in the top of the retrieved results. (Color figure online)

**Table 13.** The performance comparison of SIFT$-128D$, SIFT$-64D$, SURF$-64D$ and our RC-SIFT-64$D$(R)and RC-SIFT-64$D$(C) in the case of applying salt and pepper noise and rotation to *UKbench5* benchmark. For each query image the retrieval task is achieved if its corresponding image in the noised and rotated benchmark appears in the top of the result. The results are presented for two noise densities (i.e. 15%, 35%) for each five rotation values are applied: $40°, 135°, 215°, 250°, 300°$. The results are presented in percent.

| Method | Noise 15% | | | | | Noise 35% | | | | |
|---|---|---|---|---|---|---|---|---|---|---|
| | 40° | 135° | 215° | 250° | 300° | 40° | 135° | 215° | 250° | 300° |
| SIFT-128$D$ | **32.0** | **39.8** | **40.0** | **43.0** | 37.6 | **6.2** | **5.6** | 5.4 | **5.8** | 5.4 |
| SIFT-64$D$ | **31.8** | **39.5** | **40.0** | 42.0 | 36.8 | **6.1** | **5.6** | 5.6 | 5.2 | 5.0 |
| SURF-64$D$ | 30.7 | 39.2 | 40.2 | 42.3 | 37.1 | 5.8 | 5.5 | **5.9** | **5.8** | 5.4 |
| RC-SIFT64 (R) | 31.7 | 39.2 | **41.2** | **43.8** | **40.0** | 6.0 | **6.0** | 5.5 | **6.2** | **6.0** |
| RC-SIFT64 (C) | **32.0** | 39.3 | **41.2** | **43.0** | **40.0** | 5.8 | 5.3 | **5.7** | **6.2** | **6.2** |

noised rotated images. Table 13 describes how the performance of all proposed descriptors decrease very strongly for a fixed rotation angle when the density of the added noise is increased.

## 6.4    Combination of Blurring and Affine Transformation

To study the effect of image blurring combination with various kinds of affine transformation on the performance of the original SIFT$-128D$, SIFT$-64D$, SURF$-64D$ and RC-SIFT$-64D$, the *UKbench5* benchmark images are firstly blurred by convolving the image with Gaussian filters using three variations i.e., $\sigma^2 = 5$, $\sigma^2 = 10$ and $\sigma^2 = 15$ (the process of fileting is described in [22]). After that, the illumination of the blurred images is increased or decreased using the same values presented in Subsect. 6.3. The best Performance of near-duplicate retrieval is obtained when the Gaussian filter with $\sigma^2 = 5$ is used. This performance is below 25% for all proposed descriptors. However, when a Gaussian filter with $\sigma^2 = 10$ or $\sigma^2 = 15$ is applied the performance decreases to be not more than 16% or 13%, respectively. When a combination of image blurring and rotation change (the rotation values are: $40^o, 135^o, 215^o, 250^o, 300^o$) is applied, the successfully retrieved images are not more than 13% for all descriptors when $\sigma^2 = 5$. Whereas, the performance decreases to 8% or 4% when $\sigma^2 = 10$ or $\sigma^2 = 15$, respectively. A Combination of the Gaussian blur with the salt pepper noise retrieve successfully less than 15% of the applied query images when the density of noise is 15% and the blurring variation is $\sigma^2 = 5$. The number of retrieved images decreases to 10% or 8% when the blurring variation increases to $\sigma^2 = 10$ or $\sigma^2 = 15$, respectively. The results of combining Gaussian blur with different kinds of image affine transformations present that the performance of all proposed descriptors decreases strongly and the extracted descriptors become unstable when more blurring is added to the images.

## 7   Conclusion

In this work, we evaluated the performance of the RC-SIFT$-64D$ descriptor to solve the near-duplicate retrieval task in two cases: Firstly, for benchmarks of different size. Secondly, using the same benchmark but for different numbers of extracted features. The experiments show a slight improvement in matching results compared to the original SIFT$-128D$ when tested on various benchmark databases. Moreover, the RC-SIFT$-64D$ needs shorter time for indexing and less memory.

We also evaluate the robustness and stability of the original SIFT$-128D$, SIFT$-64D$, SURF$-64D$ and RC-SIFT$-64D$ against combinations of image affine transformations. The results show that all proposed descriptors are robust for combination of transformations with small changes. However, the stability of descriptors decreases when the amount of the combined transformations increases especially, in the case of combination with noise. When the image affine transformations are combined with blurring, the performance of all proposed descriptors decreases very strongly. So that in this case the extracted descriptors loose their robustness.

In the case of extracting variable amounts of features from the benchmark, the performance increase when the numbers of extracted features decrease. Therefore, we are going in the next step to study the factors that may help to reduce the amount of detected features but enhance the performance of descriptors in solving the near-duplicate retrieval task. Moreover, we aim to study if the RC-SIFT$-64D$ can be used in the field of human visual attention, e.g., as a more stable predictor for creating a saliency map of human gaze as discussed in a previous study [18].

# References

1. Lowe, D.: Distinctive image features from scale-invariant keypoints. J. Comput. Vis. **60**, 91–110 (2004)
2. Nistèr, D., Stewènius, H.: Scalable recognition with a vocabulary tree. In: Conference on Computer Vision and Pattern Recognition (CVPR), pp. 2161–2168 (2006)
3. Jiang, M., Zhang, S., Li, H., Metaxas, D.N.: Computer-aided diagnosis of mammographic masses using scalable image retrieval. IEEE Trans. Biomed. Eng. **62**, 783–792 (2015)
4. Jianchao, Y., Kai, Y., Yihong, G., Thomas, H.: Linear spatial pyramid matching using sparse coding for image classification. In: Proceedings IEEE Computer Society Conference on Computer Vision and Pattern Recognition (CVPR) (2009)
5. Zhang, C., Wang, S., Huang, Q., Liu, J., Liang, C., Tian, Q.: Image classification using spatial pyramid robust sparse coding. Pattern Recogn. Lett. **34**, 1046–1052 (2013)
6. Zhang, D.Q., Chang, S.F.: Detecting image near-duplicate by stochastic attribute relational graph matching with learning. In: Proceedings of the 12th Annual ACM International Conference on Multimedia (2004)
7. Chum, O., Philbin, J., Isard, M., Zisserman, A.: Scalable near identical image and shot detection. In: Proceedings of the 6th ACM International Conference on Image and Video Retrieval (CIVR), pp. 549–556 (2007)
8. Chum, O., Philbin, J., Zisserman, A.: Near duplicate image detection: min-Hash and tf-idf weighting. In: British Machine Vision Conference, pp. 50.1–50.10 (2008)
9. Auclair, A., Vincent, N., Cohen, L.D.: Hash functions for near duplicate image retrieval. In: Applications of Computer Vision (WACV), pp. 1–6 (2009)
10. Xu, D., Cham, T., Yan, S., Duan, L., Chang, S.: Near duplicate identification with spatially aligned pyramid matching. IEEE Trans. Circ. Syst. Video Technol. **20**, 1068–1079 (2010)
11. Chu, L., Jiang, S., Wang, S., Zhang, Y., Huang, Q.: Robust spatial consistency graph model for partial duplicate image retrieval. IEEE Trans. Multimedia **15**, 1982–1996 (2010)
12. Steinbach, M., Ertoz, L., Kumar, V.: The challenges of clustering high dimensional data. In: Wille, L.T. (ed.) New Vistas in Statistical Physics-Applications in Econophysics, Bioinformatics, and Pattern Recognition, pp. 273–309. Springer, Heidelberg (2004)
13. Khan, N.Y., McCane, B., Wyvill, G.: SIFT and SURF performance evaluation against various image deformations on benchmark dataset. In: Digital Image Computing Techniques and Applications (DICTA), pp. 501–506 (2011)
14. Grauman, K., Darrell, T.: The pyramid match kernel: efficient learning with sets of features. J. Mach. Learn. Res. **8**, 725–760 (2007)
15. Grauman, K., Darrell, T.: Pyramid match kernels: discriminative classification with sets of image features. In: Proceedings of IEEE International Conference on Computer Vision (ICCV), pp. 1458–1465 (2005)
16. Yang, Y., Newsam, S.: Comparing SIFT descriptors and Gabor texture features for classification of remote sensed imagery. In: Proceedings of the 15th IEEE on Image Processing, pp. 1852–1855 (2008)
17. Li, J., Qian, X., Li, Q., Zhao, Y., Wang, L., Tang, Y.Y.: Mining near duplicate image groups. Multimedia Tools Appl. **74**, 655–669 (2014). Springer Science and Business Media, New York

18. Steffen, J., Christian, H., Ahmad Alyosef, A., Tönnies, K., Nürnberger, A.: Rotational invariance at fixation points - experiments using human gaze data. In: Proceedings of the 1st International Conference on Pattern Recognition Applications and Methods, pp. 451–456 (2012)
19. Bay, H., Tuytelaars, T., Van Gool, L.: SURF: speeded up robust features. In: Leonardis, A., Bischof, H., Pinz, A. (eds.) ECCV 2006. LNCS, vol. 3951, pp. 404–417. Springer, Heidelberg (2006). doi:10.1007/11744023_32
20. Jègou, H., Douze, M., Schmid, C., Pèrez, P.: Aggregating local descriptors into a compact image representation. In: Proceedings of IEEE Computer Vision and Pattern Recognition, pp. 3304–3311 (2010)
21. Ke, Y., Sukthankar, R.: PCA-SIFT: a more distinctive representation for local image descriptors. In: Computer Vision and Pattern Recognition, no. 2, pp. 506–513 (2004)
22. Ahmad Alyosef, A., Nürnberger, A.: Adapted SIFT descriptor for improved near duplicate retrieval. In: Proceedings of the 5th International Conference on Pattern Recognition Applications and Methods, pp. 55–64 (2016)
23. Manning, C.D., Raghavan, P., Schütze, H.: Chapter 8: evaluation in information retrieval. Part of Introduction to Information Retrieval, pp. 151–175 (2009)
24. Aly, M., Welinder, P., Munich, M., Perona, P.: Towards automated large scale discovery of image families. In: Computer Vision and Pattern Recognition Second IEEE Workshop (CVPR), pp. 9–16 (2009)
25. Mikolajczyk, K., Schmid, C.: A performance evaluation of local descriptors. IEEE Trans. Pattern Anal. Mach. Intell. **27**, 1615–1630 (2005)
26. Foo, J.J., Sinha, R.: Pruning SIFT for scalable near-duplicate image matching. In: Proceedings of the Eighteenth Conference on Australasian Database, pp. 63–71 (2007)
27. Jègou, H., Douze, M., Schmid, C.: Hamming embedding and weak geometric consistency for large scale image search. In: European Conference on Computer Vision, pp. 301–317 (2008)
28. Nistèr, D., Stewènius, H.: Recognition Benchmark Images. http://www.vis.uky.edu/~stewe/ukbench/
29. Aly, M., Welinder, P., Munich, M., Perona, P.: Caltech-Buildings Benchmark. http://www.vision.caltech.edu/malaa/datasets/caltech-buildings/

# Activity Recognition for Elderly Care by Evaluating Proximity to Objects and Human Skeleton Data

Julia Richter[✉], Christian Wiede, Enes Dayangac, Ahsan Shahenshah, and Gangolf Hirtz

Department of Electrical Engineering and Information Technology,
Technische Universität Chemnitz,
Reichenhainer Straße 70, 09126 Chemnitz, Germany
julia.richter@etit.tu-chemnitz.de

**Abstract.** Recently, researchers have shown an increased interest in the detection of activities of daily living (ADLs) for ambient assisted living (AAL) applications. In this study, we present an algorithm that detects activities related to personal hygiene. The approach is based on the evaluation of pose information and a person's proximity to objects belonging to the typical equipment of bathrooms, such as sink, toilet and shower. In addition to this high-level reasoning, we developed a skeleton-based algorithm that recognises actions using a supervised learning model. Therefore, we analysed several feature vectors, especially with regard to the representation of joint trajectories in the frequency domain. The results gave evidence that this high-level reasoning algorithm can reliably recognise hygiene-related activities. An evaluation of the skeleton-based algorithm shows that the defined actions were successfully classified with a rate of 96.66%.

**Keywords:** Video analysis · 3-D image processing · Activity recognition · Machine learning · Pose estimation · High-level reasoning · Ambient assisted living

## 1 Introduction

The past years have seen increasingly rapid advances in the development of technical assistance systems that aim at supporting elderly persons at home. This application field is referred to the term AAL. Together with medical partners, we are developing an AAL system that aims at helping elderly at an early stage of dementia to continue to live longer in their familiar environment instead of moving to a nursing home. In order to optimise the caring process, we firstly propose to support elderly by reminding them to perform the daily routines which they tend to forget. Secondly, meta data about the patients' performed daily activities can be accessed through an intuitive web interface by their caregivers. By integrating this information in their patients' individual caring plans, caregivers shall be supported during their daily work. Moreover, the individual needs of every patient can be assessed and attended to in a more appropriate fashion.

© Springer International Publishing AG 2017
A. Fred et al. (Eds.): ICPRAM 2016, LNCS 10163, pp. 139–155, 2017.
DOI: 10.1007/978-3-319-53375-9_8

To this end, we recognise ADLs by integrating so-called smart sensors into the elderlies' living environment. These sensors are composed of a stereo camera and an internal processing unit. The sensors are mounted at the ceiling in the living environment and monitor ADLs – without releasing raw image data, but only meta data, e.g. the detected activities.

The meta data is being logged real-time into a database, which can be accessed by the caring personnel via a web interface. In this way, caring personnel can obtain information that has been inaccessible so far because elderly with dementia are often incapable to communicate the daily activities they already have performed. Caregivers could, as another example, better interpret a patient's uncooperative behaviour during the morning visit when they are informed by the system that the patient was very active during the night and showed a restless sleeping behaviour. Such type of information allows caregivers to better understand and interpret their patients behaviour. As a consequence, the caring process can be adapted to individual needs of the elderly. Another contribution of our system is that caregivers can now react promptly to sudden changes of their patients' behaviour and directly adapt the necessary care to the actual circumstances. Elderly at an early stage of dementia can benefit from the individualised caring plan. At the same time, they can be supported to maintain their daily routines with the help of reminding messages provided by the system.

The first part of the presented study is focussing on the high-level recognition of hygiene-related activities in a bathroom using pose and proximity information. This choice of room was mainly motivated by medical reasons. The frequency of toileting, for example, provides relevant information for diagnosing and treating incontinence. The second part presents the detection of hygiene-related activities by using a machine learning based approach that evaluates skeleton information, i.e. 3-D joint positions.

The paper is structured as follows: We present related work in Sect. 2 while focussing on different approaches for ADL recognition. Section 3 describes the modules that are related to high-level reasoning based activity recognition. The employed person detection algorithm and the pose estimation algorithm are briefly summarised in Sects. 3.1 and 3.2 respectively. In Sect. 3.3, the high-level reasoning evaluating the proximity to objects and pose information is described in detail. Section 3.4 presents the analysis method for the reasoning algorithm, whereas the results are presented and discussed in Sect. 3.5. Section 4 explains and evaluates the skeleton-based algorithm, including the aspects of sequence duration and frame skipping, the description of feature vectors as well as the obtained results accompanied by a discussion. Section 5 draws conclusions about the accuracy of both the high-level reasoning and the skeleton-based algorithm. Besides, an outline regarding further developments is given.

## 2   Related Work

A number of studies have employed different types of sensors, such as motion sensors [15] or body-worn sensors [11], to analyse daily activities or to detect emergencies. Pirsiavash and Ramanan [6] reported that ADLs can be detected by processing the first-person camera view acquired by a wearable camera. Several previous studies have attempted to monitor home activities using acoustics, including bathroom-related activities. Fogarty et al. [4] installed low-cost sensors in the water distribution structure of a home to measure water usage patterns and deduce activities especially performed in a kitchen and in a bathroom. The sensors were attached to the outside of the water pipes in the basement. These sensors consisted of microphones that provided audio signals indicating that the toilet was flushed, the shower was used or the sink was active, for example. The evaluation of these audio signals allowed the recognition of activities that are connected to water consumption. Chen et al. [3] focussed on activity monitoring in bathrooms while using omni-directional microphones. The recognised activities included washing hands, teeth brushing, flushing the toilet and urination.

In our project, we have decided to apply non-wearable optical sensors. These are wide-angle stereo cameras that are designed to be mounted at the ceiling of a room. We came to this decision because people with dementia are apt to remove wearable sensors and they often forget to put them on again. Moreover, compared to motion sensors or acoustic data, image data can provide information with a higher grade of detail.

A meaningful indicator for ADLs represents the room a person stays in [10, 15]. Richter et al. [10] gave evidence that, based on the chronological order of the rooms a person entered, several ADLs could be detected. They therefore introduced a person detection algorithm that derived the person's position in a room using a stereo camera and then assigned the corresponding room the person stayed in. If the person was localised in the sleeping room in the morning and afterwards in the bathroom for several minutes, then they deduced that the person has attended to personal hygiene. In a similar way, other activities, such as the sleeping behaviour or preparing food, can be detected.

However, for a more detailed prediction of daily activities, the evaluation of the room itself is insufficient. For this reason, Richter et al. [8] proposed a concept that deduces ADLs by analysing the proximity of a person to certain objects in the room. Moreover, pose information obtained by a machine learning based approach [9] was included. This work focussed on three objects in the bathroom, the shower, the sink and the toilet. In order to deduce whether a person is close to an object, the person is required to be localised within the room he or she currently stays in. For this purpose, person detection algorithms were required to be applied. Several studies have introduced person detection algorithms, such as Harville and Li [5], Yous et al. [19] and Richter et al. [10], which work on image data obtained by stereo sensors. The studies of Harville et al. and Yous et al. have revealed shortcomings of their stereo vision-based algorithms. Harville and Li [5] determined a point-wise mean positional error from reference data of 160 mm. Yous et al. [19] evaluated their algorithm empirically by marking detected persons by a cuboid. Moreover, false positive and false negative rates were determined. However, there has been little quantitative analysis of spatial accuracy in the mentioned studies. When utilizing a person detection algorithm, it is essential to know the grade of reliability of the obtained person's position. If the person is detected close to the toilet, for example, it is necessary to know how reliable this information is. In this way, uncertainties can be reduced so that misinterpretations can be avoided.

In order to reason about activities performed in a bathroom, the work of Richter et al. [8] applied the person detection algorithm described in [10]. The authors aimed at refining the room assignment to the assignment of objects the person is probably occupied with in a certain room. The accuracy analysis of this person detection algorithm yields that persons can be localised in a very accurate way: During their evaluation, they determined a mean error ranging from 74 mm to 87 mm [8]. Thus, the obtained position can be relied on even if the specific objects are very close to each other. On the basis of this finding, they have designed a high-level reasoning algorithm that recognises bathroom-related activities.

In addition to the high-level reasoning algorithm described in [8], we introduce a skeleton-based algorithm that recognises actions that are related to the bathroom on the one hand and that are the basis for further reasoning about bathroom-related activities on the other hand. The presented skeleton-based activity recognition was inspired by the work of Raptis and Sigal [7] and Beaudry et al. [1]. Raptis and Sigal [7] demonstrated that by only considering local discriminative key frames in a sequence, actions can be accurately classified. Similarly, this was also shown in early publications, such as in the works of Carlsson and Sullivan [2] as well as of Schindler and Van Gool [12]. Beaudry et al. [1] transformed trajectories of relevant points from the optical flow into the frequency domain. In their experiments, they used the KTH dataset [13] and showed that the classification rates are comparable to the results of the highest state-of-the-art algorithms.

Since the launch of the Kinect, researchers have the opportunity to access human skeleton data easily. Already existing approaches, such as the work of Beaudry et al. [1], can now be adapted to skeleton joint coordinates instead of using detected relevant points. Recent work already highlighted that the integration of joint data led to improved classification results. Yao et al. [17] compared the performance of low-level appearance features with pose-based features derived from joint coordinates of successive frames. They demonstrated that their pose-based features outperform appearance-based features under the same circumstances, such as same classifier and same dataset. Wang et al. [16] calculated potential and kinetic energy features of skeletal joint using key frames. The survey of Ye et al. [18] can be referred to for more details regarding the recent works in the field.

Our investigation focusses on the evaluation of different feature vectors, e.g. features obtained by a Fourier transformation on skeleton joints. The impact of different sequence durations as well as the utilisation of key frames is a further part of this study. Moreover, the influence of the frequency resolution changes were investigated.

# 3    Activity Recognition Based on Proximity to Objects and Pose Information

In this section, we present the approach introduced by Richter et al. [8] by explaining the algorithms that allow the detection of bathroom-related activities. This approach is the basis for our further developments.

## 3.1    Person Localisation

The person detection algorithm that is applied in the study of Richter et al. locates a person in a 3-D point cloud. This point cloud is derived from an image pair of a stereo sensor, which is composed of high-quality wide field of view lenses. They have a focal length of 3.5 mm, whereas the sensor has a resolution of $1360 \times 1024$ pixels. The coordinates are calculated with respect to a world coordinate system whose $x$-$y$ plane is aligned with the floor of the flat. The $z$ axis represents the height above the floor, and is not relevant in this work. After a foreground-background segmentation [21], all points belonging to the foreground are projected onto the $x$-$y$ plane. In a subsequent step, blobs are detected on the resulting projection image. The centers of blobs with a certain size are regarded as the centers of detected persons and are denoted as $p_s = (x_s, y_s)$. This center $p_s$ is used for proximity determination in a later processing step. For a better legibility, we dispense with vector signs in this paper.

## 3.2    Pose Information

In order to derive information about the general pose for the high-level reasoning in [8], Richter et al. apply an algorithm that first trains a linear classifier, i.e. a support vector machine [9]. The feature vector is a histogram that represents the distribution of points belonging to a person's surface according to their $z$ component. After training, the classifier can distinguish between standing, sitting and lying. In the work of Richter et al. [8], only sitting and standing are relevant.

## 3.3    High-Level Reasoning Using Position and Pose Information

The high-level reasoning algorithm [8] is able to detect the activities "showering", "using the toilet" and activities normally performed in front of a sink, such as "washing hands, combing and teeth brushing". An overview about this algorithm is presented in Fig. 1. The algorithm unites the person's and objects' position data as well as information about the person's general pose. First of all, the algorithm determines whether a person is close to a certain object. In their study, Richter et al. used the three objects "shower", "toilet" and "sink". For a comparison of the respective position data, stereo sensors are distributed in a test flat and extrinsically calibrated so that they share the same world coordinate system. All determined positions are specified with respect to the origin of this coordinate system. The $N$ object's positions, i.e. their centres and their expansions are stored in a look-up-table (LUT). As a result, there are $N$ entries in the LUT, whereas the expansions are used as thresholds $thresh_n$ for determining whether a person is close to a certain object.

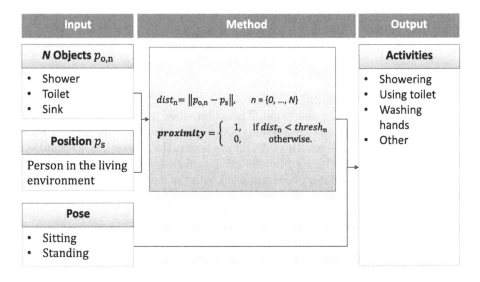

**Fig. 1.** Overview about the high-level reasoning algorithm that evaluates a person's proximity to objects and pose information.

In order to decide whether a person is close to an object, the distance $dist_n$ between the person's position $p_s = (x_s, y_s)$ and all the objects' positions $p_{o,n} = (x_{o,n}, y_{o,n})$ is compared with the corresponding expansion values $thres_n$ in the LUT, whereas $n \in \{1, 2, ..., N\}$. At this point, $N$ denotes the number of objects and $n$ the index of a specific object. A person is considered to be close to an object if the distance between their centres is smaller that the stored threshold value in the LUT. Then, the boolean variable *proximity* is 1, corresponding to *close*, otherwise it is 0, corresponding to *far*. If there are $N$ objects in the room, the following equations are applied $N$ times to check the proximity criterion.

$$dist_n = \|p_{o,n} - p_s\|, \; n \in \{1, ..., N\}, \tag{1}$$

$$proximity = \begin{cases} 1, & \text{if } dist_n < thresh_n \\ 0, & \text{otherwise.} \end{cases} \tag{2}$$

In this context, it can be stated that the smaller the room and the closer the objects are to each other, the more challenging the described assignment will be.

In a second step, the integration of pose information, i.e. standing and sitting, allows to draw conclusions about the following ADLs:

- Activities that are performed when standing in front of a sink, such as washing hands, combing and teeth brushing, if the person is close to the sink and standing.

- Using the toilet if the person is close to the toilet and sitting.
- Taking a shower if the person is close to the shower and standing.
- Other activities if none of the above mentioned activities are detected.

The listed activities are written to a First-In First-Out (FIFO) and only after a certain amount of detections within the FIFO element, the activity is forwarded as a valid system output. If none of the previously mentioned scenarios occur, we assume that the person is doing another action.

Figure 2 illustrates a scenario in the bathroom of a testing flat, where a person is attending to his or her personal hygiene. In this point cloud, the person is detected to be very close to the toilet. As the person is also determined to be sitting, we consider that the person probably uses the toilet. This view was generated during experiments. As high attention is devoted to privacy aspects, this view is not visible outside of the smart sensor in the final implementation.

### 3.4   Analysis Method

Richter et al. analysed their algorithm by recording video sequences with three persons in the bathroom of their testing flat. The probands performed a typical morning routine: After entering the bathroom they attend to his or her personal hygiene including activities, such as using the toilet, washing hands and showering. Simultaneously, the system determines the performed ADLs using the high-level reasoning algorithm. Both the recorded frames and the output of the algorithm are accompanied by timestamps and saved together to the memory. Each frame of the recorded sequence is labelled with the actual ADL afterwards. In a final step, the actual ADL as well as the system output are plotted over time to allow a comparison. In this way, it is also possible to analyse the system latency.

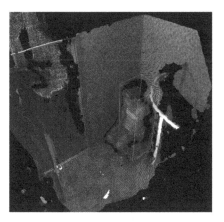

**Fig. 2.** A scenario in the bathroom of a testing flat, where a person is attending to her personal hygiene.

## 3.5    Results and Discussion

Figures 3, 4 and 5 present the results for each sequence that has been recorded in the bathroom of the testing flat. The filled bars represent the real, i.e. the manually labelled, activity, whereas the activity that was detected by the high-level reasoning algorithm is marked with a line. The ordinate shows the time in seconds, while the abscissa shows the bathroom activities.

The reasoning algorithm shows results of high quality even for the small bath room in the testing flat. From the charts, it can be seen that only few minor false detections occurred, whereas all the performed activities were correctly detected. Additionally, a comparatively small delay can be observed, which is due to the Kalman filter that was applied in the person detection algorithm. Moreover, this delay is caused because of the low-pass behaviour, induced by the FIFO, of the reasoning algorithm. In view of the present application, however, these delays do not affect the system functionality. In summary, it was found that the high-level reasoning algorithm reliably detected most of the activities that were performed in the recordings.

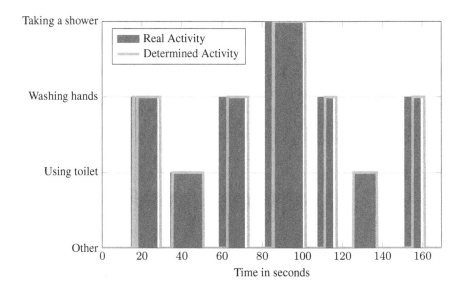

**Fig. 3.** Comparison between real and determined activity for subject 1.

Thus far, we have presented a reasoning algorithm based on proximity to objects and pose information to predict typical bathroom activities. We expect that the availability of more detailed information about patient's activities will contribute to a better reasoning and understanding of the patient's behaviour. Therefore, we further enhanced the performance of our reasoning by evaluating human skeleton joints.

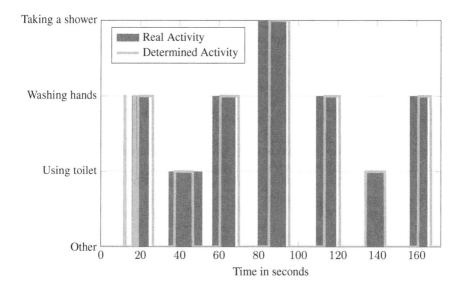

**Fig. 4.** Comparison between real and determined activity for subject 2.

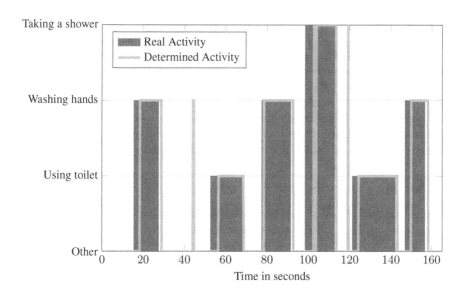

**Fig. 5.** Comparison between real and determined activity for subject 3.

## 4    System Enhancement with Skeleton-Based Activity Recognition

In this section, we present a Fourier transformation based method where we evaluate human skeleton joints to enhance the presented system.

The Kinect device is chosen for this study, because it provides skeleton joint information. The device obtains skeleton joints by processing depth data using the algorithm invented by Shotton et al. [14]. In this study, this data is the base for the proof of concept while using Kinect skeleton sequences for an off-line classification. This study, therefore provides information about the classification accuracy of the designed off-line activity recognition approach.

The presented approach focusses on recognising actions performed in a bathroom by applying a Fourier transformation on skeleton joints using only samples of a sequence instead of all captured frames.

### 4.1    Actions

In the presented approach, the aim is to classify the following actions, with the hereby assigned labels $A1$ to $A6$, that contribute to ADL recognition in a bathroom scenario:

- $A1$: Moving a hand to the mouth
- $A2$: Teeth brushing
- $A3$: Walking
- $A4$: Standing up
- $A5$: Sitting down
- $A6$: Idle

Moving the hand to the mouth can indicate that the person is cleaning the mouth with water after teeth brushing or washing the face. The action teeth brushing is recognised when a person is moving the hand in a repetitive way in front of the mouth as it is characteristic for teeth brushing. Both sitting down and standing up can be a strong clue that the toilet has been used. Walking can be considered as moving from one place in the bathroom to another, whereas the person is considered to be idle when none of the other actions is performed.

### 4.2    Sequence Duration and Frame Skipping

For this study, we recorded sequences containing skeleton data using the Kinect sensor. Each sequence shows a person performing the previously defined actions. The final dataset we used is a combination of our own recordings and the dataset provided by Yu et al. [20]. Except action $A1$, which was performed twice by 56 different persons, each action was performed twice by 28 different persons. This, except for action $A1$, results in an overall number of 56 examples per action, whereas 26 were used for training and the remaining 30 examples for testing. For action $A1$, 52 examples were used for training and 60 examples were used for testing.

The sequences were recorded at 30 fps having a total number $num_{\mathrm{BF}}$ of 256 frames, 128 frames and 64 frames per sequence. This corresponds to sequence durations $d_{\mathrm{m}}$ of approximately 8.53 s ($d_1$), 4.27 s ($d_2$) and 2.13 s ($d_3$) respectively, see Eq. 3. These different settings result from the following observation: When we measured the number of frames that the different actions were actually consuming, we realised that actions, such as sitting down and standing up, were performed in half the time we originally set with 256 frames. Therefore, we investigated the influence of a decreasing duration of the recordings from 8.53 s to 4.27 s (128 frames) and to 2.13 s (64 frames).

$$d_{\mathrm{m}} = \left\lfloor \frac{num_{\mathrm{BF}}}{30\,\mathrm{fps}} \right\rfloor , \quad num_{\mathrm{BF}} \in \{256, 128, 64\,\mathrm{frames}\}, \quad m \in \{1, 2, 3\} . \quad (3)$$

Instead of finding key frames in the style of Raptis and Sigal [7], we implemented key frame generation by frame skipping in order to reduce the number of frames for classification according to Eq. 4. Consider the sequence duration to be 256 frames, i.e. 8.53 s. In the first case, we keep all the frames ($frame_{\mathrm{keep}} = 1$), which results in 256 key frames. In the second case, when we keep every third frame ($frame_{\mathrm{keep}} = 3$), we will have 85 resulting key frames, which corresponds to a frame rate of 10 fps. When we skip five frames between two successive frames, i.e. we keep every sixth frame ($frame_{\mathrm{keep}} = 6$), we will have a resulting frame number of 42 key frames, which corresponds to a frame rate of 5 fps. For the durations of 8.53 s and 4.27 s, we performed the frame skipping in the same way as described above. This results in 128, 42 and 21 key frames for a duration of 8.53 s; and in 64, 21 and 10 key frames for a duration of 4.27 s.

$$num_{\mathrm{KF}} = \left\lfloor \frac{num_{\mathrm{BF}}}{frame_{\mathrm{keep}}} \right\rfloor . \quad (4)$$

### 4.3  Feature Vectors

In this study, we investigated the performance of different feature vectors that serve as an input to a linear one-versus-one multi-class support vector machine. The calculations include all the joints the Kinect software development kit provides except the feet and ankle joints. In this paper, the joints are numbered with the indices $k$ ranging from 1 to $K$, whereas $k = 1$ denotes the hip center *hip_center* and $K$ is the overall number of joints.

In the following, the feature vectors are presented with their denotation and the corresponding explanations.

***xyz***: For this feature vector calculation, the Cartesian joint coordinates w.r.t. the sensor are transformed and then concatenated to a feature vector. For the *hip_center*, the following transformation is applied:

$$hip\_center\_trans_{\mathrm{t}} = hip\_center_{\mathrm{t}} - hip\_center_1, \quad t \in \{1, ..., N\} . \quad (5)$$

For all $N$ frames of a recorded sequence, the hip center from the first frame $hip\_center_1$ is subtracted from the hip center positions $hip\_center_t$ of all the following frames. The idea is to detect translations of the whole skeleton, when the person walks, but as well when the person sits down or stands up. For the remaining joints $joint_k$, $k \in \{2, ..., K\}$, the following equation is valid:

$$joint\_trans_{t,k} = joint_{t,k} - shoulder\_center_{t,k} \tag{6}$$
$$t \in \{1, ..., N\}, k \in \{2, ..., K\} . \tag{7}$$

For every frame, the joints $joint_t$ are thereby translated in a coordinate system with the *shoulder_center* as origin. The main reason is to detect relative movements to the body, such as the moving hand joint while brushing teeth. The feature vector is constructed as follows: The $x$ components of the transformed hip coordinates $hip\_center\_trans_t$ are sorted frame by frame. In the same way, the $y$ and $z$ components of $hip\_center\_trans_t$ are listed and then appended to the list of the $x$ components. In the same manner, the time series of $x$, $y$ and $z$ components for the other joints are created by using $joint\_trans_{t,k}$. Afterwards, these lists are appended joint by joint.

$\mathcal{F}(xyz)$: Each coordinate of the transformed Cartesian joint coordinates of the hip $hip\_center\_trans_t$ and of the remaining joints $joint\_trans_{t,k}$, $k = \{2, ..., K\}$ can be treated as a one-dimensional time-varying signal. We applied Fast Fourier Transformation (FFT) to each of these signals. The amplitude responses for each coordinate are assembled for one joint. The feature vector is formed by concatenating the Fourier transformations joint by joint.

$xyz$, $\mathcal{F}(xyz)$: The feature vectors $xyz$ and $\mathcal{F}(xyz)$ are assembled to one feature vector by concatenating them.

$\rho\theta\phi$: This feature vector has the same structure as the feature vector $xyz$. However, the Cartesian coordinates are transformed to spherical coordinates.

$\mathcal{F}(\rho\theta\phi)$: Analogue to feature vector $\mathcal{F}(xyz)$, we applied FFT to the single spherical coordinates and assembled the amplitude responses joint by joint.

$\rho\theta\phi$, $\mathcal{F}(\rho\theta\phi)$: The feature vectors $\rho\theta\phi$ and $\mathcal{F}(\rho\theta\phi)$ are assembled to one feature vector by concatenating them.

## 4.4    Results and Discussion

From Table 1, the following relationships can be deduced: The highest classifications rates were achieved with feature vectors that combine Cartesian or spherical coordinates respectively with their corresponding Fourier transformation, i.e. $xyz$, $\mathcal{F}(xyz)$ and $\rho\theta\phi$, $\mathcal{F}(\rho\theta\phi)$. The best classification rate can be reached by using the feature $\rho\theta\phi$, $\mathcal{F}(\rho\theta\phi)$ with a duration of 8.53 s and 42 KF, which corresponds to a sampling rate of 5 fps.

With regard to Cartesian and spherical coordinates, we can deduce that neither of the two representations shows significant improvements when compared to each other.

**Table 1.** Overall classification rates in percent using defined features for different numbers of key frames (KF) and different frame rates. The three highest classification rates are highlighted within each of the three blocks. A block represents the three columns that belong to one duration.

| Feature | Overall classification rates in % | | | | | | | | |
|---|---|---|---|---|---|---|---|---|---|
| | 8.53 sec. | | | 4.27 sec. | | | 2.13 sec. | | |
| | 256 KF | 85 KF | 42 KF | 128 KF | 42 KF | 21 KF | 64 KF | 21 KF | 10 KF |
| | @30 fps | @10 fps | @5 fps | @30 fps | @10 fps | @5 fps | @30 fps | @10 fps | @5 fps |
| $xyz$ | 85.71 | 85.23 | 87.61 | 83.33 | 83.33 | 83.80 | 79.04 | 79.04 | 78.57 |
| $\mathcal{F}(xyz)$ | 87.14 | 91.42 | 91.42 | 84.76 | **89.04** | 88.09 | 80.95 | 82.85 | 83.80 |
| $xyz, \mathcal{F}(xyz)$ | 91.90 | 91.42 | 95.23 | **89.04** | 88.57 | **89.04** | **84.76** | **86.66** | **86.19** |
| $\rho\theta\phi$ | 90.47 | 90.47 | 89.52 | 81.90 | 81.90 | 83.33 | 80.00 | 78.57 | 74.28 |
| $\mathcal{F}(\rho\theta\phi)$ | 87.61 | 88.09 | 87.14 | 83.33 | 83.33 | 82.38 | 75.71 | 78.09 | 81.42 |
| $\rho\theta\phi, \mathcal{F}(\rho\theta\phi)$ | **95.71** | **96.19** | **96.66** | 87.61 | 87.61 | 87.61 | 80.95 | 83.80 | 83.80 |

The different number of key frames does not show a strong effect on the classification rates. However, we can state the classification results do not decrease with a smaller number of frames. In some cases, the classification rates even improved slightly. Consequently, in order to save processing power, the algorithm should preferably be used with only a small number of key frames.

Moreover, it is obvious that the classification rates decrease with smaller duration of the captured sequence (8.53 to 2.31 s) for the same number of key frames. This could be due to two reasons: Firstly, with a longer duration, more information about the activity can be obtained, whereas for shorter durations, less information is available for processing. The second reason could be that with shorter durations and for the set number of key frames and sampling rate, the frequency resolution decreases, i.e. the distance between coefficients in the spectrum is higher. This can cause a loss of information about the frequency behaviour of the activity. In order to explore these aspects, we padded the sequences with a duration of 4.2 s and 2.13 s with zeros so that a number of 256 frames is reached, and performed the Fourier transformation again. By doing this, we achieve the same frequency resolution for different sequence durations. The results in Table 2 reveal that by setting the same frequency resolution for all three duration scenarios, the classification rates do not show a substantial change. Several of the classification rates decrease whereas others increase slightly. This indicates that the decrease of classification rates for shorter durations as shown in Table 1 is not caused by the lower frequency resolution, but rather by the smaller extract of the recorded action.

For the best feature vector, i.e. $\rho\theta\phi, \mathcal{F}(\rho\theta\phi)$, we calculated the confusion matrix, see Table 3. The evaluation confirms that all actions could be classified accurately.

**Table 2.** Overall classification rates in percent using zero padding for the durations 4.27 s and 2.13 s. By doing this, the same frequency resolution is achieved for the different durations.

| Feature | Overall classification rates in % | | | | | | | | |
|---|---|---|---|---|---|---|---|---|---|
| | 8.53 sec. | | | 4.27 sec. | | | 2.13 sec. | | |
| | 256 KF @30 fps | 85 KF @10 fps | 42 KF @5 fps | 128 KF @30 fps | 42 KF @10 fps | 21 KF @5 fps | 64 KF @30 fps | 21 KF @10 fps | 10 KF @5 fps |
| $xyz$ | 85.71 | 85.23 | 87.61 | 83.33 | 83.33 | 83.80 | 79.04 | 79.04 | 78.57 |
| $\mathcal{F}(xyz)$ | 87.14 | 91.42 | 91.42 | 88.09 | 86.19 | 88.09 | 82.85 | 80.00 | 72.38 |
| $xyz, \mathcal{F}(xyz)$ | 91.90 | 91.42 | 95.23 | 88.09 | 84.76 | 84.28 | 82.38 | 81.42 | 79.04 |
| $\rho\theta\phi$ | 90.47 | 90.47 | 89.52 | 81.90 | 81.90 | 83.33 | 80.00 | 78.57 | 74.28 |
| $\mathcal{F}(\rho\theta\phi)$ | 87.61 | 88.09 | 87.14 | 88.09 | 87.14 | 88.09 | 80.47 | 83.80 | 75.23 |
| $\rho\theta\phi, \mathcal{F}(\rho\theta\phi)$ | 95.71 | 96.19 | 96.66 | 86.66 | 86.66 | 85.71 | 81.90 | 81.90 | 76.19 |

**Table 3.** Confusion matrix for feature vector $\rho\theta\phi, \mathcal{F}(\rho\theta\phi)$ with 256 frames (8.53 s) and 42 key frames. The overall classification rate is 96.66%.

| Action | Prediction | | | | | |
|---|---|---|---|---|---|---|
| | $A1$ | $A2$ | $A3$ | $A4$ | $A5$ | $A6$ |
| $A1$ | **57** | 2 | 1 | 0 | 0 | 0 |
| $A2$ | 1 | **28** | 0 | 0 | 0 | 1 |
| $A3$ | 0 | 0 | **30** | 0 | 0 | 0 |
| $A4$ | 0 | 0 | 0 | **30** | 0 | 0 |
| $A5$ | 0 | 0 | 2 | 0 | **28** | 0 |
| $A6$ | 0 | 0 | 0 | 0 | 0 | **30** |

## 5   Conclusions and Future Work

The main goal of this study was to investigate an activity recognition approach with the aim of refining ADL recognition for activities typically performed in a bathroom. We therefore presented an existing high-level reasoning algorithm that determines ADLs based on proximity to objects and pose information. In addition to this, we introduced a skeleton data-based action recognition algorithm that perspectively, after integrating it into the existing system, can enhance the performance of this AAL system. In contrast to the high-level reasoning algorithm, this new approach employs machine learning techniques.

The evaluation of the reasoning algorithm gave evidence that activities normally performed in front of a sink, such as "washing hands, combing, teeth brushing, etc.", "showering" and "using the toilet" could be accurately detected. Since these test had been conducted in a small room, it is likely that the algorithm will show good results in larger rooms as well. This aspect will be evaluated in future tests. Further research will enhance the algorithm by including more objects and other rooms with the aim of recognising further ADLs, such as "preparing food", "washing up" or "cooking". Besides detecting the proximity to locally fixed objects, an extension to moving objects by using object detection algorithms is sensible as well. Further work needs to be done in order to integrate the presented skeleton-based action recognition into the current high-level reasoning system. This implies, inter alia, to convert the off-line classification of pre-recorded skeleton sequences with known length to an on-line version. This study yielded that basic actions, such as "teeth brushing" or "rising an object to the mouth", can be reliably recognised using spherical coordinates of skeleton joints and their Fourier transformation as a feature vector. Thus, we added new activities that can be recognised by our AAL system. Future work will focus on the evaluation of the order of performed actions and draw further conclusions about bathroom activities. For example, sitting down and standing up could be used to better judge the likelihood whether the person used the toilet.

In this study, we used the algorithm of Shotton et al. [14] in combination with the Kinect to obtain skeletal data. For our AAL application, however, we plan to adapt this algorithm to work with data derived by stereo sensors, so that it can be integrated in our existing stereo sensor system.

In addition to the above mentioned work, we intend to install the designed system in real living environments. To achieve this, we will continue working together with local housing associations and our partners from care facilities. Ensuring a high quality of care for people with dementia should be a priority for our society. With regard to the lack of caring personnel and the increasing number of elderly, technical support systems could contribute to achieve that by reminding patients and providing care-related information to the caring staff. However, although these modern developments can be beneficial for all involved persons, we should not forget that such technologies can and shall never replace human closeness and care.

**Acknowledgements.** This project is funded by the European Social Fund (ESF). We furthermore would like to express our thanks to all the persons who contributed to this project during the recordings.

# References

1. Beaudry, C., Péteri, R., Mascarilla, L.: Action recognition in videos using frequency analysis of critical point trajectories. In: 2014 IEEE International Conference on Image Processing (ICIP), pp. 1445–1449. IEEE (2014)
2. Carlsson, S., Sullivan, J.: Action recognition by shape matching to key frames. In: Workshop on Models versus Exemplars in Computer Vision, vol. 1, p. 18 (2001)
3. Chen, J., Kam, A.H., Zhang, J., Liu, N., Shue, L.: Bathroom activity monitoring based on sound. In: Gellersen, H.-W., Want, R., Schmidt, A. (eds.) Pervasive 2005. LNCS, vol. 3468, pp. 47–61. Springer, Heidelberg (2005). doi:10.1007/11428572_4
4. Fogarty, J., Au, C., Hudson, S.E.: Sensing from the basement: a feasibility study of unobtrusive and low-cost home activity recognition. In: Proceedings of the 19th Annual ACM Symposium on User Interface Software and Technology, pp. 91–100. ACM (2006)

5. Harville, M., Li, D.: Fast, integrated person tracking and activity recognition with plan-view templates from a single stereo camera. In: Proceedings of the 2004 IEEE Computer Society Conference on Computer Vision and Pattern Recognition, CVPR 2004, vol. 2, pp. II-398. IEEE (2004)
6. Pirsiavash, H., Ramanan, D.: Detecting activities of daily living in first-person camera views. In: 2012 IEEE Conference on Computer Vision and Pattern Recognition (CVPR), pp. 2847–2854. IEEE (2012)
7. Raptis, M., Sigal, L.: Poselet key-framing: a model for human activity recognition. In: Proceedings of the IEEE Conference on Computer Vision and Pattern Recognition, pp. 2650–2657 (2013)
8. Richter, J., Christian, W., Dayangac, E., Heß, M., Hirtz, G.: Activity recognition based on high-level reasoning: an experimental study evaluating proximity to objects and pose information. In: Fifth International Conference on Pattern Recognition Applications and Methods, ICPRAM, Rome, pp. 415–422 (2016)
9. Richter, J., Christian, W., Hirtz, G.: Mobility assessment of demented people using pose estimation and movement detection. In: Fourth International Conference on Pattern Recognition Applications and Methods, ICPRAM, Lisbon, pp. 22–29 (2015)
10. Richter, J., Findeisen, M., Hirtz, G.: Assessment and care system based on people detection for elderly suffering from dementia. In: Consumer Electronics Berlin (ICCE-Berlin), ICCE Berlin 2014. IEEE Fourth International Conference on Consumer Electronics, pp. 59–63. IEEE (2014)
11. Scanaill, C.N., Carew, S., Barralon, P., Noury, N., Lyons, D., Lyons, G.M.: A review of approaches to mobility telemonitoring of the elderly in their living environment. Ann. Biomed. Eng. 34(4), 547–563 (2006)
12. Schindler, K., Van Gool, L.: Action snippets: how many frames does human action recognition require? In: IEEE Conference on Computer Vision and Pattern Recognition, CVPR 2008, pp. 1–8. IEEE (2008)
13. Schüldt, C., Laptev, I., Caputo, B.: Recognizing human actions: a local SVM approach. In: Proceedings of the 17th International Conference on Pattern Recognition, ICPR 2004, vol. 3, pp. 32–36. IEEE (2004)
14. Shotton, J., Girshick, R., Fitzgibbon, A., Sharp, T., Cook, M., Finocchio, M., Moore, R., Kohli, P., Criminisi, A., Kipman, A., et al.: Efficient human pose estimation from single depth images. IEEE Trans. Pattern Anal. Mach. Intell. 35(12), 2821–2840 (2013)
15. Steen, E.E., Frenken, T., Frenken, M., Hein, A.: Functional assessment in elderlies homes: early results from a field trial. In: Lebensqualität im Wandel von Demografie und Technik (2013)
16. Wang, Y., Sun, S., Ding, X.: A self-adaptive weighted affinity propagation clustering for key frames extraction on human action recognition. J. Vis. Commun. Image Represent. 33, 193–202 (2015)
17. Yao, A., Gall, J., Fanelli, G., Van Gool, L.J.: Does human action recognition benefit from pose estimation? In: BMVC, vol. 3, p. 6 (2011)
18. Ye, M., Zhang, Q., Wang, L., Zhu, J., Yang, R., Gall, J.: A survey on human motion analysis from depth data. In: Grzegorzek, M., Theobalt, C., Koch, R., Kolb, A. (eds.) Time-of-Flight and Depth Imaging. LNCS, vol. 8200, pp. 149–187. Springer, Heidelberg (2013). doi:10.1007/978-3-642-44964-2_8
19. Yous, S., Laga, H., Chihara, K., et al.: People detection and tracking with world-z map from a single stereo camera. In: The Eighth International Workshop on Visual Surveillance, VS 2008 (2008)

20. Yu, G., Liu, Z., Yuan, J.: Discriminative orderlet mining for real-time recognition of human-object interaction. In: Cremers, D., Reid, I., Saito, H., Yang, M.-H. (eds.) ACCV 2014. LNCS, vol. 9007, pp. 50–65. Springer, Heidelberg (2015). doi:10.1007/978-3-319-16814-2_4

21. Zivkovic, Z.: Improved adaptive Gaussian mixture model for background subtraction. In: Proceedings of the 17th International Conference on Pattern Recognition, ICPR 2004, vol. 2, pp. 28–31. IEEE (2004)

# Real-Time Swimmer Tracking
# on Sparse Camera Array

Paavo Nevalainen[1]([⊠]), M. Hashem Haghbayan[1], Antti Kauhanen[2],
Jonne Pohjankukka[1], Mikko-Jussi Laakso[1], and Jukka Heikkonen[1]

[1] Department of Information Technology, University of Turku, 20014 Turku, Finland
{ptneva,mohhaq,mikko-jussi.laakso,jukka.heikkonen}@utu.fi
[2] Sport Academy of Turku Region, Kaarinankatu 3, 20500 Turku, Finland
antti.kauhanen@turku.fi

**Abstract.** A swimmer detection and tracking is an essential first step
in a video-based athletics performance analysis. A real-time algorithm is
presented, with the following capabilities: performing the planar projection of the image, fading the background to protect the intimacy of other
swimmers, framing the swimmer at a specific swimming lane, and eliminating the redundant video stream from idle cameras. The generated
video stream is a basis for further analysis at the batch-mode. The geometric video transform accommodates a sparse camera array and enables
geometric observations of swimmer silhouette. The tracking component
allows real-time feedback and combination of different video streams to
a single one. Swimming cycle registration algorithm based on markerless
tracking is presented. The methodology allows unknown camera positions and can be installed in many types of public swimming pools.

**Keywords:** Athletics · Swimming · Body motion tracking · Camera
calibration · Background subtraction · Video processing · Silhouette
registration · Movement cycle registration

## 1 Introduction

Video systems in swimming coaching have three contradictory requirements.
They should be economical to implement and operate while they have to produce
adequate visualization and there should be some analysis capacity for real-time
numerical feedback. This paper proposes a pipeline of video processing algorithms which are implemented already or intended to be implemented in the
near future. The pipeline aims in early reduction of the video data amount to
be stored and computed. It consists of a fast planar projection based on camera
calibration, background filtering and swimmer tracking. The planar projection
makes it possible to combine different camera streams to a single one and gives a
solid physical framework for further analysis. Since swimming speed is an essential feature, speed accuracy estimation is included. The calibration based on
planar projection is not conventional one, thus a comparison to an ideal pinhole
model with existing camera placement uncertainty is documented.

© Springer International Publishing AG 2017
A. Fred et al. (Eds.): ICPRAM 2016, LNCS 10163, pp. 156–174, 2017.
DOI: 10.1007/978-3-319-53375-9_9

The background filtering serves as a preparatory step in athlete tracking but also protects the intimacy of the general public. This is needed since the site (Impivaara Swimming Center, Turku, Finland) is open to the general public while the coaching sessions take place. Video recordings are used as handouts for athletes and are sometimes used as a public scientific resource.

The swimmer tracking helps to record only the swimmer and his immediate surroundings (bubble clouds, wave form at the water surface). This amounts to storing only 2% of the original raw video data. The registration and vectorization of the swimmer silhouette in real time remains as a possible further development. This paper documents one silhouette vectorization algorithm implemented as a post-processing step. The vectorization quality has been compared to cycle registration and to the horizontal velocity records of the real-time swimmer tracking. The final aim of our project is an establishment of a video database with combination of bio-mechanical indicators, silhouette dynamics, silhouette state vector, time stamps for automatically and manually detected events and handout videos.

Starting a new site for swimming analysis requires usually considerable resources and our economical approach should be of interest to any swimming coach considering a basic computerized real-time feedback at a local site.

The rest of the paper is organized as follows. Section 2 is a short presentation about the contemporary research. Section 3 describes the site and the computing equipment. Section 4 presents an economical and simple calibration method based on planar projection. Also a brief comparison to mono-camera and ideal pin-hole model is provided. Section 5 is a proposal for real-time markerless tracking of the swimmer and it may have relevance to the swimming research community. The swimming cycle registration based on markerless tracking is proposed in Sect. 6. Section 7 has conclusions and discussion about the possible future developments.

## 2   Literature Review

The oldest approach in swimming tracking uses mechanical wire. [16] reports about measuring the force in the wire while some object is dragged behind, another method is measuring the swimmer speed directly using the wire. The mechanical method is used especially to verify the video installments.

Currently, performance analysis is based on video analysis, see e.g. a review in [15]. A typical approach is:

- to produce video stream from multiple cameras
- and detect marked or nonmarked anchor points of the body, and
- combine the trace information to a biomechanical model and visualization tools.

Several commercial tools are available, see e.g. a summary at [15]. Known examples are Dartfish [14] and Sports Motion [13].

A combination of other sensors are used in coaching, e.g. wearable accelerometers with data cache for whole length of the pool [5] and pressure pads [19] at hands.

A typical mature system is documented at [21]. An excellent analysis of real-time and post-session coaching feedback is provided. Swedish Center for Aquatic Studies has AIM (Athletes in Motion) system capable of combining views from submerged and above-water cameras, see [11]. The calibration process is close to our approach although they use striped poles while our approach is based on chessboard pattern. AIM has been developed by Chalmers and Lund Universities.

Another possibility is the virtual camera technique. A synthetical moving viewpoint is computed between adjacent cameras. The view between two stationary cameras can be interpolated as in [9]. This approach is possible in our site to eliminate the projection error between neighboring cameras. A concise method for virtual view generation is described in [22].

Video analysis without markers [10] simplifies the coaching sessions. It is much less intrusive and provides a smooth coaching process. This is also our approach. Although [10] aims for 3D body capture, their arrangement cannot be applied to a sparse camera-row easily. Our approach is more modest, 2D silhouette capture and for swimming style analysis.

A real-time human silhouette detection system is documented in [3]. In their application, the background is stable and can be recorded before the session. Our environment has potential moving objects (non-athlete swimmers sharing the same site during athletic sessions). The aim of [3] is to estimate the center of the body of elderly people at domestic conditions. This data is then used to activity detection and classification.

Silhouette-based gait detection is the topic of [4]. The technique divides the standing or walking target to subparts for analysis. The approach is directly applicable on our field. The added difficulty comes from presence of bubbles and light scattering underwater.

The trend in research is towards 3D visualization and biomechanical models. Analysis of the recorded data is quite developed but there is room for improvement what comes to quick performance feedback by understandable performance measures.

There are mature systems which already serve the coaching activities well. The implementation seems to be rather involved requiring technical assistance, set-up times and high initial and running costs. Our approach is economical, we seek a non-intrusive rudimentary implementation with basis for further improvement.

## 3    The Site and the System Description

### 3.1    The Site

We had only 3 cameras allocated for the project. The cameras cover 18 m over the 25 m pool length. Cameras had to be placed to underwater window sills at

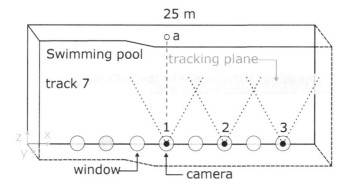

**Fig. 1.** The general layout of the site seen from above. The tracking plane of lane 7 is emphasized. Orientation point $a$ of camera 1 is depicted.

depth of 560 mm at the pool side. A fourth camera can be added to the grey dot depicted in Fig. 1 in the future. This setup will cover the whole 25 m pool length.

The tracking plane is positioned c. 200 mm aside towards the cameras from the centerline of the swimming lane 7. The distance has been chosen so that it approximates the dimensions of the pelvis of an average-sized adult male and female.

The image mapping constructs pixel intensities directly in relation to the tracking plane. This method does not use any camera model, camera locations nor orientations. The World Coordinate System (WCS) is depicted in Fig. 1. Axis $y$ stands for depth. The horizontal length $x$ is oriented from the turning point towards the starting point.

### 3.2   System

The system consists of:

- one 2-core 3.2 GHz 64 bits computer with. 6 TB of disc space
- 3 permanently placed 50 fps cameras (Basler acA2000-50gc GigE). The image size is 750 × 2044. Pixel size at the tracking plane is fixed to $4 \times 4\,mm^2$. The frame time difference $\Delta t = 50^{-1}\,s$ will be used later in this paper.

The uncompressed data from three cameras amounts to about 1 GB for a 10 s clip. All cameras are synchronized so that they capture images at the same time. The time stamps are stored in the video files and they can be used in determining how to stitch the video streams together.

## 4   Planar Projection

Stereo and mono camera calibration methods adapted to underwater conditions are the usual choice for the calibration phase. However, on this site the actual camera locations were not precise preventing the mono-camera calibration and the camera array was too sparse for stereo camera calibration. Instead, a direct planar calibration was used, where pixels and their WCS position on the projection plane $G$ (see Figs. 1 and 2) were sampled in such a quantity, that the rest of the pixel mapping was accomplished by interpolation.

A preliminary measurement process delivering the direct geometric mapping from pixels to global positions is documented first. The method can be categorized as an ad hoc approach answer to two demands: sparsely placed camera array and real-time video transformation. Nearest reference is [17], which uses a camera model and requires the co-planarity of the camera image plane and the tracking plane. Our method requires no camera model but can optionally use one as an interpolant.

The calibration data was gathered by floating a calibration checkerboard along the surface at the tracking plane and recording its horizontal position at each picture. The chessboard had buoys at the top and weight at the bottom. The global position $x_0$ of the board was measured within 10 mm accuracy std.

Next, a quick computation scheme for target pixel intensities will be derived. It is divided to a pre-computation of constants of Eq. 5 and real-time computation of Eq. 6. The following treatise refers to Fig. 2. Informally, one needs mappings $F_s$ and $F_t$ describing the relation from source image $\mathcal{I}_s$ and target image $\mathcal{I}_t$ to WCS. Then a functional composition $F_t \circ F_s^{-1}$ will provide the proper source image intensity $I_t(p_t)$ for a target image pixel $p_t$. The following definitions use somewhat unconventional definition of image as a set of pixels and a function $I(.)$ for pixel intensities because the pixel set $P_s$ of source image is not a conventional rectangle, but merely a general subset of the full frame.

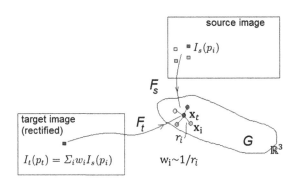

**Fig. 2.** Relation of source image, target image and the projection plane.

The tracking plane $G = \{\mathbf{g} \in \mathbb{R}^3 \mid (\mathbf{g} - \mathbf{g}_G) \cdot \mathbf{n}_G = 0\}$ is defined by the unit normal vector $\mathbf{n}_G$ aligned with WCS $z$ axis and one plane point $\mathbf{g}_G = (0, 0, z_G)^T$, where $z_G = 5050$ mm in case of lane 7. See Fig. 1, which depicts the tracking plane $G$ and the WCS axes $x, y, z$. A source image $\mathcal{I}_s = (P_s, I_s)$ has a set $P_s$ of pixels with intensities $I_s(p)$, $p = (i, j, 1)^T \in P_s$, where $i$ and $j$ are conventional image pixel grid indices. Whereas source image intensities $I_s$ are given, the target image $\mathcal{I}_t = (P_t, I_t)$ requires the geometrically rectified intensity map $I_t$ to be computed. One can cover a conveniently chosen part of the plane $G$ with the following mapping $F_t : P_t \rightarrow G$:

$$F_t(p_t) = \begin{pmatrix} 0 & \gamma & x_{min} \\ \gamma & 0 & y_{min} \\ 0 & 0 & z_G \end{pmatrix} p_t, \tag{1}$$

where $\gamma = 4.0\,mm$ is an arbitrarily chosen scale factor and $g_{min} \in G$ defined in Eq. 2 is a camera view specific upper left corner (UL) point. The lower right (LR) corner point $g_{max}$ has similar definition Eq. 3. Corner points UL and LR, and the target image pixel sizes $n_t, m_t \in \mathbb{N}_+$ can be chosen freely as long as they lead to connected combined image and fulfill the constraint posed by Eq. 3:

$$g_{min} = (x_{min}\,y_{min}\,z_G)^T \tag{2}$$
$$g_{max} = F_t\left((n_t,\,m_t,\,1)^T\right). \tag{3}$$

The target image size $P_t = [1, n_t] \times [1, m_t] \in \mathbb{N}_+^2$ and the UL position $g_{min}$ are free parameters, which are to be fixed by a practical choice so that the mapping quality is good and all camera views combine to a continuous total view. Figure 3 shows a choice made for all three camera views.

A camera board has a checkerboard pattern. Each corner pixel $p$ on the checkerboard on each calibration image and its corresponding global position $g \in G$ form a measurement pair $(p, \mathbf{g}) \in U$, where $p \in P_U \subset P_s$. The calibration data set $U$ is cumulated over 37 calibration images with c. 3000 measurement pairs. The pixel samples $P_U$ of $U$ cover only a part of the source image pixels $P_s$ whereas the end result of the direct plane calibration have to map whole source image onto the tracking plane using $F_s : P_s \rightarrow G$. In that sense this is an interpolation problem.

[2] uses two bilinear functions $f_x, f_y : \mathbb{R}^2 \rightarrow \mathbb{R}$ to map WCS coordinates $x$ and $y$ separately:

$$F_s(p_s) = (f_x(p_s), f_y(p_s), z_G)^T. \tag{4}$$

Specifically, for the calibration measurement set $U$ it holds that $(p, \mathbf{g}) \in U \rightarrow F_s(p) \approx \mathbf{g}$. It is worth to mention that the above mapping $F_s$ could have been any relatively smooth function with some sort of regularization control.

## 4.1 Pre-computation and Post-computation

The initial problem of finding the target image content has been accomplished, except it is inconvenient to inverse $F_s$. Instead, one needs to match source image pixels to target image on the projection plane. There are many possibilities to this. One can e.g. use Shepard interpolation as in [2] and pre-compute the necessary neighborhood sets and weights. In that respect, $k = 4$ nearest neighbors of a point $F_t(p_t)$ amongst the point set $F_s(P_s)$ are used. The interpolation neighborhood matrix $M \in \mathbb{N}_+^{|P_t| \times k}$ and neighborhood weight matrix $W \in \mathbb{R}_+^{|P_t| \times k}$ can be pre-computed and used for the real-time pixel intensity computation:

$$I_t(p_t) = \sum_{i=1}^{k} W_{p_t i} \, I_s(M_{p_t i}), \, p_t \in P_t. \tag{5}$$

A special case $k = 1$ leads to the nearest neighbor approximation, which is very fast and requires no weights $W$. By denoting $C = M_{p_t 1}$ as a table of the nearest neighbor pixels of $p_t \in P_t$ at $P_s$, one can write the final real-time transformation as:

$$\mathcal{I}_t := (P_t, I_s(C(P_t))) \tag{6}$$

The pre-computed image mapping is simple and efficient enough for to be used in real-time. The usage of the projection plane $G$ also makes it possible to combine each camera signal accurately to one single video, as demonstrated in Fig. 3. What comes to implementation, two image intensities rest on two memory blocks, and $C$ is an index array.

## 4.2   Error Analysis

The measurement set $U$ has c. 150 mm vertical gap and c. 50 mm average horizontal distance between points. This requires the interpolant to have rather high penalty for non-smoothness.

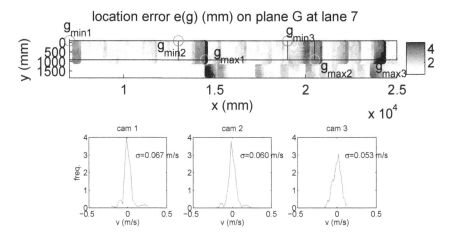

**Fig. 3.** Above: Error of the interpolation as a difference $e(p) = \|F_s(p) - \mathbf{g}\|, (p, \mathbf{g}) \in U$ at the calibration data set $U$. Also the choice of target image views $P_t$ for each camera depicted. The color bar shows errors within the tracking area limited by view frames $(\mathbf{g}_{min\,i}, \mathbf{g}_{max\,i}), i = 1, 2, 3$. Maximum error outside the tracking area is 9 mm. Below: Speed error for a point moving at $v = 1.6 m/s$. Velocity distribution is measured to all directions. The std. is 0.07 m/s. (Color figure online)

The measurement errors for each observation $(p, \mathbf{g}) \in U$ were as follows:

- pixel $p$ detection was done with Matlab *detectCheckerBoard.m* function, theory of which is contained in [6]. The pixel detection error is $p \approx (1, 1)$ std.
- the mechanical placement accuracy of a measured point $\mathbf{g}$ is $\Delta\mathbf{g} \approx (10, 10, 10 + 0.01\, y)^T$ mm as an approximate std.

The so called back-projection (see e.g. [18]) plot shows that the geometric mapping error is un-biased but not Gaussian. The back-projection map is not included to this paper. The final accuracy of $F_s(p_s)$ is much better than of initial data $U$. The geometric mapping error measure $e(p)$ is given in Eq. 7 and depicted at Fig. 3:

$$e(p) = \|\mathbf{g} - F_s(p)\|, \ (p, \mathbf{g}) \in U \tag{7}$$

Since the sample set $U$ is of rather good quality and since the function $F_s$ is rather smooth, the error stays almost constant even if the tuning of the interpolation is subjected to cross-validation over subsets of $U$. The positioning error with std. $\sigma_g \approx 1.8$ mm (see [2]) is largest in occasional points at the border and grows rapidly when extrapolating. The border areas are seldom occupied by a swimmer, though, and the problem is more of aesthetical nature. The border error can be eliminated in the future by applying a different interpolant instead of one in Eq. 4. Future calibrations will use a laser device for the horizontal position measurement, hopefully improving the visible errors occurred during the recording of the calibration data set $U$ revealed by Fig. 3.

The speed error with the result std. $\sigma_v \approx 0.07 m/s$ has been estimated by assuming an observable point moving at a typical velocity $v = 1.6\, m/s = 8$ pixels/frame. An observation cannot occur at a shorter time interval than one frame, thus more pessimistic finite differences must be used in estimation. Instead of usual chain differentiation of velocity $v(p) = F_s'(p)\dot{p}\gamma$ where pixel speed $\dot{p}$ is assumed to be constant $\dot{p} = 8/\Delta t$, one needs to employ a probabilistic variables $p \sim \mathcal{U}(P_s)$ and $\dot{p}\Delta t \sim \mathcal{N}(8, 2 \times 0.5^2)$, where variance term comes from two additive measurements of consecutive frames with std. 0.5 pixels ($\approx 2$ mm according to [2]). The resulting distribution function of speed $v$ is presented at Fig. 3:

$$v \sim \frac{F_s(p_r) - F_s(p)}{r}\dot{p}\gamma, \ \|p_r - p\| = r = 8, \ p_r \in P_s. \tag{8}$$

Both the location and speed errors are somewhat larger in practise, since there are always observation or registration errors involved.

### 4.3  Effect of Inaccurate Camera Placement to Traditional Calibration

Alternative calibration methods were evaluated and tested. The best alternative was mono-camera model used as an interpolant of planar projection Eq. 4, but even that failed to reach the same accuracy as direct planar calibration. The traditional mono-camera calibration was next and the stereo-calibration worst - this was due to small overlapping of the camera view cones, see Fig. 1.

The traditional camera calibration requires an arrangement where the camera locations and orientations are known or can be measured accurately. Unfortunately, the geometric accuracy (see Table 1) technically possible on the site was not enough for traditional calibration to be competitive. Since the approaches differ so much, it is good to have a criterion for required camera positioning accuracy before mono or stereo calibration become an option.

A treatise of the application of mono-camera model combined with the plane projection follows. There are three distinct parts: pinhole camera model, coordinate transformation due the camera orientation, and projection to the tracking plane.

**Pinhole Camera Model** $Q_1^{-1}$ : $P \to S_2 \subset \mathbb{R}^3$ from pixels $P$ to camera-related projection directions on the unit sphere $S_2$. Literature has the camera model defined usually the opposite way as $Q_1$ but uses the concept in both meanings. We used Matlab Toolbox [20] and an underwater calibration chessboard.

**Camera Orientation** of a camera $c$ is defined by WCS orientation matrix $\mathbf{N}_c = \{\mathbf{n}_i\}, \mathbf{n}_i \in S_2, i = 1..3$, to be derived on the next page. The WCS coordinate base is denoted by $\{\mathbf{e}_i\}$ and camera coordinate base by $\{\mathbf{e}_{ci}\}, i = 1..3$, see Figs. 4 and 1.

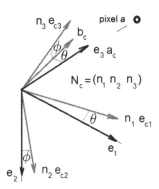

**Fig. 4.** The WCS orientation base $\mathbf{E}$, an orientation $\mathbf{E}_c$ specific to a camera $c$ and column vectors of the coordinate transform matrix $\mathbf{N}_c$ depicted at the camera $c$. The pixel $a$ is the global $z$ axis marker for camera $c$.

The orientation is measured by an alignment marker pixel $a \in P$, which indicates the global $z$ axis direction in the camera coordinates, see Figs. 1 and 4. The following definitions are needed:

$$\mathbf{a}_c = Q_1^{-1}(a, \boldsymbol{\alpha}) \text{ (geom. axis 3 in camera coords)}$$
$$\mathbf{P}_{c23} = \mathbf{e}_{c3}\, \mathbf{e}_{c3}^T + \mathbf{e}_{c2}\, \mathbf{e}_{c2}^T \text{ (a projection matrix)}$$
$$\mathbf{b}_c = (\mathbf{P}_{c23}\, \mathbf{a}_c)^0 \text{ (projecting } \mathbf{a}_c \text{ to } (\mathbf{e}_{c2}, \mathbf{e}_{c3}) \text{ plane),} \tag{9}$$

where $\boldsymbol{\alpha}$ are the pinhole model parameters achieved e.g. by Matlab toolbox [20] session and $(.)^0$ denotes vector normalization.

The above definition relies on the fact that cameras have been placed very carefully to have very small rotational error around the optical axis. This was achieved by special horizontal reference markers at the opposite side of the pool. The achieved optical axis rotation accuracy is c. $0.1 \times 10^{-3}\, rad$.

A rotation matrix $\mathbf{R_{uv}}$ rotates a unit vector $\mathbf{u}$ to a unit vector $\mathbf{v}$:

$$\mathbf{R_{uv}} = \mathbf{H(u + v)H(u)}, \tag{10}$$

where the so called line reflection matrix $\mathbf{H(u)} = -\mathbf{I} + 2\mathbf{u}\mathbf{u}^T/\|\mathbf{u}\|_2$, see e.g. [24]. The rotation matrix $\mathbf{R_{uv}}$ has the following necessary properties: $\mathbf{H(u,u)} = \mathbf{I}$, $\mathbf{H(u,v)u} = \mathbf{v}$, $\mathbf{w}{\cdot}\mathbf{u} = \mathbf{w}{\cdot}\mathbf{v} = 0 \rightarrow \mathbf{R(u,v)w} = \mathbf{w}$, $\mathbf{R(u,(u+v)^0)}^2 = \mathbf{R(u,v)}$.

A coordinate transformation matrix $\mathbf{N}_c$ can now be based on the definitions of Eq. 9 and two consecutive Euler rotations $\mathbf{R}_\phi$ and $\mathbf{R}_\theta$ depicted in Fig. 4:

$$\mathbf{R}_\phi = \mathbf{R}_{\mathbf{e}_{c3}\,\mathbf{b}_c} \tag{11}$$

$$\mathbf{R}_\theta = \mathbf{R}_{\mathbf{b}_c\,\mathbf{a}_c} \tag{12}$$

$$\mathbf{N}_c = (\mathbf{R}_\theta\,\mathbf{R}_\phi)^{-1}. \tag{13}$$

The final transformation from coordinates specific to a camera $c$ to WCS equals:

$$\mathbf{n} = \mathbf{N}_c\,\mathbf{n}_c \tag{14}$$

where $\mathbf{n}_c = Q_1^{-1}(p,\boldsymbol{\alpha}), p \in P$ is the camera-specific orientation of a pixel $p$ from the previously addressed pinhole model.

**Projection:** The final map $Q_2 : \mathbf{n} \rightarrow \mathbf{g} \in G$ is the projection to the tracking plane $G$:

$$l = (z_G - g_c \cdot \mathbf{e}_3)/(\mathbf{n} \cdot \mathbf{e}_3) \;\; \text{(projection length)} \tag{15}$$

$$g = g_c + l\mathbf{n} \in G \;\; \text{(on-line condition)} \tag{16}$$

The projection length $l$ has been solved from the in-plane condition: $g \in G \rightarrow g \cdot \mathbf{e}_3 - z_G = 0$.

The total mapping $F_{sM} = Q_2 \circ \mathbf{N}_c \circ Q_1^{-1} : P \rightarrow G$ for mono-camera has been now constructed. By assuming normal distributed measurements of Table 1 and normal distributed pinhole model parameters from Matlab Toolbox [20], one gets the pixel placement accuracies listed in Table 2. Due to nonlinearities in Eqs. 9–16, the resulting distribution of tracking plane position $g \in G$ is not normal distributed. Std. values are used in Table 2 to make comparisons possible. Computations were done by python Sympy package [23] by coding Eqs. 9–16 with their corresponding variance terms (first five values) of Table 1. $\Delta\mathbf{N}$ at the last row of Table 1 stands for the maximum angular orientation variance at $\mathbf{e}_3$ direction and it is a result, not an input for the computation.

The ideal camera placement is exact, whereas the current accuracy is within 10 mm (std.). The ideal camera model is impossible but it can be modelled with Eqs. 9 by assuming the pixel specific local direction vector $\mathbf{a}_c$ be given. All existing camera models are between the ideal one and the Matlab Toolbox, which is based on [6,12].

**Table 1.** Error terms (1 std.) of the mono-camera calibration.

| Term | Explanation | Value |
|------|-------------|-------|
| $\Delta a$ | Direction pixel error | 2.0 pix. |
| $\Delta p$ | Chessboard pixel error | 0.9 pix. |
| $\Delta z_G$ | Tracking plane position | 10 mm |
| $\Delta g_c$ | Real global camera position (*) | (10,10,15) mm |
| $\Delta g_c$ | Ideal global camera position (**) | (0,0,0) mm |
| $\Delta \mathbf{N}$ | Global camera orientation | $0.4 \times 10^{-3}$ rad |

The accuracy is clearly limited by the camera placement if direct calibration is not utilized. Also, the limiting factor in direct calibration is horizontal position measurement, which can be improved in the future.

**Table 2.** The effect of camera placement inaccuracy. The projection calibration, an ideal and a real mono-camera model accuracy $\Delta g$ compared. Values are std. in vertical and horizontal direction.

| Method | No pos. error (*) (mm) | Real pos. error (**) (mm) |
|--------|------------------------|---------------------------|
| Direct calibration | 2.0 ... 7.0 | 2.0 ... 7.0 |
| Ideal camera model | 3.2 ... 5.2 | 13 ... 15 |
| Mono-calibration | 3.4 ... 10 | 15 ... 21 |

## 5   Silhouette Tracking

The real-time swimmer tracking method is based on finding the horizontal pixel translation which minimizes the intensity difference of two sequential images. There are three caveats in this simplistic approach though:

1. The background does not move and causes a strong slow-down bias. In our chosen approach, we use the difference of two consecutive images as a basis of further computations, thus eliminating this effect.
2. Hands are constantly propelling the swimmer forwards with a backwards stroke aside the body. This causes similar slow-down bias, but only locally. This error source can be eliminated by dividing the observed area to e.g. $3 \times 6$ subparts as depicted in Fig. 5 and taking the average of the majority translational shift observed.
3. The resulting speed or horizontal position as a function of time cannot be directly associated to any particular body part. This problem can be solved only by a secondary tuning at the post-processing phase.

The swimmer tracking succeeds in keeping the swimmer at focus, thus enabling the reduction of the video record size. Bio-mechanical indicators e.g. accurate speed of head, pelvis etc. have to be measured at the post-processing phase. The inexact preliminary speed measurement does not prevent later additive corrections!

The algorithm to find the horizontal shift $u^* \in \mathbb{R}$ (in pixels) between two sequential images $\mathcal{I}_1$ and $\mathcal{I}_2$ is presented. It is based on minimizing the following simple and fast dissimilarity measure:

$$d_A(\mathcal{I}_1, \mathcal{I}_2) = \sum_{p \in P_1 \cap P_2} |I_1(p) - I_2(p)|/|P_1 \cap P_2|, \tag{17}$$

where the summation is applied to the overlapping part of two images or sub-images after a possible horizontal translation explained later in the text. Since the athlete silhouette has a slight shape change between two frames, the dissimilarity minimization is made by sub-images, and only the most similar sub-image pairs will contribute to the decision about the effective horizontal translation happening between frames. This limited set of sub-images is detected by their deep local minima of $d_A(.,.)$ measured by the second order derivative (or its approximant) at the minimum. The algorithm is designed to be of $O(|P_t|)$ (constant time for a pixel), and it allows simple task-parallelization per each sub-window, although it is intended to a single processor. It requires one parameter, the a priori value $u_0$ for the vertical speed.

Following conventions have been used in the algorithm: $\mathcal{I}(\Delta i)$ is an image $\mathcal{I}$ shifted by $\Delta i$ pixels horizontally along the current swimming direction. $\mathcal{I}_{i,k}$ is a $k$'th sub-image of an image $\mathcal{I}_i$. $g(u)$ is a low-order polynomial fit to a data set $\{(i, f(i))| i \in$ a short interval $\subset \mathbb{N}_+\}$, e.g. of degree 2 or 3.

**Data**: Sequential images $\mathcal{I}_1$ and $\mathcal{I}_2$, and the previous horizontal translation
$\quad\quad u_0 \in \mathbb{R}_+$
**Result**: Translation $u^* \in \mathbb{R}_+$ which best fits sub-images $\mathcal{I}_{.k}$
**forall** $\mathcal{I}_{1,k} \subset \mathcal{I}_1$ **do**
$\quad\quad f(i) \equiv_{df} d_A\left(\mathcal{I}_{1,k}(i), \mathcal{I}_{2,k}(0)\right)$ see Eq. 17;
$\quad\quad i_0 \leftarrow \lfloor u_0 \rfloor$ truncated value;
$\quad\quad i_k \leftarrow \underset{i \in [i_0 - \Delta u, i_0 + \Delta u]}{\text{argmin}} f(i)$ optim. evaluations based on local convexity;
$\quad\quad g(u) \equiv_{df} f(i)$ interpolated to continuous $u \in \mathbb{R}_+$;
$\quad\quad u_k \leftarrow \underset{u \in [i_k - \Delta u, i_k + \Delta u]}{\text{argmin}} g(u)$;
$\quad\quad w_k \leftarrow \frac{d^2 g}{du^2}(u_k)$ approximative or exact second derivative;
**end**
$u^* \leftarrow \text{mean}_{k \in [1,4]} u_k$ when ordered by descending $w_k$;

**Algorithm 1.** Finding the optimal horizontal pixel shift between two images.

There is a faster, non-interpolating and more inexact version of this algorithm using discrete value $i_k$ directly and approximating the importance $w_k$ of a sub-image $k$ by the second order difference $D^2(.)$ with a unit step size:

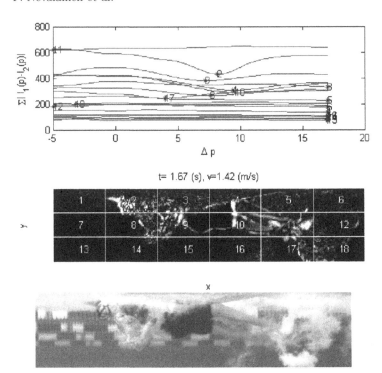

**Fig. 5.** Above: Horizontal pixel translation $\Delta p$ in for the optimal match at each subwindow. Middle: Division of the camera view to sub-images $\mathcal{I}_{i,k}$, $k \in [1, 18]$. The background is neutral since the difference of two sequential images has been used for the swimmer tracking. Below: Original video frame (G channel of the RGB signal).

$$u_k \leftarrow \qquad\qquad i_k$$
$$w_k \leftarrow f(i_0 - 1) - 2f(i_0) + f(i_0 + 1) = D^2(i_0). \tag{18}$$

A typical speed of a female backstroke swimmer is $u_0 = 8...9$ pixels/frame ($\approx 1.5...1.7$ m/s). This speed is specific to abilities of the athlete, swimming style, sex and age. It serves as an initial guess for the Algorithm 1 when first two images are being processed. The later image pairs will use the latest observed $u^*$ as the initial guess. Two sequential images are then divided to sub-image pairs $\mathcal{I}_{.k}$, which have in some cases unique optimal horizontal shift $u_k$ (in pixels). The shifts are made in exact pixels $i$, and usually only few trials are needed to find the best pixel fit $i_k = \arg\min_i f(i)$ when starting the search from the truncated initial guess $i_0 = \lfloor u_0 \rfloor$. The continuous counterpart $g(u)$ of the image fit function $f(i)$ is evaluated only locally at a short interval, and a low-order 2nd or 3rd order polynomial is enough to estimate the quality of the fit. Figure 5 depicts more generic splines in order to illustrate the regularity of the local fit.

If two sub-images resemble each other, they provide more useful information about the translation than other sub-image pairs, and should be preferred. The quality of the fit is reflected by the depth of the local minimum. E.g. sub-images

2,4,9 and 10 in Fig. 5 indicate the horizontal speed best and are chosen to form the averaged horizontal shift $u^*$.

The momentary horizontal velocity of the swimmer is now $v(t) = u^*(t)\,\gamma/\Delta t$ where $\gamma = 4\,mm$ is the geometric pixel size from Eq. 1 and time $t = i\Delta t$ for an image $\mathcal{I}_i$. Figure 6 depicts the resulting tracking speed. Individual strokes can be registered but the plot would require some smoothing. The current system has a real-time velocity plot based on marker-tracking, but at the moment it seems to have poorer quality. There are further improvements possible to Algorithm 2 in order to produce better tracking speed plot, but similar improvements can be achieved in the post-processing phase, as well. E.g. a Kalman filtering employed in [2] can be used.

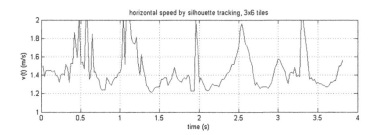

**Fig. 6.** Silhouette tracking. Above: Horizontal translation for the optimal match at each subpart. Below: Division of the camera view to subparts $I_k,\ k \in [1, 18]$. The background is black since the difference image is being used for the swimmer tracking.

## 6   Swimming Cycle Registration

There are several alternatives for swimming cycle registration. One can detect pikes from the velocity curve of Fig. 6 and produce a visual dissimilarity graph based on velocity histories of each swimming stroke as in [2].

Another alternative is silhouette vectorization, which would provide access to the state space of the performance. This approach would open the door to fully nonlinear dynamics analysis.

Third alternative uses the translational sub-image dissimilarity demonstrated in Algorithm 2 to find the closest match from the frames which exist approximately one estimated cycle duration away from the current frame. This gives a continuous cycle duration measure. The next matching frames form a detectable path in the image dissimilarity matrix $D$ depicted in Fig. 7. The definition of the dissimilarity matrix $D$ follows:

$$D = \{d(\mathcal{I}_i, \mathcal{I}_j)\}_{i,j \in 1..n}, \tag{19}$$

where $n$ is the number of video images over one length of the pool. There are several possibilities for the dissimilarity $d(.,.)$ which will be explored next. Figure 5 has been produced by matching whole images with only horizontal

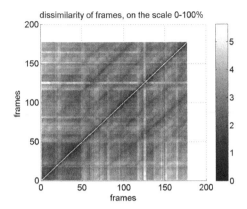

**Fig. 7.** Translational dissimilarity of full frames in a swimming performance with a gliding phase (frames 1–50) and backstroke phase (frames 51...). Exceptional events are clearly visible. 100% dissimilarity means difference between completely black and completely white images.

translation allowed, while left part of Fig. 6 has been produced using sub-images and fully free translation. An algorithm implementing a dissimilarity measure $d_B(\mathcal{I}_i, \mathcal{I}_j)$ based on extent of minimizing free translational shift is presented next. The previous notion of horizontally shifted image $\mathcal{I}(i)$, $i \in \mathbb{N}_+$ is now extended to completely free translation $\mathcal{I}(\mathbf{i})$, $\mathbf{i} \in \mathbb{N}^2$. Also, previous interval search for minimization is substituted by square area search over $\mathrm{box}((u_x, u_y), \Delta u) = [u_x - \Delta u, u_x + \Delta u] \times [u_y - \Delta u, u_y + \Delta u] \subset \mathbb{R}^2$, $u_x, u_y \in \mathbb{R}$. A similar $\mathrm{box}(\mathbf{i}, \Delta u) \subset \mathbb{N}^2, \mathbf{i} \in \mathbb{N}^2$ has a corresponding discrete domain.

---

**Data**: Two images or sub-images $\mathcal{I}_i$ and $\mathcal{I}_j$ and the search limit $\Delta u = 2$.
**Result**: the image dissimilarity $d_{ij} = d_B(\mathcal{I}_i, \mathcal{I}_j)$ based on the norm of
        minimizing 2D pixel translation.
$f(\mathbf{i}) \leftarrow d_A\left(\mathcal{I}_1(\mathbf{i}), \mathcal{I}_2(0)\right)$, $\mathbf{i} \in \mathbb{N}^2$ ;
$\mathbf{i}^* \leftarrow \underset{\mathbf{i} \in \mathrm{box}(0, \Delta u)}{\mathrm{argmin}} \; f(\mathbf{i})$ optimizing evaluations based on local convexity;
$g(\mathbf{u}) \equiv_{def} f(\mathbf{i})$ interpolated to continuous $\mathbf{u} \in \mathbb{R}^2$;
$\mathbf{u}^* \leftarrow \underset{\mathbf{u} \in \mathrm{box}(\mathbf{i}^*, \Delta u)}{\mathrm{argmin}} \; g(\mathbf{u})$;
$d_{ij} \leftarrow \|\mathbf{u}^*\|_2$

**Algorithm 2.** Finding the extent of an optimal pixel shift between two images.

---

Another variant of Algorithm 2 defines dissimilarity $d_C(.,.)$ with translations **u** restricted to horizontal shift only. Yet another dissimilarity applies to whole images only:

$$d_D(\mathcal{I}_i, \mathcal{I}_j) = \underset{\mathcal{I}_{1k} \in \mathcal{I}_1}{\mathrm{mean}} \, d_C(\mathcal{I}_{1k}, \mathcal{I}_{2k}), \qquad (20)$$

which has a possible variant restricting the value set used for averaging by a similar interpolation and weight strategy as used in Algorithm 1.

The following cycle duration algorithm assumes the dissimilarity matrix $D$ partially pre-computed using the dissimilarity variant $d_D(.,.)$ of Eq. 20 so that it contains the path of best matching future frames. This can be ensured by computing a diagonal stripe of $D$ using some practical margin width, say 15 frames. The algorithm then traces the path to produce the cycle duration plot on right side of Fig. 8. First, the best matching future frame $\mathcal{I}_{i+j_i}$ is searched for a frame $\mathcal{I}_i$. Then, an interpolation function $g(.,.)$ is based on a modest sample set of neighboring dissimilarity values of $D$ to find a best matching cycle duration $u_i$. This algorithm produces interpolated frames $\mathcal{I}_{i+u_i}$, $u_i \in \mathbb{R}_+$, from where the cycle duration $T(t) = u_i \Delta t$, $t = i\Delta t$ can be derived:

**Data**: Dissimilarity matrix $D$, image index $i$ and the previous cycle duration
$\quad\quad u_{i-1} \in \mathbb{R}_+$ or a priori $u_0$
**Result**: Duration $u_i \in \mathbb{R}_+$ which best fits frame pairs $\mathcal{I}_i, \mathcal{I}_{i+u_i}$.
**forall** $i \in 1...n$ **do**
$\quad j_0 \leftarrow \lfloor u_{i-1} \rfloor$;
$\quad j_i \leftarrow \underset{j}{\operatorname{argmin}}\, D_{ij}$ starting the search from $j_0$;
$\quad g(v, u) \leftarrow D_{ij}$ interpolated to continuous $u, v \in \mathbb{R}$ using earlier computed
$\quad\quad$ values of $D$ in a neighborhood box$((i, j_k), \Delta u)$;
$\quad u_i \leftarrow \underset{j_k - \Delta u \leq u \leq j_k + \Delta u}{\operatorname{argmin}}\, g(i, u)$ a local search with $\Delta u = 3$;
**end**

**Algorithm 3.** Finding the optimal match over the swimming cycle duration.

As before, a faster and less accurate version can be provided by bypassing the interpolation step: $u_i \leftarrow j_i$. It seems, that a practical choice is to use the

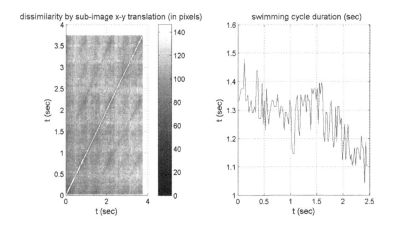

**Fig. 8.** Left: The frame dissimilarity matrix $D$ based on the dissimilarity measure $d_D(.,.)$. Both horizontal and vertical translations are allowed. See Algorithm 3. The cycle length shows as diagonal stripes. Right: The cycle duration based on the dissimilarity matrix at left. The graph shows how the athlete shifts from slide mode to stroke mode. The graph contains noise, since the interpolation routine of Algorithm 3 was by-passed for quality assurance.

latter less accurate model and use e.g. Kalman smoothing (see e.g. [7]) with a physically feasible point mass and damping coefficients to produce the final velocity plot of Fig. 6 and the cycle period graphs depicted in Fig. 8.

# 7  Conclusions

A principal objective of the system is to provide immediate trainer feedback. To achieve that in the early phase of the project, the implementation has been based on the following principles:

- The analysis is limited to a monotonic stage of the athletic performance. Detection of stage changes like from gliding to stroking remains a future problem. At the moment the analysis is triggered manually. Several swimming styles are supported, though.
- A real-time marker tracking has been implemented. It will be substituted by the markerless tracking in the future.
- The geometric mapping of images, markerless silhouette tracking, seamless combination of 3 camera signals, tracked velocity visualization and cycle regularity visualization are real-time processes, which will quite likely perform at 25 fps (for every second frame).
- silhouette capture and vectorization, separate legs-vs-hands stroke analysis, biomechanical analysis and various problems about clustering athletes by their swimming technique are all left as a post-processing step.

As the the execution order of the Algorithm 1 is $O(1)$ per pixel, and there are c. 100 processor instructions per pixel, it is estimated that the execution time will be very small and that the intended computing equippement [1] can handle the tracking of 25 fps video input. This means two processors will be capable of real-time geometric transform [2] and tracking task with c. 25 fps i.e. every second frame are subjected to computation and every second frame is interpolated with a simple instruction set. The computation tasks are inherently parallel, facilitating upgrade to system speed 50 fps.

A simple pipeline of real-time swimmer tracking, which requires relatively modest computational arrangements and is able to provide immediate coaching feedback, has been presented. The analysis is limited to projective 2D plane, and it is our hope that it will be capable of silhouette vectorization and further bio-mechanical analysis. The camera calibration process adapts to the economically feasible sparse camera array arrangement and to poor camera placement accuracy. A procedure to decide between conventional camera calibration and direct projection plane calibration has been presented.

At the moment the system performance has not been validated against contemporary techniques and technologies. A new approach avoiding many pitfalls of computationally heavy operations has been proposed, and further improvements depend on the feedback of the research community with long traditions and experience on this field. The concrete micro-array chip implementation of algorithms will proceed later, and we remain optimistic that the result will have practical value.

The current system will be upgraded by a fourth camera at the location indicated by a grey circle in Fig. 1. The video monitoring would then span whole the pool length.

There are several alternatives to algorithms used to track the swimming speed. One can preprocess or vectorize the images by FFT etc. methods. These alternatives were not of interest now, since we aim at a video database of large number of athletic sessions with correctly focused video frames in order to cumulate samples for further application of Machine Learning methodologies, e.g. for swimming gait evaluation.

The geometric mapping of the video image is fast while it maintains enough signal quality for markerless tracking algorithms. An interesting alternative interpolant for the mapping $F_s$ of Eq. 4 is the best available water-plexiglass-air camera model as described in [8]. The same model has been applied e.g. in [11]. The model requires more tuning parameters than e.g. [20]; camera position and orientation, plexiglass thickness and refraction coefficient, and the air length between camera and plexiglass must be fit by mean square error minimization of Eq. 7. This approach will be attempted in the future.

We hope that proposals and demonstrations of this paper stir up some interest in the research community. We aim at a combination of simple and feasible methods, and the current set of algorithms lacks only a high-quality silhouette registration with a simple real-time version to detect swimming phase changes and events.

**Acknowledgements.** The project is a joint venture of University of Turku IT department and Sports Academy of Turku region and it has been funded by city of Turku, National Olympic Committee, Finnish Swimming Federation, Urheiluopistosäätiö and University of Turku. Machine Technology Center Turku Ltd. participated to instrumentation.

# References

1. ARM Cortex-A7 Processor (2016). http://www.arm.com/products/processors/cortex-a/cortex-a7.php
2. Nevalainen, P., Kauhanen, A., Raduly-Baka, C., Heikkonen, J.: Video based swimming analysis for fast feedback, In: Proceedings of International Conference on Pattern Recognition Applications and Methods (ICPRAM2016), pp. 457–466 (2016)
3. Christodoulidis, A., Delibasis, K.K., Maglogiannis, I.: Near real-time human silhouette and movement detection in indoor environments using fixed cameras. In: Proceedings of the 5th International Conference on Pervasive Technologies Related to Assistive Environments (PETRA 2012) (2012)
4. Choudhury, S.D., Tjahjadi, T.: Silhouette-based gait recognition using Procrustes shape analysis and elliptic Fourier descriptors. Pattern Recogn. **45**, 3414–3426 (2012)
5. Dadashi, F., Millet, G., Aminian, K.: Inertial measurement unit and biomechanical analysis of swimming: an update. Sportmedizin **61**, 21–26 (2013)

6. Zhang, Z.: A flexible new technique for camera calibration. IEEE Trans. Pattern Anal. Mach. Intell. **22**, 1330–1334 (2000)
7. Hartikainen, J., Seppänen, M., Särkkä, S.: State-space inference for non-linear latent force models with application to satellite orbit prediction. In: CoRR (2012)
8. Sedlazeck, A., Koch, R.: Perspective and non-perspective camera models in underwater imaging – overview and error analysis. In: Dellaert, F., Frahm, J.-M., Pollefeys, M., Leal-Taixé, L., Rosenhahn, B. (eds.) Outdoor and Large-Scale Real-World Scene Analysis. LNCS, vol. 7474, pp. 212–242. Springer, Heidelberg (2012). doi:10.1007/978-3-642-34091-8_10
9. Makoto, H.S., Kimura, M., Yaguchi, S., Inamoto, N.: View interpolation of multiple cameras based on projective geometry. In: International Workshop on Pattern Recognition and Understanding for Visual Information (2002)
10. Ceseracciu, E.: New frontiers of Markerless Motion Capture: Application to Swim Biomechanics and Gait Analysis. Padova University, Padua (2011)
11. Haner, S., Svärm, L., Ask, E., Heyden, A.: Joint under and over water calibration of a swimmer tracking system. In: Proceedings of the International Conference on Pattern Recognition Applications and Methods, pp. 142–149 (2015)
12. Heikkilä, J., Silven, O.: A four-step camera calibration procedure with implicit image correction. In: Proceedings of IEEE Conference on Computer Vision and Pattern Recognition, pp. 1106–1112 (1997)
13. Sportsmotion: A motion analysis system (2011–2015). http://www.sportsmotion.com/
14. Dartfish: Dartfish video analysis tool (2011–2015). http://www.sportmanitoba.ca/page.php?id=116
15. Kirmizibayrak, J., Honorio, J., Xiaolong, J., Russell, M., Hahn, J.K.: Digital analysis and visualization of swimming motion. Int. J. Virtual Real. **10**, 9–16 (2011)
16. Bideau, B.: Biomechanics and medicine in swimming IX. In: Chatard, J.-C. (ed.) IXth International World Symposium on Biomechanics and Medicine in Swimming, pp. 52–53 (2003)
17. Luo, H.-G., Zhu, L.-M., Ding, H.: Camera calibration with coplanar calibration board near parallel to the imaging plane. Sens. Actuators A: Phys. **132**, 480–486 (2006)
18. Kannala, J., Heikkilä, J., Brandt, S.S.: Geometric camera calibration. In: Wiley Encyclopedia of Computer Science and Engineering (2008)
19. Bottoni, A., et al.: Technical skill differences in stroke propulsion between high level athletes in triathlon and top level swimmers. J. Hum. Sport Exerc. **6**(2), 351–362 (2011)
20. Bouguet, J.Y.: Camera calibration toolbox for Matlab (2008)
21. Mullane, S.L., Justham, L.M., West, A.A., Conway, P.P.: Design of an end-user centric information interface from data-rich. Procedia Eng. **2**, 2713–2719 (2010)
22. Martin, N., Roy, S.: Fast view interpolation from stereo: simpler can be better. In: Proceedings of 3DPTV 2008. The Fourth International Symposium on 3-D Data Processing, Visualization and Transmission, Georgia Institute of Technology, Atlanta, GA, USA (2008)
23. Rocklin, M., Terrel, A.R.: Symbolic statistics with SymPy. Comput. Sci. Eng. **14**(3), 88–93 (2012)
24. Dorst, L., Fontijne, D., Mann, S.: Geometric Algebra for Computer Science: An Object-Oriented Approach to Geometry (2007)

# Fundamentals of Nonparametric Bayesian Line Detection

Anne C. van Rossum[1,2,3]([✉]), Hai Xiang Lin[1,2,3], Johan Dubbeldam[1,2,3], and H. Jaap van den Herik[1,2,3]

[1] Distributed Organisms B.V., Rotterdam, The Netherlands
anne@dobots.nl, {h.x.lin,j.l.a.dubbeldam}@tudelft.nl,
h.j.vandenherik@law.leidenuniv.nl
[2] Delft University of Technology, Delft, The Netherlands
[3] Leiden University, Leiden, The Netherlands

**Abstract.** Line detection is a fundamental problem in the world of computer vision. Many sophisticated methods have been proposed for performing inference over multiple lines; however, they are quite ad-hoc. Our fully Bayesian model extends a linear Bayesian regression model to an infinite mixture model and uses a Dirichlet Process as a prior. Gibbs sampling over non-unique parameters as well as over clusters is performed to fit lines of a fixed length, a variety of orientations, and a variable number of data points. Bayesian inference over data is optimal given a model and noise definition. Initial computer experiments show promising results with respect to clustering performance indicators such as the Rand Index, the Adjusted Rand Index, the Mirvin metric, and the Hubert metric. In future work, this mathematical foundation can be used to extend the algorithms to inference over multiple line segments and multiple volumetric objects.

**Keywords:** Bayesian nonparametrics · Line detection

## 1 Introduction

The task of line detection in point clouds has hitherto been performed through rather ad-hoc methods. Two of the most familiar methods have been RANSAC and Hough. The RANSAC [3] method iteratively tests a hypothesis for line parameters. A set of points is selected out of all existing points and a line is fitted through them. Points that are not in this set, but fit the same line (according to a predefined loss function), are added to this set. If the fit, considering all points, is not sufficiently adequate, this process is reiterate till a predefined performance level is obtained. The Hough transform [15] is not an iterative method, but it deterministically maps all points in the image space to curves in the so-called Hough space. The Hough space has typically two dimensions: slope and intercept. A line is characterized in Hough space by a curves intersecting at the same point. Each point defines a unique slope and intercept. By rasterizing the Hough space

© Springer International Publishing AG 2017
A. Fred et al. (Eds.): ICPRAM 2016, LNCS 10163, pp. 175–193, 2017.
DOI: 10.1007/978-3-319-53375-9_10

in a grid, each grid cell represents a vote from points for that particular range of slopes and intercepts. The estimate for the line parameters is obtained by finding the grid cell that receives votes from the maximum number of points. The line is extracted by finding the unique maximum in the Hough space.

The ad-hoc nature of these methods leads to three main problems. First, the Hough or RANSAC models incorporate assumptions on the nature of the noise that are not well-defined. The loss function in RANSAC only indirectly corresponds with a noise distribution. The rasterization in the Hough transform defines the noise in an even more indirect manner. It is not possible to incorporate knowledge about the nature of the noise in a formal manner. Second, even though line detection can be expanded on from detection of a single line to that of multiple lines, this requires nontrivial adaptations [6,12,27] without guarantees about optimality. For example, a Hough transform extension for multiple lines might iteratively remove the maximum of the most salient line with its corresponding points. However, this maximum is not unique, and the result can depend on the order in which the maxima are removed. Third, extensions of the methods to inference over hierarchical structures, such as squares or volumetric forms, are difficult to implement. The inference methods are not independent from the form of the model or would not scale well with increasing model dimensions (for example, to find the maximum in an $N$-dimensional Hough space scales with the resolution to the power of $N$).

Bayesian methods [9] solve the first problem. They define noise properly and perform optimal inference with the type of noise considered. A Bayesian model is defined by a likelihood function and a prior. Bayes' rule gives then the unique, optimal method to combine the likelihood with the prior to obtain the posterior. This method is optimal from the viewpoint of information processing [26]. Note that in this context, it is the inference procedure that is optimal. Optimality does not reflect correctness of the likelihood functions or the postulated prior.

Nonparametric Bayesian models (introduced in Sect. 2) solve the second and third problems of inference over multiple objects and extensions to hierarchical forms. The detection task of multiple lines might seem a rather straightforward problem. A proper definition, however, allow extensions of line estimation to application domains beyond computer vision. In robotics depth sensors generate large point clouds of data that are difficult to process in its raw form. Compression of this data into lines, planes, and volumetric objects [17] is of paramount importance to accelerate the inference in, for example, simultaneous localization and mapping [25] procedures. A method that is able to infer multiple lines simultaneously can be extended to perform inference over multiple planes and objects. The Bayesian approach will allow for setting intriguing priors, such as a prior that describes a prevalence for certain horizontal and vertical angles in man-made environments compared to more natural scenes (as seems to be the case for the number of unique objects [24].

In this chapter we provide context and expand on our previous work [23], a nonparametric Bayesian method to perform inference over the number of lines and over the fitting of points on that line. To be able to do so, we establish the theory on nonparametric Bayesian models and will use it to extract multiple lines from 2D point clouds simultaneously.

The course of this chapter is as follows. In Sect. 2 we discuss Bayesian nonparametrics. In Sect. 3 we describe the Dirichlet Process. In Sect. 4 we propose our Infinite Line Mixture Model. In Sect. 5 the results of the Infinite Line Mixture Model. Section 6 ends with conclusions.

## 2    Bayesian Nonparametrics

The basic idea behind using Bayesian methods to extract certain features from supplied data, such as to which line a certain point belongs, is as follows. Observations provide information about the unknown density from which they came, as well information about the distribution of these densities.

The difficulty consists of (1) finding a correct prior distribution that can be matched with the observed data and (2) finding a computationally feasible algorithm to update the prior distribution after each observation or after each number of observations.

In a Bayesian model we are performing inference over parameters $\theta$ and observations $w$ by Theorem 1.

**Theorem 1.** *Laplace (1812) published the first modern formulation of Bayes' theorem:*

$$p(\theta \mid w) = \overbrace{p(w|\theta)}^{\text{likelihood}} \; \overbrace{p(\theta)}^{\text{prior}} \underbrace{p(w)^{-1}}_{\text{constant}} \propto p(w|\theta)p(\theta) \tag{1}$$

In summary, the posterior is proportional to the product of the prior and likelihood, with the proportionality defined through the normalizing constant, also known as evidence. When observations on a line are conditionally independent and identically distributed (i.i.d.), this can be written as a product:

$$p(\theta \mid w) = \prod_j p(\theta \mid w_j). \tag{2}$$

Let us now define not only a single model, but a set of models (in our particular case, lines parameterized by $\theta_i$) to which our observations belong. These lines are not ordered: the index of a line is irrelevant to the inference process. A process in which data is generated from objects that are exchangeable is called an exchangeable process. The joint distribution $p(\theta) = p(\theta_1, \ldots, \theta_N)$ over these objects is the same for all permutations (indicated by the permutation vector $\rho$) of these objects.

$$p(\theta_1, \ldots, \theta_N) = p(\theta_{\rho(1)}, \ldots, \theta_{\rho(N)}) \tag{3}$$

If we have a fixed number $N$ of such models, we can describe our problem as a finite mixture model:

$$p(\theta) = \sum_{i=1}^{N} \pi_i p(\theta_i) \tag{4}$$

Here $p(\theta_i)$ is a (not necessarily unique) probability density function. The $\pi_i$ are called mixture weights and sum up to one: $\sum_{i=1}^{N} \pi_i = 1$. A countably infinite mixture model is defined by allowing $N = \infty$. An uncountable mixture, or compound probability distribution is defined through a probability density function $p(\pi)$:

$$p(\theta) = \int p(\pi)p(\theta \mid \pi)d\pi = \int p(\theta \mid \pi)dP(\pi), \qquad (5)$$

with $dP(\pi) = p(\pi)d\pi$. De Finetti [10] showed in 1935 that a joint probability distribution $p(\theta)$ for exchangeable objects $\theta_i$ admits a representation as a mixture (see Theorem 2).

**Theorem 2.** *De Finetti (1935) If the random variables $\theta_i$ are exchangeable the joint probability distribution $p(\theta)$ can be represented as*

$$p(\theta) = \int \prod_{i=1}^{N} p(\theta_i|G)dP(G), \qquad (6)$$

*for a hidden random variable $G$ where $P(G)$ is a distribution on measures.*

An important example of such a distribution on measures is the Dirichlet Process which we will consider in the next section.

There are three reasons why the preceding theory is interesting. First, $p(\pi)$ can be seen as a prior with respect to weights and be defined with additional hyperparameters in a hierarchical manner. Second, $p(\theta|\pi)$ is integrated over all possible $P(\pi)$, even if this space is an infinite measure space. Third, De Finetti shows in Eq. 6 that there are only $N$ clusters after all in Eq. 5.

## 3   Dirichlet Process

If the number of lines is known in advance, the only task is to assign points to lines and fit the points to these lines. If the number of lines is not known, it still can be fixed apriori. Models that fix the number of lines apriori are k-means clustering and Gaussian mixture models. The number of clusters, $k$, determines the number of clusters in advance.

Owing to the exchangeability property and De Finetti's theorem we know that it is possible to establish $p(\theta)$ and to have automatically the number $k$ inferred. Let us introduce now the Dirichlet Process (DP). The Dirichlet Process is a distribution over probability measures. It is defined as follows.

$$G \sim DP(\alpha, H) \qquad (7)$$

Equation 7 is a shorthand notation to describe that for every (finite measurable) partition $\{\Theta_1, \ldots, \Theta_k\}$ of the parameter set $\Theta$, the random distribution $G(\Theta_i)$ is a Dirichlet distribution with base distribution $H$ and concentration parameter $\alpha$.

$$\{G(\Theta_1), \ldots, G(\Theta_k)\} \sim Dir(\alpha H(\Theta_1), \ldots, \alpha H(\Theta_k)) \qquad (8)$$

Note that although $H$ is continuous, $\Theta_i$ is a partition with multiple parameters $\theta_j$ that are non-unique.

Below we discuss how the Dirichlet Process can be used for inference over objects. We describe the Dirichlet Mixture Model in Sect. 3.1 that uses a Dirichlet Process as prior over the lines. Then we describe two possible inference methods. Gibbs sampling over parameters in Sect. 3.2 and Gibbs sampling over clusters in Sect. 3.3.

### 3.1 Dirichlet Mixture Model

To generate observations from a Dirichlet Process requires three steps (see [1,7,18]): (1) The Dirichlet Process $DP$ with hyperparameters $\alpha$ and $H$ generates an almost surely discrete distribution $G$, (2) the distribution $G$ generates non-unique parameters $\theta_i$, (3) the parameters $\theta_i$ generate through the likelihood function $F$ the observations $w_i$.

$$G \sim DP(\alpha, H)$$
$$\theta_i \sim G$$
$$w_i \sim F(\theta_i)$$

Alternatively, this can be visualized as a graphical model (see Fig. 1).

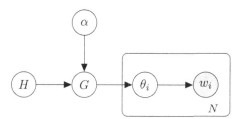

**Fig. 1.** The Dirichlet Mixture Model. The almost surely discrete distribution $G$ is generated by the Dirichlet Process from $H$ and $\alpha$. The parameters $\theta_i$ are generated from $G$. Plate notation [5] is used to visualize that there are multiple ($N$) parameters and observations. Each observation $w_i$ gets assigned its non-unique line $\theta_i$.

The Dirichlet Process can be used as a prior for a mixture model thanks to the fact that the distribution $G$ is almost surely discrete[1].

---

[1] In particular, if $H$ is a Uniform distribution $U(0,1)$ and $\alpha$ is small, the distribution $G$ will have positive probability for only a few discrete $\theta$ values in the $[0,1]$ range. The parameters $\theta_i$ are only sampled from the discrete distribution $G$ rather than from the continuous distribution $H$. The values with positive probability correspond to individual, discrete, clusters.

If we integrate over $G$ we sample the parameters directly from the base distribution $H$.

$$p(w, \theta|\alpha, H) = \overbrace{p(w \mid \theta)}^{F} \underbrace{p(\theta \mid H, \alpha)}_{\text{DP prior}} \tag{9}$$

Equivalently, we can introduce parameters $z$ as assignment parameters with $z_i = k$ denoting that observation $w_i$ belongs to parameter $\theta_k$ and by describing the DP prior as a product of a multinomial and Dirichlet distribution.

$$p(w, z, \theta|\alpha, H) = \underbrace{p(w \mid z, \theta)}_{F(\theta)} \underbrace{p(\theta \mid H)}_{H(\lambda)} \underbrace{p(z \mid \pi)}_{\text{multinomial}} \underbrace{p(\pi \mid \alpha)}_{\text{Dirichlet}} \tag{10}$$

For individual assignment variables this can be described as in Eq. 11.

$$p(w, z, \theta|\alpha, H) = p(w \mid z, \theta)p(\theta \mid H) \prod_{i=1}^{N} \underbrace{p(z_i \mid \pi)}_{\text{multinomial}} \underbrace{p(\pi \mid \alpha)}_{\text{Dirichlet}} \tag{11}$$

There are $N$ observations $w_i$, each tied to a partition $\pi$ over parameters $\theta_k$ of which there are infinitely many generated from $H(\lambda_0)$.

For a proper understanding of the Dirichlet Mixture Model we establish three concepts, the Dirac measure, the stick-breaking construction, and the Pólya urn.

**Dirac Measure.** The definition of the Dirac measure $\delta_{\theta_j}(\Theta_k)$ is as follows. Given a set $\Theta_k$ with a $\sigma$-algebra over subsets of $\Theta$, then the Dirac measure is defined as in Eq. 12.

$$\delta_{\theta_j}(\Theta_k) = 1_{\Theta_k}(\theta_k) = \begin{cases} 1, & \text{if } \theta_j \in \Theta_k \\ 0, & \text{if } \theta_j \notin \Theta_k \end{cases} \tag{12}$$

**Stick-Breaking Construction.** A stick-breaking construction $\pi \sim GEM$ (named after Griffiths, Engen, and McCloskey [8]) of the Dirichlet Process generates first weights $\phi$ according to a Beta distribution and deterministically through $f_1$ defines $\pi$ through $\phi$.

$$p(w, z, \theta|\alpha, H) = p(w \mid z, \theta)p(z \mid \pi) \overbrace{\prod_{k} f_2(p(\Theta_k \mid H), \underbrace{f_1(\pi, p(\phi \mid \alpha))}_{\text{GEM}}))}^{\text{stick-breaking construction}} \tag{13}$$

The function $f_2$ defines $p(\theta \mid H)p(\pi, \alpha)$ through a countably infinite sum of weights.

$$f_2 = p(\Theta_k \mid \pi, \alpha, H) = \sum_{j=1}^{\infty} \pi_j \delta_{\theta_j}(\Theta_k) \tag{14}$$

The stick-breaking construction shows explicitly how the weights for each partition come from an infinite sum.

**Pólya Urn.** A closed form for the conditional probabilities is known since Blackwell and MacQueen [2] as a weighted combination of the base distribution $H$ and the other parameters $\theta_i$.

$$p(\theta_{n+1} \in \Theta_k \mid \theta_1, \ldots, \theta_n, H, \alpha) = \frac{1}{\alpha + n} \left( \alpha H(\Theta_k) + \sum_{i=1}^{n} \delta_{\theta_i}(\Theta_k) \right) \quad (15)$$

The base distribution $H$ generates discrete parameter values from the continuous range $\Theta_k$. The sum of Dirac measures describes the nonzero possibility of sampling the same discrete parameter multiple times. The probability for $\theta_{n+1}$ is proportional to a weighted combination of these new and old probability distributions with $\alpha$ defining the weight of new samples (see Fig. 2). Pólya urn sampling is also known as the Chinese Restaurant Process.

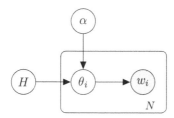

**Fig. 2.** In Pólya urn sampling $\theta_i$ is directly sampled from $H$ in an iterative manner without first sampling distributions $G$ (compare with Fig. 1).

### 3.2   Gibbs Sampling of Parameters

Gibbs sampling requires the conditional probabilities for every random variable involved [13]. Gibbs sampling, just like other Markov chain Monte Carlo methods, generates a sequence of correlated samples. Subsequently, if necessary, the Maximum A Posteriori estimation of a value can be found through estimating the mode (most common occurring value) of a parameter. Below we discuss how to do perform (1) Gibbs sampling in general, (2) Gibbs sampling of a new parameter, and (3) Gibbs sampling of an existing parameter.

**Sampling.** The derivation of the conditional probabilities of parameters with respect to the remaining parameters has been extensively described in the literature [20]. Such a derivation uses an important property of the Dirichlet Process, namely that it is the conjugate prior of the multinomial distribution. Owing to conjugacy the Eqs. 16–18 have closed-form descriptions.

Escobar [7] has shown that the conditional distribution of a parameter given the other parameters and the data forms a mixture.

$$p(\theta_i \mid \theta_{-i}, w_i) \propto \alpha H_i(\theta_i) \int F(w_i, \theta) dH(\theta) + \sum_{j \neq i} F(w_i, \theta_j) \delta_{\theta_j}(\theta_i) \quad (16)$$

The notation $\theta_{-i}$ describes the set of all parameters $\Theta$ with $\theta_i$ excluded. In Eq. 16 we sample (1) a new or (2) an existing parameter.

**Sampling a New Parameter.** The probability of sampling a new parameter is defined by the product of the posterior density of a new parameter given the prior and a single observation with that of the likelihood integrated over all possible parameters generated by the base distribution.

$$p(\theta_{i,new} \mid \theta_{-i}, w_i) \propto \alpha H_i(\theta_{new}) \int F(w_i, \theta) dH(\theta) \tag{17}$$

$H_i(\theta_{new})$ is the posterior density of $\theta_{new}$ given $H$ as prior and a single observation $w_i$, namely $H_i(\theta_{new}) \propto F(w_i, \theta_{new}) H(\theta_{new})$. Conjugacy plays a role between $F$ and $H$ in the integral, as well as for the update for a single observation.

**Sampling an Existing Parameter.** The probability of sampling an existing parameter is defined directly through the likelihood $F(w_i, \theta_j)$.

$$p(\theta_{i,existing} \mid \theta_{-i}, w_i) \propto \sum_{j \neq i} F(w_i, \theta_j) \delta_{\theta_j}(\theta_i) \tag{18}$$

Equations 16–18 can be used to perform inference directly with all (non-unique) parameters $\theta_i$ tied to observations $w_i$. Details on inference will be provided in Sect. 4.

### 3.3   Gibbs Sampling of Clusters

The preceding method iterates over the observations (and the non-unique lines that the observations are assigned to). Alternatively, we can iterate only over the lines themselves. The derivation [20] leads to an update for the component indices that only depends on the number of data items per cluster, the parameter $\alpha$, and the data at hand. Below we briefly describe (1) sampling a new cluster and (2) sampling an existing cluster.

**Sampling a New Cluster.** A new cluster is sampled with a probability that depends on $\alpha$ and the total number of data items. This is described in Eq. 19.

$$p(c_{i,new} \mid c_{-i}, \alpha) \propto \frac{\alpha}{\alpha + n - 1} \int F(w_i \mid \theta_i) dH(\theta) \tag{19}$$

We can obtain an analytical expression for the integral in Eq. 19 as a consequence of the fact that $F$ and $H$ are conjugate.

**Sampling an Existing Cluster.** An existing cluster is sampled with a probability that depends on the number of items in that cluster (except the data item at hand). This is expressed in Eq. 20.

$$p(c_{i,existing} \mid c_{-i}, w_i, \alpha, \theta) \propto \frac{n_{c,-i}}{\alpha + n - 1} F(w_i \mid \theta_i) \tag{20}$$

The model to which we will apply Gibbs sampling as described in Eqs. 19 and 20 will be proposed in Sect. 4.

## 4 Infinite Line Mixture Model

The proposed model extends Bayesian linear regression to multiple lines using a Dirichlet Process as a prior for the partitioning of points over lines and the number of lines overall. A suitable name for this model might be the "Infinite Line Mixture Model". This name follows the naming convention for other models in the nonparametric Bayesian literature [11,14,22]. In particular, "infinite" means that there are a potentially infinite number of lines to be inferred (see Fig. 3).

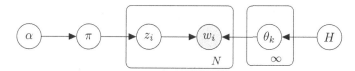

**Fig. 3.** The Infinite Line Mixture Model, a Bayesian linear regression model for multiple lines. The Dirichlet Process is defined with hyperparameters $\alpha$ and $H$. At the left the concentration parameter $\alpha$ generates the partitions $(\pi_1, \ldots, \pi_k)$. The partitions use assignment parameters $z_i$ to denote which observation $w_i$ (with index $i$) belongs to which cluster $k$. The cluster is defined through the parameter set $\theta_k$. The parameters are sampled from the base distribution $H$.

The course of this section is as follows. In Sect. 4.1 we briefly describe the Bayesian linear regression model. In Sect. 4.2 we describe the conjugate prior for the Bayesian linear regression model. In Sect. 4.3 we outline the implementation of Gibbs sampling of parameters. In Sect. 4.4 we show the implementation of Gibbs sampling of clusters.

### 4.1 Bayesian Linear Regression Model

We resume the Bayesian linear regression model for a single line [4], where the line is assumed to contain Gaussian noise. The individual points $i$ are normally distributed.

$$y_i \sim \mathcal{N}(x_i \beta, \sigma^2) \tag{21}$$

The coordinate (column) vector $\beta$ maps the (row) vector with independent variables $x_i$ to the dependent variable $y$. The noise is normally distributed with standard deviation $\sigma$ in the $y$ coordinates. In a computer vision task with images, it holds that $x_i = [1, x_{value}]$ and $y_i = y_{value}$. We transform the x-coordinate to be able to infer the intersect, $\beta[0]$, and slope, $\beta[1]$.

All observations assigned to the same line have a likelihood function that corresponds to a normally distributed random variable with $y$ and $X$ as parameters.

$$p(y \mid X, \beta, \sigma^2) \propto \sigma^{-n} \exp\left(-\frac{1}{2\sigma^2}(y - X\beta)^T(y - X\beta)\right) \tag{22}$$

Equation 22 can also correspond to multiple points. The dependent variable $y$ becomes a column vector of values. The individual observations $X$ becomes a matrix with on each row an observation. The coordinate vector $\beta$ and the standard deviation $\sigma$ are considered to be the same for all observations related to the line.

### 4.2  Conjugate Prior for the Bayesian Linear Regression Model

The conjugate prior has the form of Eq. 22 which can be composed out of a separate prior for the standard deviation $p(\sigma)$ and the conditional probability of the line coefficients given the standard deviation $p(\beta \mid \sigma^2)$.

$$p(\sigma^2, \beta) = p(\sigma^2)p(\beta \mid \sigma^2) \tag{23}$$

The standard deviation $\sigma$ is sampled from an Inverse-Gamma (IG) distribution.

$$p(\sigma) \propto (\sigma^2)^{-(\nu_0/2+1)} \exp(-\frac{1}{2\sigma^2}\nu_0 s_0^2) \tag{24}$$

This is an $IG(a, b)$ with $a = \nu_0/2$ and $b = 1/2\nu_0 s_0^2$. The conditional probability with respect to the line coefficients has a normal distribution as prior.

$$p(\beta \mid \sigma^2) \propto \sigma^{-n} \exp\left(-\frac{1}{2\sigma^2}(\beta - \mu_0)^T \Lambda_0(\beta - \mu_0)\right) \tag{25}$$

**Sufficient Statistics.** Due to conjugacy of the distribution (25) we have a simplified description for updating the parameters at once, given a set of observations. The sufficient statistics are updated [19] according to Eq. 26.

$$\begin{aligned}
\Lambda_n &= (X^T X + \Lambda_0) \\
\mu_n &= \Lambda_n^{-1}(\Lambda_0 \mu_0 + X^T y) \\
a_n &= a_0 + n/2 \\
b_n &= b_0 + 1/2(y^T y + \mu_0^T \Lambda_0 \mu_0 - \mu_n^T \Lambda_n \mu_n)
\end{aligned} \tag{26}$$

Let us collect (1) the hyperparameters $\Lambda_0, \mu_0, a_0, b_0$ in $\lambda_0$, (2) the updated hyperparameters $\Lambda_n, \mu_n, a_n, b_n$ in $\lambda_n$, (3) our observation $(X, y)_k$ in $w_k$, and (4) all $n$

observations in $w$. The update for the sufficient statistics (Eq. 26) can then be summarized as:

$$\lambda_n = U_{ss}(\lambda_0, w). \tag{27}$$

**Sample from the NIG Distribution.** Using $\lambda_0$ defined above, we can find our base distribution $H$ as a Normal-Inverse-Gamma distribution (NIG).

$$H(\theta_k) = NIG(\theta_k; \lambda_0) \tag{28}$$

When we sample from a Normal-Inverse-Gamma distribution, the standard deviation is obtained using the Gamma distribution with $a_0$ and $b_0$ as hyperparameters and the line coefficients are drawn from a Normal distribution.

$$\sigma_k = \tau_k^{-1/2} \qquad \tau_k \sim \mathcal{G}(a_0, b_0)$$
$$\mu_k \sim \mathcal{N}(\mu_0, \sigma^2 \Lambda_0^{-1}) \tag{29}$$

For sampling from the NIG posterior, observations up to the last one, $w_n$, have to be incorporated.

$$H_n(\theta_k) = NIG(\theta_k; \lambda_n) \tag{30}$$

Here $\lambda_n$ is updated through the sufficient statistics update $U_{ss}$.

## 4.3    Implementation of Gibbs Sampling of Parameters

The implementation for Gibbs sampling of parameters closely follows Sect. 3.2. The individual steps are described in detail in Algorithm 1 (with the likelihood and prior for the data unspecified). This Gibbs algorithm is known as Algorithm 1 [20]. Let us define Eq. 16 in five distinct equations (Eqs. 31–35).

$$p(\theta_i \mid \theta_{-i}, w_i) \propto r_i H_i(\theta_i) + \sum_{j \neq i} L_{i,j} \delta(\theta_j - \theta_i) \tag{31}$$

$$r_i = \alpha Q(w_i, \lambda_0) = \alpha \int_\Theta F(w_i, \theta) dH(\theta) \tag{32}$$

$$H_i(\theta_i) \propto H(\theta_i) F(w_i, \theta_i) \tag{33}$$

$$L_{i,j} = F(w_i, \theta_j) \tag{34}$$

$$p(\theta_{new}) = \frac{r_i}{r_i + \sum_{j \neq i} L_{i,j}} \tag{35}$$

We can use these equations to sample all parameters $\theta_i$.

In Algorithm 1 we perform a loop in which for $T$ iterations each $\theta_i$ belonging to observation $w_i$ is updated in sequence. First, the likelihood $L_i$ for all $\theta_{-i}$ given $w_i$ is calculated. Second, the posterior predictive for $w_i$ given the hyperparameters $p(w_i \mid \phi_0)$ is calculated. The fraction with the Dirichlet Process concentration parameter $\alpha$ subsequently defines whether (1) $\theta_i$ will be sampled

---

**Algorithm 1.** Gibbs sampling over parameters $\theta_i$.

---

1: **procedure** GIBBS ALGORITHM $1(w, \lambda_0, \alpha)$    ▷ Accepts points $w$, hyperparameters
    $\lambda_0, \alpha$ and returns $k$ line coordinates
2:    **for all** $t = 1 : T$ **do**
3:        **for all** $i = 1 : N$ **do**
4:            $r_i = \alpha Q(w_i, \lambda_0)$        ▷ Predictive of $w_i$ given hyperparameters (Eq. 32)
5:            **for all** $j = 1 : N, j \neq i$ **do**
6:                $L_{i,j} = F(w_i, \theta_j)$    ▷ Likelihood for a line given observation (Eq. 34)
7:            **end for**
8:            $p(\theta_{new}) = \frac{r_i}{r_i + \sum_{j \neq i} L_{i,j}}$        ▷ New parameter probability (Eq. 35)
9:            $u \sim U(0, 1)$
10:           **if** $p(\theta_{new}) > u$ **then**        ▷ Sample with probability $p(\theta_{new})$
11:               $\lambda_n = U_{ss}(w_i, \lambda_0)$    ▷ Update sufficient statistics with $w_i$ (Eq. 27)
12:               $\theta_i \sim NIG(\theta_i; \lambda_n)$        ▷ Sample $\theta_i$ from NIG (Eq. 30)
13:           **else**
14:               $\theta_i$ sampled from existing clusters        ▷ Sample existing cluster
15:           **end if**
16:       **end for**
17:   **end for**
18:   **return** summary on $\theta_k$ for $k$ lines
19: **end procedure**

---

from a new cluster or (2) one of the existing clusters will be sampled. If a new cluster is sampled, the sufficient statistics are updated with information on $w_i$ and thereafter $\theta$ is sampled from a Normal-Inverse-Gamma distribution with the updated hyperparameters. If an existing cluster is sampled, the sufficient statistics are known and do not need to be updated.

### 4.4    Implementation of Gibbs Sampling of Clusters

Directly sampling over the clusters is known as Algorithm 2 [20]. Rather than updating each $\theta_i$ per observation $w_i$, an entire cluster $\Theta_k$ is updated. In Algorithm 1 the update of a cluster would require a first observation to generate a new cluster at $\theta_j$ and then moving all observations of the old cluster $\theta_i$ to $\theta_j$. The so-called mixing is slow: it takes a long time for the MCMC chain to visit all high probability regions in the state space. Algorithm 2 presents a sequence of events in a single move. This speeds up mixing in the Gibbs sampling algorithm.

The likelihood for all observations corresponding with parameter $\Theta_k$ is a product of the likelihood function for (i.i.d.) individual observations.

$$L_{i,k} = F(w_i, \Theta_k) = \prod^{m_k} F(w_i, \theta_k) = m_k F(w_i, \theta_k) \tag{36}$$

In sampling over clusters we also need to be able to remove observations from a cluster. This requires an update for the sufficient statistics in which an observation is removed.

$$\Lambda_0 = (\Lambda_n - X^T X)$$
$$\mu_0 = \Lambda_0^{-1}(\Lambda_n \mu_n - X^T y)$$
$$a_0 = a_n - n/2 \tag{37}$$
$$b_0 = b_n - 1/2(y^T y + \mu_0^T \Lambda_0 \mu_0 - \mu_n^T \Lambda_n \mu_n)$$

We summarize this by a "downdate".

$$\lambda_0 = U_{ss}^{-1}(\lambda_n, w) \tag{38}$$

Here $U_{ss}^{-1}$ is the inverse operation of the mapping $U_{ss}$ defined previously.

Algorithm 2 follows the same procedure in excluding $w_i$ from calculating the likelihood. This requires a so-called "downdate" from the corresponding sufficient statistics. In Algorithm 2 after all observations have been iterated over and assigned to the corresponding cluster $k$, an outer loop iterates over all clusters to obtain new parameters $\theta$ from the NIG prior.

---

**Algorithm 2.** Gibbs sampling over clusters $c_k$.

---

1: **procedure** GIBBS ALGORITHM 2$(w, \lambda_0, \alpha)$          ▷ Accepts points $w$ and
        hyperparameters $\lambda_0$ and $\alpha$, returns $k$ line coordinates
2:     **for all** $t = 1 : T$ **do**
3:         **for all** $i = 1 : N$ **do**
4:             $c = \text{cluster}(w_i)$      ▷ Get cluster $c$ currently assigned to observation $w_i$
5:             $\lambda_c = U_{ss}^{-1}(w_i, \lambda_c)$       ▷ Downdate sufficient statistics with $w_i$ (Eq. 38)
6:             $m_c = m_c - 1$                              ▷ Adjust cluster size $m_c$
7:             $r_i = \alpha Q(w_i, \lambda_0)$       ▷ Predictive of $w_i$ given hyperparameters (Eq. 32)
8:             **for all** $k = 1 : K$ **do**
9:                 $L_k = m_k \, F(w_i, \Theta_k)$    ▷ Likelihood for cluster $k$ given $w_i$ (Eq. 36)
10:             **end for**
11:             $p(\theta_{new}) = \frac{r_i}{r_i + \sum_k L_k}$      ▷ New parameter probability (Eq. 35)
12:             $u \sim U(0, 1)$
13:             **if** $p(\theta_{new}) > u$ **then**               ▷ Sample with probability $p(\theta_{new})$
14:                 $\lambda_c = U_{ss}(w_i, \lambda_0)$    ▷ Update sufficient statistics with $w_i$ (Eq. 27)
15:                 $\theta_i \sim NIG(\theta_i; \lambda_n)$              ▷ Sample $\theta_i$ from NIG (Eq. 30)
16:             **else**
17:                 $c$ sampled from existing clusters
18:                 $\lambda_c = U_{ss}(w_i, \lambda_c)$  ▷ Restore sufficient statistics with observation $w_i$
19:             **end if**
20:             $m_c = m_c + 1$                                ▷ Increment cluster size $m_c$
21:         **end for**
22:         **for all** $k = 1 : K$ **do**
23:             $\Theta_k \sim NIG(\lambda_k)$                          ▷ Sample $\Theta_k$ from NIG
24:         **end for**
25:     **end for**
26:     **return** summary on $\Theta_k$ for $k$ lines
27: **end procedure**

---

## 5   Results

The Infinite Line Mixture Model as explained in Sect. 4 can be applied to the problem of fitting an infinite number of lines through a point cloud in two dimensions. Here we touch upon three issues. The first issue concerns our model only pertaining to lines rather than line segments. Of course, when testing the model, we consider only finite length lines. Hence in practice we consider line segments. However, by choosing the line length to be much larger than the typical standard deviation of points on the line, the theory developed here is adequately applicable as is witnessed by the numerical results. A second issue that we dealt with in the algorithms, is the correlation between samples that arise from Gibbs sampling. We counter this with subsampling to drastically remove the amount of correlation. The third issue rises from the Bayesian methodology. The Bayesian inference method does not yield a single result (a point estimate), but a distribution. For example, given a point cloud, our model assigns probability $p = 0.2$ to two lines (and line parameters) and $p = 0.7$ to three lines (and line parameters). To obtain a single point estimate out of our Bayesian model, we chose the median values for all the parameters involved. This is the Maximum A Posterior estimate to our clusters.

### 5.1   Clustering Performance

The results are measured using conventional metrics for clustering performance. For example, the Rand Index describes the accuracy of cluster assignments [21].

$$R = \frac{a + b}{a + b + c + d} \tag{39}$$

Here $a$ numbers the pair of points that belong to the same cluster, both at ground truth as well as after the inference procedure. Likewise $b$ numbers the pair of points that belong to different clusters in both sets. The values $c$ and $d$ describe discrepancies between the ground truth and the results after inference. A Rand Index of one ($R = 1$) means that there have been no mistakes.

The clustering performance is separate from the line estimation performance for three reasons. First, if the points are not properly assigned, the line will not be estimated correctly. Due to the fact that line estimation has this secondary effect, the line estimation performance is not taken into account. Second, some lines will generate only a single, or very few points. We can extract point assignments, but line coefficients are impossible to derive in this case. It would lead to introducing a threshold for the number of points per cluster. Third, the performance would require (artificially) weighting the clustering performance with the line estimation performance.

The performance of Algorithm 1 can be seen in Fig. 4 and is rather disappointing. On average the inference procedure agrees upon the ground truth for 75% of the cases considering the Rand Index. This seems rather okay, but if we adjust for chance as with the Adjusted Rand Index, the performance drops to only having 25% correct!

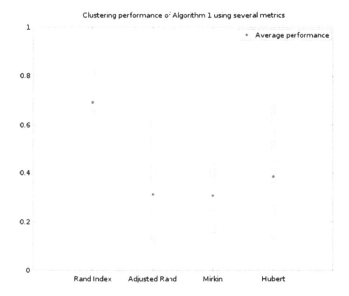

**Fig. 4.** The performance of Algorithm 1 with respect to clustering is measured using the Rand Index, the Adjusted Rand Index, the Mirvin metric, and the Hubert metric. A figure of 1 means perfect clustering for all metrics, except Mirvin's where 0 denotes perfect clustering.

This demonstrates that faster mixing Gibbs sampling is essential, as implemented in Algorithm 2. Sampling over clusters leads to excellent performance measures as can be seen from Fig. 5. Apparently updating entire clusters at once with respect to their parameter values leads at times to perfect clustering, bringing the performance metrics close to their optimal values.

We conclude Sect. 5.1 by two considerations on Algorithm 1. First, the lack of performance is caused by slower mixing: it requires more time to reach the steady state distribution. However, also when allowing it ten times the number of iterations of Algorithm 2, it still does not reach the same performance levels. Our second consideration is that a line seems to form isolated, local regions of high probability in the sample space. This makes it difficult for individual points to postulate slightly changed line coordinates.

## 5.2 Two Examples

We next illustrate the performance of the inference process by discussing two examples. The first example, Fig. 6 shows the assignment after a single Gibbs step in Algorithm 1. One can clearly see that a single line of points from the top left to the lower right corner (orange and red points) is composed of two clusters. This is caused by the fact that only single points can be transferred between clusters by Algorithm 1. It therefore takes a very long time before all points in one cluster are transferred to a new cluster. This could be repaired by

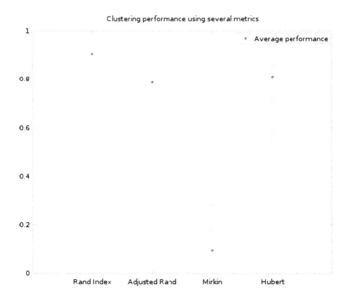

**Fig. 5.** The performance of Algorithm 2 with respect to clustering is measured using the Rand Index, the Adjusted Rand Index, the Mirvin metric, and the Hubert metric. A figure of 1 means perfect clustering for all metrics, except Mirvin's where 0 denotes perfect clustering.

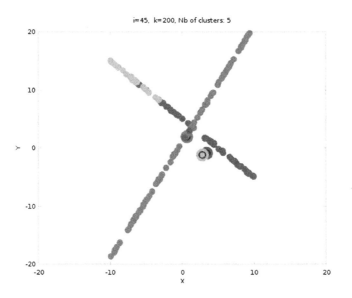

**Fig. 6.** One of the Gibbs steps in the inference of two particular lines. The points are more or less distributed according to the lines, but one line exists out of two large clusters. The line coordinates are visualized by a double circle. The x-coordinate is the y-intercept of the line, the y-coordinate is the slope.

adding split-merge steps to Algorithm 1. Algorithms for taking such steps into account have been reported in the literature [16].

The second example is visualized in Fig. 7. It shows that a single point as an outlier is not a problem for our method. A single point might throw off Bayesian linear regression, but because there are multiple lines to be estimated in our Infinite Line Mixture Model, this single point is assigned its own line.

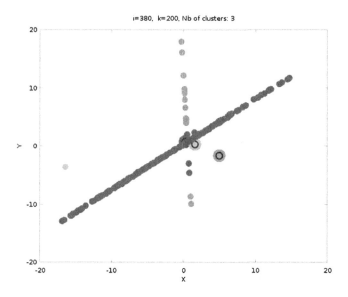

**Fig. 7.** The assignment of a line to a single point. There are three clusters found, rather than only the obvious two.

On top of that, the extension to more points as outliers would, of course, require us to postulate a distribution for these outlier points as well. A uniform distribution might, for example, be used in tandem with the proposed model. This however would lead to a non-conjugate model and hence different inference methods.

## 6   Conclusions

We have seen that the Infinite Line Mixture Model proposed in Sect. 4 extends the well known Bayesian linear regression model to an infinite number of lines using a Dirichlet Process as prior. From the theoretical considerations and the practical observations we may conclude that a nonparametric Bayesian model does indeed allow for inference over multiple lines simultaneously with the fitting of a single line. Since the Infinite Line Mixture Model is a Bayesian model - in contrast to ad-hoc methods such as the Hough transform or RANSAC - it optimally [26] determines the number of lines given the model and noise definition.

Rather than only defining a theoretical optimal model, we have the model also tested on point data. The results given in Sect. 5 show high values for difference performance metrics for clustering, such as the Rand Index, the Adjusted Rand Index, and other metrics. Inference in our Bayesian model is implemented by two types of algorithms. Algorithm 1 iterates over all observations and suffers from slow mixing. The individual updates make it hard to reassign large number of points at the same time. Algorithm 2 iterates over entire clusters. This allows updates for groups of points leading to much faster mixing. Here we note, that even optimal inference results so now and then in occasional misclassifications. The reason is that the dataset is generated by a random process. Hence, occasionally two lines are generated with almost the same slope and intercept. Points on these lines are impossible to assign to the proper line.

The essential contribution of this chapter is to introduce a fully Bayesian method called the Infinite Line Mixture Model. In essence, the model infers lines given point clouds and lays down the mathematical foundation for future research. Regarding future work, there are two ways in which the postulated model can to be extended for full-fledged inference in computer vision as required in robotics. The first way extends the extend the point clouds from 2D to 3D and extend from the extraction of lines to that of planes. This is quite a trivial extension that does not change anything of the model except for the dimension of the data points. The second way is as follows. Somehow a prior needs to be incorporated to restrict the lines of infinite length, to that of line segments of finite length. To restrict points on the lines to a uniform distribution of points over a line segment, a symmetric Pareto distribution can be used as prior (for the end points). This would subsequently allow for a hierarchical model in which these end points are in their turn part of more complicated objects. We are convinced that the described nonparametric Bayesian models and their application as an Infinite Line Mixture Model form a proper mathematical foundation for more complex computer vision problems.

# References

1. Antoniak, C.E.: Mixtures of Dirichlet processes with applications to Bayesian nonparametric problems. Ann. Stat. **2**, 1152–1174 (1974)
2. Blackwell, D., MacQueen, J.B.: Ferguson distributions via Pólya urn schemes. Ann. stat. **1**, 353–355 (1973)
3. Bolles, R.C., Fischler, M.A.: A RANSAC-based approach to model fitting and its application to finding cylinders in range data. IJCAI **1981**, 637–643 (1981)
4. Box, G.E.P., Tiao, G.C.: Bayesian Inference in Statistical Analysis, vol. 40. Wiley, Hoboken (2011)
5. Buntine, W.L.: Operations for learning with graphical models. JAIR **2**, 159–225 (1994)
6. Chen, H., Meer, P., Tyler, D.E.: Robust regression for data with multiple structures. In: Proceedings of the 2001 IEEE Computer Society Conference on Computer Vision and Pattern Recognition, CVPR 2001, vol. 1, pp. I–1069. IEEE (2001)
7. Escobar, M.D., West, M.: Bayesian density estimation and inference using mixtures. J. Am. Stat. Assoc. **90**(430), 577–588 (1995)

8. Ewens, W.J.: Population genetics theory-the past and the future. In: Lessard, S. (ed.) Mathematical and Statistical Developments of Evolutionary Theory, pp. 177–227. Springer, Heidelberg (1990)
9. Fienberg, S.E., et al.: When did Bayesian inference become "Bayesian"? Bayesian Anal. **1**(1), 1–40 (2006)
10. de Finetti, B.: Foresight its logical laws its subjective sources. In: Kyburg, H.E., Smokler, H.E. (eds.) Studies in Subjective Probability, pp. 93–158. Wiley, New York (1992)
11. Gael, J.V., Teh, Y.W., Ghahramani, Z.: The infinite factorial hidden Markov model. In: Advances in Neural Information Processing Systems, pp. 1697–1704 (2008)
12. Gallo, O., Manduchi, R., Rafii, A.: CC-RANSAC: fitting planes in the presence of multiple surfaces in range data. Pattern Recogn. Lett. **32**(3), 403–410 (2011)
13. Geman, S., Geman, D.: Stochastic relaxation, Gibbs distributions, and the Bayesian restoration of images. IEEE Trans. Pattern Anal. Mach. Intell. **6**, 721–741 (1984)
14. Ghahramani, Z., Griffiths, T.L.: Infinite latent feature models and the Indian buffet process. In: Advances in Neural Information Processing Systems, pp. 475–482 (2005)
15. Hough, P.V.: Method and Means for Recognizing Complex Patterns, December 1962. https://www.google.com/patents/US3069654, patent US 3069654 A
16. Jain, S., Neal, R.M.: A split-merge Markov chain Monte Carlo procedure for the Dirichlet process mixture model. J. Comput. Graph. Stat. **13**(1) (2004)
17. Kwon, S.W., Bosche, F., Kim, C., Haas, C.T., Liapi, K.A.: Fitting range data to primitives for rapid local 3D modeling using sparse range point clouds. Autom. Constr. **13**(1), 67–81 (2004)
18. MacEachern, S.N., Müller, P.: Estimating mixture of Dirichlet process models. J. Comput. Graph. Stat. **7**(2), 223–238 (1998)
19. Minka, T.: Bayesian linear regression. Technical report, Citeseer (2000)
20. Neal, R.M.: Markov chain sampling methods for Dirichlet process mixture models. J. Comput. Graph. Stat. **9**(2), 249–265 (2000)
21. Rand, W.M.: Objective criteria for the evaluation of clustering methods. J. Am. Stat. Assoc. **66**(336), 846–850 (1971)
22. Rasmussen, C.E.: The infinite Gaussian mixture model. NIPS **12**, 554–560 (1999)
23. van Rossum, A.C., Lin, H.X., Dubbeldam, J., van den Herik, H.J.: Nonparametric Bayesian line detection - towards proper priors for robotic computer vision. In: Proceedings of the 5th International Conference on Pattern Recognition Applications and Methods, pp. 119–127 February 2016
24. Sudderth, E.B., Jordan, M.I.: Shared segmentation of natural scenes using dependent Pitman-Yor processes. In: Advances in Neural Information Processing Systems, pp. 1585–1592 (2009)
25. Vasudevan, S., Gächter, S., Nguyen, V., Siegwart, R.: Cognitive maps for mobile robots - an object based approach. Robot. Auton. Syst. **55**(5), 359–371 (2007)
26. Zellner, A.: Optimal information processing and Bayes's theorem. Am. Stat. **42**(4), 278–280 (1988)
27. Zhang, W., Kôsecká, J.: Nonparametric estimation of multiple structures with outliers. In: Vidal, R., Heyden, A., Ma, Y. (eds.) WDV 2005–2006. LNCS, vol. 4358, pp. 60–74. Springer, Heidelberg (2007). doi:10.1007/978-3-540-70932-9_5

# Computing the Number of Bubbles and Tunnels of a 3-D Binary Object

Humberto Sossa[1(✉)] and Hermilo Sánchez[2]

[1] Instituto Politécnico Nacional-CIC, Av. Juan de Dios Bátiz S/N,
Gustavo a Madero, 07738 Mexico City, Mexico
hsossa@cic.ipn.mx
[2] Centro de Ciencias Básicas, Universidad Autónoma de Aguascalientes,
Aguascalientes, Mexico
hsanchez@correo.uaa.mx

**Abstract.** We present two formulations and two procedures that can be used for computing the number of bubbles and tunnels of a 3-D binary object. The first formulation is useful to determine the number of bubbles of an object, while the second one can be used to calculate the number of tunnels of an object. Both formulations are formally demonstrated. Examples are provided to numerically validate the functioning of both formulations. On the other hand, the first procedure allows obtaining the number of bubbles and tunnels of a 3-D object while the second procedure allows computing the number of bubbles and tunnels of several 3-D objects. Examples with 3-D images are provided to illustrate the utility and validity of the second procedure.

## 1 Introduction

Many fabricated or natural objects might have bubbles (voids or cavities) and/or tunnels (holes); washers, nuts, some varieties of French or Swiss cheese, sponges, bread, some kind of stones, bones, are some the examples one can mention.

In many image analysis applications, computing the number of bubbles and tunnels for a 3-D object could be an important. It could help in the: (1) analysis of 3-D microstructures of human trabecular bones in relation to its mechanical properties, as is performed in [1], (2) determination of the quantitative morphology and network representation of soil pore structure [2], (3) unambiguous classification of complex microstructures by their three-dimensional parameters applied to graphite in cast iron [3], and (4) analyse the connectivity of the trabecular bone in identifying the deterioration of the bone structure [4].

In this chapter we first introduce two mathematical expressions that allow computing, separately, the number of bubbles (voids) and tunnels (holes) of a 3-D object. These two propositions are also formally demonstrated. In each case, examples are added to numerically validate each proposition. Second, we describe two general methods. The first method allows determining, in two steps, the number of bubbles and tunnels of a 3-D object with both cavities and holes. The second method, permits, in three steps, to accomplish the same task but for several objects into the same image. This chapter is an extended version of the material presented in [42].

© Springer International Publishing AG 2017
A. Fred et al. (Eds.): ICPRAM 2016, LNCS 10163, pp. 194–211, 2017.
DOI: 10.1007/978-3-319-53375-9_11

The rest of the chapter is organized as follows. In Sect. 2, the problem to be faced is stated. In Sect. 3, several related pioneering and recent related methods to compute the Euler number of a digital 3-D image (object) are described. Next, in Sect. 4, several basic definitions that will facilitate the reading of the rest of the paper will be provided. After that, in Sect. 5, the proposed two expressions will be presented and demonstrated. Simple examples are added to numerically validate the operation of each equation. Section 6 will be focused to present and explain the functioning of two general methods for determining the number of bubbles and tunnels of 3-D objects. Section 7 will be devoted to present several examples to show the functioning and applicability of the proposals. In short, Sect. 8 will be oriented to show present the conclusions and directions for further research concerning this investigation.

## 2 Problem Statement

The problem to be solved in the content of this investigation could be stated as follows: Given a 3-D object composed of face-connected voxels, determine its number of bubbles and tunnels. One way to provide a solution to this problem could be by first computing the Euler number of the 3-D object.

In 3-D, as it known, the Euler number relates the number of bubbles and tunnels of the object. One expression commonly used for this is the following [5, 6]:

$$e = 1 - b_1 + b_2 \qquad (1)$$

In this case, $b_1$ is the number of tunnels or holes of the object and $b_2$ is its number of bubbles, cavities or voids [7, 8].

It is not difficult to see that Eq. (1) is the simplification of the more general formulation:

$$e = b_0 - b_1 + b_2 \qquad (2)$$

where $b_0$ represents the number of objects in the 3-D binary image $I(x, y, z)$ (for short $I$) under study. If in the 3-D image there is only one object, then $b_0 = 1$. Just to remember, $b_0$, $b_1$ and $b_2$ are the first three Betti numbers used to distinguish topological spaces based on the connectivity of $n$-dimensional *simplicial* complexes [9].

The careful reader can rapidly see that Eq. (1) exhibits two problems:

1. The two Betti numbers $b_1$ and $b_2$ cannot be obtained by computing local features of the 3-D object. They cannot be got through computing the local features such as the number of vertices or edges. In other words, the computation of Eq. (1) cannot be broken into subtasks, meaning that Eq. (1) cannot be used to compute local measures.
2. Both numbers $b_1$ and $b_2$ are part of the same equation. Thus, these two numbers cannot be computed directly from Eq. (1). Indeed, if a 3-D object has both bubbles and tunnels, number $b_2$ will add up a 1 to Eq. (1) for each bubble found; in the other hand, number $b_1$ will subtract a 1 to Eq. (1) for each tunnel found. Thus, an 3-D object with exactly the same number of bubbles and the same number of tunnels

will not alter the Euler number of the object because Eq. (1) will produce a 1, due to $(-b_1 + b_2)$ will cancel each other. In Sect. 5, we will explore how to solve these two problems.

## 3   Related Work

Different methods to compute the Euler number of a 3-D digital object (image) have been reported in literature. In this section, we will briefly describe some of the most important of these works that allow computing this number. We will first describe some of the oldest methods, next we will do the same for some of the most representative recent methods, reported in literature.

### 3.1   Pioneers Works

One of the first methods introduced to the world to determine the Euler number of a 3-D object was the work reported in [10, 11], but it was only applicable for 6-connectivity. In the first paper, the authors first define what they name "differentials", based on these differentials, the authors present a set of processing algorithms. The algorithms described in [11] can be used for labelling, counting, and computing connected objects in binary three dimensional arrays.

In [12], the authors study the 3-D surface Euler number of a polyhedron based on the Gauss-Bonnet theorem of differential geometry.

Another influential work can be found in [5]. In this work, the authors report several methods to compute the Euler number of a discrete digital image in both 2-D and 3-D.

In [13], the authors introduce an approach to computing the Euler characteristic of a three dimensional digital image by computing the change in numbers of black components, tunnels and cavities in a $3 \times 3 \times 3$ neighbourhood of an object (black) point due to its deletion. In the paper, the authors describe how a parallel implementation of the method is possible using the concept of sub-field [14].

In short, in [15], the authors describe an algorithm for computing the Euler number of a 3D digital image using the topological parameters computed by so called algorithm *topo–para*. This algorithm allows the change in the numbers of object components, tunnels and cavities in the $3 \times 3 \times 3$ neighborhood of a transformed object point (non-object point). Non-object points are obtained by means of a particular transformation acting on the original object points.

Other interesting works concerning the calculation of the Euler number of a 3-D image can be found in [16, 17].

### 3.2   Recent Works

To begin with this section, in [18, 19], the authors make use of the first two Betti numbers $b_0$ and $b_1$ of Eq. (1) for analysing the shape of a 3-D object by first labelling the points of its skeleton into four types of interest points: boundary, branching, regular and arc points. Authors label skeleton points according to the intersection of its maximal ball with the object. In order to add tolerance to the process, the radius of maximal balls is slightly increased. In a second step, authors make use of reversibility of the skeleton to deduce a labelling of the whole object.

In [20], authors present several fundamental properties of the topological structure of a 3-D digitized picture including the concept of neighbourhood and connectivity among volume cells (voxels) of 3-D digitized binary pictures defined on a cubic grid. They also introduce the concept of simplicial decomposition of a 3-D digitized object. Following this, the authors present two algorithms for calculating the Euler number (genus) of a 3-D figure. The first algorithm is based on the computation of value of a polynomial of binary variables representing voxel densities. The second algorithm utilizes local pattern matching. Both are performed only on the information of a $(2 \times 2 \times 2)$ local space. This method for obtaining the Euler number is an extension of the method for two-dimensional figures given by Gray in [10]. Lobregt et al. [21] derived a similar method for only the 6-connectivity and 26-connectivity cases basing upon a closed netted surface model and the following equation:

$$n - d + f = 2 - q \tag{3}$$

where $n$ is the number of so-called nodal points of the net, $d$ is the number of edges, $f$ is the number of faces of the net and $q$ is the so called connectivity number.

In [22], authors combine integral and digital geometry to develop a method for efficient simultaneous calculation of the intrinsic volumes of sets observed in binary images including surface area, integral of mean curvature, and Euler number. To make this rigorous, the concepts of discretization with respect to an adjacency system and complementarity of adjacency systems are introduced.

In [6], authors describe a method to compute the Euler feature of a 3-D image based on two definitions of foreground run and neighbour number. Following the definitions of foreground run and neighbour number in 2D image [23, 24], they redefine the concepts of foreground run and neighbour number in 3D image. Based on these two concepts, they propose their formula for locally computing the Euler Number of 3D image.

In [25], authors first provide a detailed description of the basics of three-dimensional digital image processing. They then talk about geometric properties of 3D images and 3D image processing fundamentals. After introducing localized processing (filtering), they show how 2D image processing methods are extended to 3D images. Next, they go to the core portion of the research where they first begin with a definition of connectivity and define some fundamental concepts such as topology preservation conditions, Euler numbers, and path and distance functions, thereby leading to some important properties; at the end authors use the ideas developed to present algorithms for processing connected components (for example, labelling, surface/axis thinning, distance transformation, and of course Euler number computation in 3-D.

In short in [26], the authors introduce two equations to compute the Euler of a 3-D object. By using the relationship between contact and enclosing surface concepts, as well as the relationships between vertices, edges and enclosing surfaces, authors derive an algorithm for obtaining the Euler feature of a 3-D object. It is worth mentioning that this is the most similar work reported in literature to the investigation reported in this chapter.

Other interesting approaches to obtain the Euler number of a 3-D object of 3-D image can be found in [27–30].

## 4   Definitions

In this section several concepts are defined, these are provided, in the one hand, to help the reader to easily follow the idea behind the proposal. On the other hand, these concept and definition are used to derive and prove the formal theorems baseline of the proposed methods that allow computing the number bubbles and tunnels of a 3-D object.

**Definition 1** (**voxel**). In three dimensions, in the case of a regular grid, a voxel is defined as the cubic unit that makes part of any 3-D object.

**Definition 2** (**connectivity among voxels**). Let $v_1$ and $v_2$ to voxels. If $v_1$ and $v_2$ share a face, then both voxels are *face connected*; otherwise, if $v_1$ and $v_2$ are connected by an edge or by a corner, then they are said to be *edge-connected* or *corner-connected*; respectively, else, $v_1$ and $v_2$ they are said to be not connected.

Figure 1 illustrates the four cases provided in Definition 2: Fig. 1(a) shows two voxels connected by a face (*face connected voxels*). In Fig. 1(b) the two voxels are connected by an edge (*edge connected voxels*), while in Fig. 1(c) the same two voxels appear connected by a corner (*corner connected voxels*). In short, Fig. 1(d) depicts the two voxels but disconnected. From now on, in this chapter we will work only with objects connected by faces. We thus have the following definition:

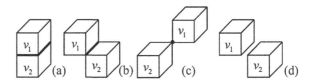

**Fig. 1.** (a) Two voxels connected by a face. (b) Two voxels connected by an edge. (c) Two voxels connected by a corner. (d) Two not connected voxels.

**Definition 3.** Any 3-D object $O_n$ composed of $n$ face-connected voxels is any connected region of voxels where all its voxels are only connected by faces.

Figures 2(a) and (b) show two objects connected by their faces. The first object is composed of five voxels, while the second object is composed of seven voxels, respectively.

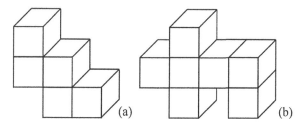

**Fig. 2.** (a) Object composed of five face-connected voxels; (b) Object composed of seven face-connected voxels.

**Definition 4.** Let $O_n$ a face-connected 3-D. The faces common to the $n$ pixels of $O_n$, this the faces interconnecting the $n$ pixels of $O_n$ will be called *contact faces*.

For example, the object shown in Fig. 2(a) has four contact faces, while the object illustrated in Fig. 2(b) has six contact faces.

**Definition 5.** A *tetra-voxel* is an arrangement of four object voxels as illustrated in Figs. 3(a), (b) or (c). Let $nt$ be the number of tetra-voxels that can be found in a 3-D binary image by a simple scanning image method.

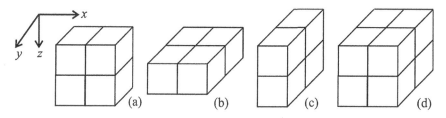

**Fig. 3.** A tetra-voxel in (a) $x$ direction, (b) $y$ direction, and (c) $z$ direction. (d) An octo-voxel.

**Definition 6.** An *octo-voxel* is an arrangement of eight object voxels as shown in Fig. 3(d). Let $no$ be the number of octo-voxels that can be found in a 3-D binary image by a simple scanning image method.

Due to these two definitions are very important in what follows, let us consider the following four objects given in Fig. 4, composed of 7, 10, 9 and 12 voxels, respectively. If we look at the first two objects, we see that they do not any tetra-voxel or octo-voxel. Now, if we observe at the third object, we note that it contains two tetra-voxels, both in the $x$ direction. Finally, the fourth object, has one octo-voxel and seven tetra-voxels.

**Definition 7 (Bubble or cavity).** Any connected component of 0-voxels that is not connected to the frame of an image is called a *bubble or cavity*.

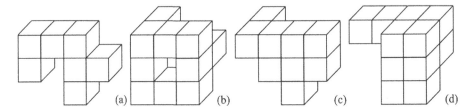

**Fig. 4.** (a) and (b) Two objects with no tetra-voxels or octo-voxels; (c) An object with two tetra-voxels; (d) An object with one octo-voxel and seven tetra-voxels.

**Definition 8 (Simply connected, multiply connected).** A connected component of 1-voxels with any holes and cavities is said to be *simply connected*, otherwise, it is said to be *multi connected.*

**Definition 9 (Tunnel or hole).** Let $O_n$ any multi connected object. A tunnel or a hole is any connected component of 0-voxels that passes through $O_n$ that diminishes the Euler number $e$ of $O_n$ by 1, according to Eq. (1).

To end up with this section, let us consider the following definition of a Connected Component Labelling Algorithm.

**Definition 10.** A Connected Component Labelling Algorithm (CCLA) applied over an image $I$ assigns a label l to each connected component found in the image $I$, according to the metric used.

For complete details about Connected Component Labelling Algorithms, the reader is refereed to [31–41].

## 5   The Proposal: Theoretical Part

Let $O_n$ a 3-D object composed of $n$ face connected voxels for which we want to determine its number of bubbles and it number of tunnels. Let $nc$, $nt$, and $no$, the number of the contact faces, number of tetra-voxels, and number of octo-voxels of $O_n$, respectively.

### 5.1   Number of Bubbles of a 3-D Object

Suppose we want to compute the number of bubbles of an object $O_n$ with no tunnels. For this we propose to use the following:

**Theorem 1.** Let $O_n$ be any connected 3-D binary object composed of $n$ face-connected voxels, the number of bubbles (voids) $nb$ of $O_n$ is given as:

$$nb = (n - nc + nt - no) - 1. \tag{4}$$

**Proof.** By mathematical induction on the number of voxels of $O_n$, for the base case: $O_1$ consisting of a single voxel, $nc = nt = no = 0$, values satisfying Eq. (4).

Induction step: let us assume that Eq. (4) holds for $O_n$. Let $nc'$, $nt'$, and $no'$ be the number of contact faces, number of tetra-voxels and number of octo-voxels, respectively, of object $O_{n+1}$ that is obtained by adding one voxel to $O_n$. Let $NC$, $NT$ and $NO$ be the corresponding numbers for this new voxel. We have that:

$$nc' = nc + NC \tag{5}$$

$$nt' = nt + NT \tag{6}$$

$$no' = no + NO \tag{7}$$

We have to show that Eq. (4) holds for $O_{n+1}$, i.e.

$$nb' = (n + 1 - nc' + nt' - no') - 1. \tag{8}$$

But this equation can be rewritten as follows:

$$\begin{aligned} nb' &= (n + 1 - nc - NC + nt + NT - no - NO) - 1 \\ &= (n - nc + nt - no) - 1 - NC + NT - NO + 1. \end{aligned} \tag{9}$$

This equation simplifies to:

$$nb' = nb - NC + NT - NO + 1. \tag{10}$$

which we know is true.    ∎

To numerically validate Eq. (10), let us consider the 3-D object composed of 25 voxels as shown in Fig. 5(a), with no bubbles and with the central voxel and the central voxel from the upper face missing. For this object, $nb = (26 - 45 + 20) - 1 = 0$ bubbles. Now, suppose that a new voxel is appended to this object as shown in Fig. 5 (b) in such a way that a bubble is obtained. In this case we have that $NC = 4$, $NT = 4$, $NO = 0$ and

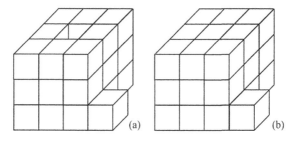

(a)          (b)

**Fig. 5.** (a) Object composed of 26 voxels with no bubbles. (b) Object with one bubble after appending a voxel to the object shown in Fig. 5(a).

$$nb' = (n + 1 - nc' + nt' - no') - 1 = (27 - 49 + 24 - 0) - 1 = 1,$$

Also

$$nb' = nb - NC + NT - NO + 1 = 0 - 4 + 4 - 0 + 1 = 1.$$

## 5.2   Number of Tunnels of a 3-D Object

Suppose now we want to compute the number of tunnels of an object $O_n$ with no bubbles. For this we propose to use the following:

**Theorem 2.** Let $O_n$ be any connected 3-D binary object composed of $n$ face-connected voxels, the number of tunnels (holes) $nh$ of $O_n$ is given as:

$$nh = 1 - (n - nc + nt - no). \tag{11}$$

**Proof.** Let us again proceed with the proof by mathematical induction on the number of voxels of $O_n$. For the base case: $O_1$ consisting of a single voxel, therefore, we have $nc = nt = no = 0$, values which satisfy Eq. (11).

For the induction step, let us assume that Eq. (11) holds for $O_n$. Let $nc'$, $nt'$, and $no'$ be the number of contact faces, number of tetra-voxels and number of octo-voxels, respectively, of object $O_{n+1}$ that is obtained by adding one voxel to $O_n$. Let $NC$, $NT$, and $NO$ be the corresponding numbers of this new voxel. We have that:

$$nc' = nc + NC \tag{12}$$

$$nt' = nt + NT \tag{13}$$

$$no' = no + NO \tag{14}$$

It must be shown that Eq. (11) holds for $O_{n+1}$, i.e.

$$nb' = 1 - (n + 1 - nc' + nt' - no'). \tag{15}$$

But this equation can be rewritten as follows:

$$\begin{aligned} nb' &= 1 - (n + 1 - nc - NC + nt + NT - no - NO) \\ &= 1 - (n - nc + nt - no) + NC - NT + NO - 1. \end{aligned} \tag{16}$$

This equation simplifies to:

$$nb' = nb + NC - NT + NO - 1. \tag{17}$$

which again we know is true.                                                      ∎

To numerically validate this last equation, let us consider the 3-D object composed of 15 voxels as shown in Fig. 6(a), with no tunnels. For this object, $nh = 1 - (15 - 14 + 0) = 0$ tunnels. Now, suppose that one new voxel is appended to this object as shown in Fig. 6(b) in such a way that a tunnel is obtained. In this case we have that $NC = 2$, $NT = 0$, $NO = 0$, and

$$nb' = 1 - (n + 1 - nc' + nt' - no') = 1 - (16 - 16 + 0 - 0) = 1$$

Also

$$nb' = nb + NC - NT + NO - 1 = 0 - 2 + 0 - 0 - 1 = 1$$

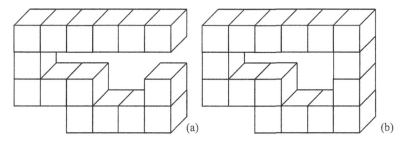

(a)    (b)

**Fig. 6.** (a) Object composed of 15 voxels with no tunnels. (b) Object with one tunnel after appending a voxel to the object shown in Fig. 6(a).

## 6 The Proposal: Practical Part

In this section we present two general procedures for determining the number of bubbles and tunnels of 3-D objects. As we will see, the first procedure is only useful for the case of images with only one object. The second procedure can be used with images containing several objects. Both procedures are only useful for the case of face-connected objects.

### 6.1 Procedure to Compute the Number of Bubbles and Tunnels of a 3-D Object

Suppose we want to determine the number of bubbles and tunnels of an object $O_n$. To do this we proceed in two steps as follows. During the first step we obtain the number of bubbles of the object. For this, we utilize of a Connected Component Labelling Algorithm (CCLA) (Definition 10). During the second step we obtain the number of tunnels of $O_n$ by applying Eq. (11). More in detail, given a 3-D image $I(x, y, z)$ of object $O_n$ as shown for example in Fig. 7(a):

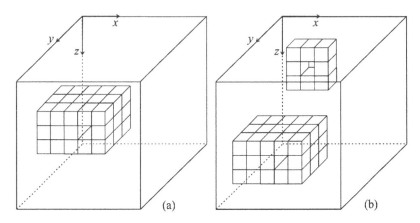

**Fig. 7.** (a) Image with one objet composed of 41 voxels used to test the functioning of the described procedure in Sect. 6.1 (b) Image with two objects to test the operation of the procedure described in Sect. 6.2.

### First Step (Determination of Number of Bubbles):

1. Apply over $I(x, y, z)$ any CCLA over the regions composed of 0-voxels. An adapted version of the algorithm reported in (Gonzalez, 2002), for the 3-D case, can be used for this. Due to bubbles are composed of 0-voxels (Definition 7), this algorithm will output value $ncc$. This variable corresponds to the number of bubbles plus 1. This 1 is obtained because the image background is also labelled; an extra label is generated.
2. Compute the number of bubbles of object $O_n$, $nb$, as $ncc - 1$.

### Second Step (Determination of the Number of Tunnels):

1. Apply Eq. (11) over image $I(x, y, z)$. If the object has bubbles and tunnels, this application will produce the number of tunnels minus the number of bubbles $nh\_b$. Refer to Eq. (1).
2. Add to the result obtained in the last step to get the number of tunnels $nh = nh\_b + nb$ of object $O_n$.

To numerically validate the above described procedure, let us consider the image with one object as shown in Fig. 7(a), composed of 41 voxels, one bubble and one tunnel. The bubble is the 0-voxel in the centre of the second vertical slice of 1-voxels of the object (left to right). The tunnel is composed of the three 0-voxels along the fourth vertical slice of 1-voxels of the object (left to right).

The first step of the above described algorithm, outputs: $nb = 1$, while the second step outputs $nh = nh\_b + nb = 1 - (41 - 76 + 36 + 0) + 1 = 0 + 1$, as desired. Note that $nh\_b = 0$ due to the object has one bubble and one tunnel, that according to Eq. (1) they cancel each other.

## 6.2    Procedure to Compute the Number of Bubbles and Tunnels of a Set of 3-D Objects

Suppose now are given an image $I(x, y, z)$ of $b_0$ voxelized objects; for each of these $b_0$ objects we would like to compute their numbers of bubbles and tunnels, respectively.

In this case we would need to apply a similar procedure as described in Sect. 6.1 with an additional step. We proceed into three steps as follows:

1. Apply any CCLA over image $I(x, y, z)$ to obtain $b_0$ labelled connected 3-D objects.
2. Apply the first step of the procedure described in Sect. 6.1 to each labelled connected region $R_i, i = 1, 2, \ldots, b_0$. For each $R_i$ we obtain its number of bubbles $nb_i, i = 1, 2 \ldots, b_0$.
3. Apply the second step of the same procedure described in Sect. 6.1 to obtain the number of tunnels of each object.

To numerically validate the above described procedure, let us consider Fig. 7(b) with two objects as depicted. As can be seen, the first object is composed of 8 voxels and one tunnel and the second one integrated of 41 voxels, with one bubble and one tunnel (the same object of Fig. 7(a)).

The first step of the above described procedure provides as a result two labels (two connected 3-D regions).

Next, for each labelled (region), the second step obtains $nb_1 = 0$ and $nb_2 = 1$, respectively. Finally, the third step outputs $nh_1 = nh\_b_1 + nb_1 = 1 - (8 - 8 + 0 - 0) + 0 = 1 + 0 = 1$ for the first object and $nh_2 = nh\_b_2 + nb_2 = 1 - (41 - 76 + 36 + 0) + 1 = 0 + 1 = 1$, for the second object, as desired.

# 7    Results and Discussion

In this section we report four experiments to validate the performance of our proposal. We only validate the correct functioning of the general algorithm described in Sect. 6.2.

## 7.1    First Experiment

During the first experiment, we utilized five 3-D images of size $100 \times 100 \times 100$ voxels. Each image was designed to have a different number of objects as established in row two of Table 1. Each time an object was added to the image it was added manually to have a control over the number of its number of bubbles and holes. Rows 3 and 4 of Table 1 show the correct number of bubbles and tunnels (Correct $nb$ and Correct $nh$) of each object of each image, respectively.

The procedure described in Sect. 6.2 was applied to each of the five images. It was programed in Java NetBeans with the Processing Applet in a desktop computer with a Core i7 model 2600 processor with 8 Gb of RAM. Rows 5 and 6 depict the computed number of bubbles and tunnels for object of each image, respectively. From these rows note also that in all cases, as expected, the correct values, $nb$ and $nh$, for each object were produced by the procedure. The average time to compute the number of bubbles

**Table 1.** Results obtained by the application of the procedure described in Sect. 6.2 to the five selected images.

| Image number | 1 | 2 | 3 | 4 | 5 |
|---|---|---|---|---|---|
| Number of objects per image | 1 | 2 | 3 | 4 | 5 |
| Correct $nb$ | 2 | 2,1 | 2,1,3 | 2,1,3,2 | 2,1,3,2,5 |
| Correct $nh$ | 1 | 1,3 | 1,3,1 | 1,3,1,2 | 1,3,1,2,4 |
| Computed $nb$ | 2 | 2,1 | 2,1,3 | 2,1,3,2 | 2,1,3,2,5 |
| Computed $nh$ | 1 | 1,3 | 1,3,1 | 1,3,1,2 | 1,3,1,2,4 |

and tunnels of each of the $b_0$ objects in image $I$ was 29.6 ms. It is worth mentioning that most of time is consumed by the connected component algorithms.

### 7.2 Second Experiment

During this experiment, we studied if the number of object-voxels influenced computation time when the total procedure was applied over an image. For this, we automatically generated a set of images with an increasing number of object-voxels. We defined a variable ($nv$) telling how many object-voxels will appear in the image. When $nv = 0.0$, it meant that the corresponding image will have only background voxels, for $nv = 0.05$, it meant that 5% of the generated voxels will belong to objects, and so on. Each time we increased variable $nv$ by 0.05. For each value of variable $nv$ we generated 10 images. We took the average time to fully process the whole set of 210 images and computed the average time. With the exception of the first case, in average the time consumed by the connected component algorithm was of 25.5 ms.

### 7.3 Third Experiment

In this case, we demonstrated the applicability of our method when applied to objects of various shapes and complexities. Figure 8 shows six of these objects: (a) a sphere, (b) a vase, (c) a torus, (d) a cheese, a (e) dragon and (f) a bookcase. Second and third row of Table 2 show the true values of number of bubbles and tunnels of each the six objects, while fourth and fifth rows show the computed values. As expected it can be seen that in all six cases, the computed values coincide with the true values. The average time to obtain the desired results was of 55.5 ms.

### 7.4 Fourth Experiment

In this last experiment we demonstrated the robustness of our method to image transformations such as translations, rotations and scale changes. For this, we took one of the objects (in this case one of the objects with a cheese shape) and translated, rotated and scaled inside its image. Four of these transformed versions are depicted in Fig. 9. Again, in all cases, one can easily verify that if the object remains face connected after being transformed, the desired number of tunnels was correctly computed.

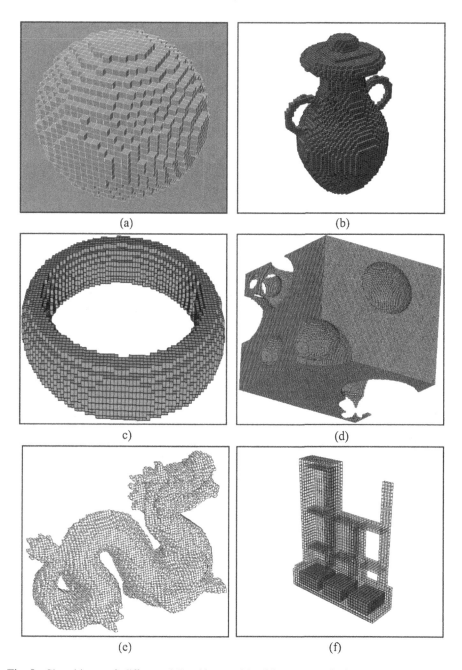

**Fig. 8.** Six objects of different 3-D objects with different complexity to demonstrate the applicability of the proposal. (a) a sphere, (b) a vase, (c) a torus, (d) a cheese, a (e) dragon and (f) a bookcase.

**Table 2.** Results obtained by the application of the procedure to the six objects of Fig. 8.

| Object | Sphere | Base | Torus | Cheese | Dragon | Bookcase |
|---|---|---|---|---|---|---|
| Correct number of bubbles $nb$ | 0 | 0 | 0 | 1 | 0 | 3 |
| Correct number of tunnels $nh$ | 0 | 2 | 1 | 4 | 1 | 2 |
| Computed number of bubbles $nb$ | 0 | 0 | 0 | 1 | 0 | 3 |
| Computed number of tunnels $nh$ | 0 | 2 | 1 | 4 | 1 | 2 |

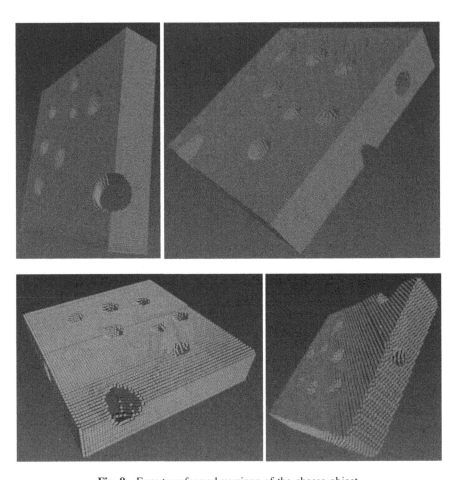

**Fig. 9.** Four transformed versions of the cheese object.

## 8   Conclusions and Directions for Further Research

In this paper we have presented two theoretical formulations and two general procedures to obtain the number of voids and tunnels of 3-D objects. The first formulation ((Eq. (4)) allows computing the number of bubbles of a 3-D binary face connected object; the second formulation (Eq. (11))) is useful for determining the number of tunnels. Both formulations have been formally demonstrated. In both cases, numerical examples were provided to numerically validate both equations.

Based on the above two formulations, two general procedures have been proposed. The first procedure, fully described in detail in Sect. 6.1, permits calculating the number of bubbles and tunnels of a 3-D binary face connected object from its binary image. The second general procedure, completely explained in Sect. 6.2, allows to produce the same but for any number of voxelixed objects. Experimental results with images of objects of different sizes and complexities have been given to show the applicability of both procedures.

Through the experiments, we have observed that the time spent in milliseconds expended by our method is reduced making it to be used in real time applications.

Considering that in some real applications, images of larger dimensions can be found, we propose for further work to implement our proposed procedure described in Sect. 6.2 into a FPGA or a GPU processor to determine how much processing times can be reduced.

**Acknowledgements.** H. Sossa thanks COFAA-IPN, SIP-IPN and CONACYT under Grants 20151187, 20161126, 155014 and 65 (Frontiers of Science), respectively, for the economic support to carry out this research. E. Sánchez thanks the Centro de Ciencias Básicas of the Universidad Autónoma de Aguascalientes for the support.

# References

1. Uchiyama, T., Tanizawa, T., Muramatsu, H., Endo, N., Takahashi, H.E., Hara, T.: Three-dimensional microstructural analysis of human trabecular bone in relation to its mechanical properties. Bone **25**(4), 487–491 (1989)
2. Vogel, H.J., Roth, K.: Quantitative morphology and network representation of soil pore structure. Adv. Water Resour. **24**, 233–242 (2001)
3. Velichko, A., Holzapfel, C., Siefers, A., Schladitz, K., Mücklich, F.: Unambiguous classification of complex microstructures by their three-dimensional parameters applied to graphite in cast iron. Acta Mater. **56**, 1981–1990 (2008)
4. Roque, W.L., de Souza, A.C.A., Barbieri, D.X.: The Euler-Poincaré characteristic applied to identify low bone density from vertebral tomographic images. Rev. Bras. Reumatol. **49**(2), 140–152 (2009)
5. Kong, T.Y., Rosenfeld, A.: Digital topology: introduction and survey. Comput. Vis. Graph. Image Process. **48**, 357–393 (1989)
6. Lin, X., Xiang, Sh., Gu, Y.: A new approach to compute the Euler number of 3D image. In: 3rd IEEE Conference on Industrial Electronics and Applications, pp. 1543–1546 (2008)
7. Lee, C.N., et al.: Winding and Euler numbers for 2D and 3D digital images, CVGIP: graph. Models Image Process. **53**(6), 522–537 (1991)
8. Lee, C.N.: Holes and genus of 2D and 3D digital images, CVGIP: graph. Models Image Process. **55**(1), 20–47 (1993)
9. Delfinado, C.J.A., Edelsbrunner, H.: An incremental algorithm for Betti numbers of simplicial complexes on the 3-sphere. Comput. Aided Geom. Des. **12**(7), 771–784 (1995)

10. Gray, S.B.: Local properties of binary images in two and three dimensions. IEEE Trans. Comput. **C-20**(5), 551–561 (1971)

11. Park, C.M., Rosenfeld, A.: Connectivity and genus in three dimension, TR-156. Computer Vision Laboratory, Computer Science Center, University of Maryland, College Park, MD (1971)

12. Lee, C.N., Rosenfeld, A.: Computing the Euler number of a 3d image. In: Proceedings of the IEEE First International Conference on Computer Vision, pp. 567–571 (1987)

13. Saha, P.K., Chaudhuri, B.B.: A new approach to computing the Euler characteristic. Pattern Recogn. **28**(12), 1955–1963 (1995)

14. Golay, M.J.E.: Hexagonal parallel pattern transformations. IEEE Trans. Comput. **C-18**, 733–740 (1969)

15. Saha, P.K., Chaudhuri, B.B.: 3D Digital topology under binary transformation with applications. Comput. Vis. Image Underst. **63**(3), 418–429 (1996)

16. Morgenthaler, D.G.: Three-dimensional digital topology: the genus. TR-980 November 1980. Computer Vision Laboratory Computer Science Center B University of Maryland (1980)

17. Morgenthaler, D.G., Rosenfeld, A.: Surfaces in three-dimensional digital images. Inf. Control **51**, 227–247 (1981)

18. Bonnassie, A., et al.: Shape description of three-dimensional images based on medial axis. In: Proceedings of the 2001 International Conference on Image Processing, pp. 931–934 (2001)

19. Bonnassie, A., Peyrin, F., Attali, D.: A new method for analyzing local shape in three-dimensional images based on medial axis transformation. IEEE Trans. Syst. Man Cybern. Part B Cybern. **33**(4), 700–705 (2003)

20. Toriwaki, J., Yonekura, T.: Euler number and connectivity indexes of a three dimensional digital picture. Forma **17**, 183–209 (2002)

21. Lobregt, S., Verbeek, P.W., Groen, F.C.A.: Three-dimensional skeletonization: principle and algorithm. IEEE Trans. Pattern Anal. Mach. Intell. **PQMI-2**(1), 75–77 (1980)

22. Schladitz, K., Ohser, J., Nagel, W.: Measuring intrinsic volumes in digital 3d images. In: Kuba, A., Nyúl, László, G., Palágyi, K. (eds.) DGCI 2006. LNCS, vol. 4245, pp. 247–258. Springer, Heidelberg (2006). doi:10.1007/11907350_21

23. Lin, X.Z., Sha, Y., Ji, J.W., Wan, J.B.: Image Euler number calculating for intelligent counting. In: Proceedings of the Seventh International Conference on Electronic Measurement and Instruments (ICEMI 2005), August 2005, vol. 8, pp. 642–645. International Academic Publishers/World Publishing Corporation, Beijing (2005)

24. Lin, X.Z., Wang, Y.M., Sha, Y., Ji, J.W.: A new approach to compute the Euler number of 2d image. In: Proceedings of the International Conference on Sensing, Computing and Automation (ICSCA 2006), May 2006, pp. 1376–1379. Watam Press, Chongqing (2006)

25. Toriwaky, J., Yoshida, H.: Fundamentals of Digital Image Processing. Springer, Heidelberg (2009)

26. Sánchez, H., Sossa, H., Braumann, U.-D., Bribiesca, E.: The Euler-Poincaré formula through contact surfaces of voxelized objects. J. Appl. Res. Technol. **11**, 55–78 (2013)

27. Yonekura, T., Toriwaki, J., Fukumura, T., Yokoi, S.: Topological properties of three-dimensional digitized picture data (1)-connectivity and Euler number. Paper of the Professional Group on Pattern Recognition and Learning, The Institute of Electronics and Communication Engineers of Japan (IECE, Denshi-Tsushin Gakkai), PRL80-1 (1980). (in Japanese)

28. Yonekura, T., Toriwaki, J., Fukumura, T., Yokoi, S.: On connectivity and Euler Number of three dimensional digitized binary pictures. Trans. IECE **E63**(11), 815–816 (1980). Japan

29. Yonekura, T., Yokoi, S., Toriwaki, J., Fukumura, T.: Connectivity and Euler number of figures in the digitized three-dimensional space. Trans. IECE (Denshi-Tsushin Gakkai Ronbunshi), Japan, **J65-D**(1), 80–87 (1982). (in Japanese)
30. Ohser, J., Nagel, W., Schladitz, K.: The Euler number of discretised sets – surprising results in three dimensions. Image Anal. Stereol. **22**, 11–19 (2003)
31. Gonzalez, R., Woods, R.: Digital Image Processing. Prentice Hall, Upper Saddle River (2002)
32. He, L., Chao, Y., Suzuki, K.: A survey of labeling algorithms. In: Joint 4th International Conference on Soft Computing and Intelligent Systems and 9th International Symposium on advanced Intelligent Systems (SCIS & ISIS 2008), September 17–21, 2008, Nagoya University, Nagoya, Japan, pp. 1293–1298 (2008)
33. He, L., Chao, Y., Suzuki, K.: An efficient first-scan method for label-equivalence-based labeling algorithms. Pattern Recogn. Lett. **31**, 28–35 (2010)
34. He, L., Chao, Y., Suzuki, K.: A run-based one and a half scan connected-component labeling algorithm. Int. J Pattern Recogn. Artif. Intell. **24**, 557–579 (2010)
35. Sutheebanjard, Ph., Premchaiswadi, W.: Efficient scan mask techniques for connected components labeling algorithm. EURASIP J. Image Video Process. **2011**, 14 (2011). doi:10.1186/1687-5281-2011-14
36. Lifeng, H., Xiao, Z., Suzuki, K.: A new two-scan algorithm for labeling connected components in binary images. In: Proceedings of the World Congress on Engineering, London, UK, 4–6 July 2012, pp. 1141–1146 (2012)
37. Grana, C., Montangero, M., Borghesani, D.: Optimal decision trees for local image processing algorithms. Pattern Recogn. Lett. **33**, 2302–2310 (2012)
38. He, L.F., Chao, Y., Suzuki, K.: An algorithm for connected-component labeling, hole labeling and Euler number computing. J. Comput. Sci. Technol. **28**, 468–478 (2013)
39. Lifeng, H., Xiao, Z., Yuyan, C., Suzuki, K.: Configuration-transition-based connected-component labeling. IEEE Trans. Image Process. **23**, 943–951 (2014)
40. Lifeng, H., Yuyan, C.: A very fast algorithm for simultaneously performing connected-component labeling and Euler number computing. IEEE Trans. Image Process. **24**, 2725–2735 (2015)
41. Chang, W.-Y., Chiu, Ch.-Ch., Yang, J.-H.: Block-based connected-component labeling algorithm using binary decision trees. Sensors **15**, 23763–23787 (2015)
42. Soosa, H.: On the computation of the number of bubbles and tunnels of a 3-d binary object. In: 5th International Conference on Pattern Recognition Applications and Methods, Roma, Italia, 24–26 February, pp. 17–23 (2016)

# Raindrop Detection on a Windshield Based on Edge Ratio

Junki Ishizuka and Kazunori Onoguchi[✉]

Hirosaki University, 3 Bunkyo-cho, Hirosaki, Aomori, Japan
`onoguchi@eit.hirosaki-u.ac.jp`

**Abstract.** This paper proposes the method for detecting raindrops with various shapes on a windshield from an in-vehicle monocular camera. Since raindrops on a windshield gives various bad influence to video-based automobile applications, for example obstacle detection and lane estimation, a driving safety support system or an automatic driving vehicle needs to understand the state of the raindrop which adheres on a windshield. Previous works are considered on isolated spherical raindrops, but raindrops on a windshield show various shapes, such as a band-like shape. The proposed method can detect raindrops regardless of the shape. In the daytime, the difference of the blur between the surrounding areas are checked for raindrop detection. The ratio of the edge strength extracted from two kinds of smoothed images is used as the degree of the blur. At night, bright areas in which the intensity does not change so much are detected as raindrops.

## 1 Introduction

Recently, video cameras are installed in many vehicles for safe driving. Since most of in-vehicle cameras are usually set behind the front windshield of a vehicle, raindrops which adhere on a windshield decrease the visibility and cause false detection in various video-based automobile applications. For example, detecting the preceding vehicle in the center of Fig. 1 is difficult because it blurs by adherent raindrops on a windshield. Therefore, the raindrop detection on the windshield is important issue for a driving safety support system or an autonomous vehicle.

In spite of a background, raindrops on a windshield are easily recognized by a person. However, it's difficult issue to detect a raindrop automatically in an image taken through a windshield because a blurred background is seen through a raindrop. Moreover, a camera usually focuses on a windshield. This makes raindrop detection more difficult because the appearance of a raindrop also blurs.

A lot of researches about raindrop detection have been conducted. Garg and Nayer [1] detected rain streaks in video sequences from intensity property of rain streaks for the first time. Although various methods including snow detection [2] have been proposed since then, these methods cannot be applied to raindrop detection on a windshield because they use the property of a falling raindrop.

© Springer International Publishing AG 2017
A. Fred et al. (Eds.): ICPRAM 2016, LNCS 10163, pp. 212–229, 2017.
DOI: 10.1007/978-3-319-53375-9_12

**Fig. 1.** Raindrops on a front windshield of a vehicle.

The device using an IR sensor has been produced for raindrop detection on the windshield. This device is used for activating a windshield wiper automatically. However, it sometimes makes a wiper malfunction since the detection area covered by an IR sensor is too narrow to cover driver's visibility. It's desirable to detect raindrops on the windshield by an in-vehicle camera since a camera can observe wide area.

Kurihata et al. [3] detected raindrops on the windshield by the subspace method. The template of a raindrop, so called eigendrops, is created by PCA. This method showed the good results only in a homogeneous area, such as in the sky, but a lot of false detection appears in the strong textured area. They improved this drawback by integrating detection results obtained in several frames [4]. Halimeh and Roser [5] detected raindrops using the geometric-photometric model which represented the refractive property of an adherent raindrop. This model assumed that the shape of a raindrop on a windshield was a section of a sphere. Sugimoto et al. [6] extended this assumption to a spheroid section. Liao et al. [7] detected raindrops from three characteristics that a raindrop exists in the similar location in the video frames for a period of time, its shape is close to an ellipse and it is bright. Nashashibi et al. [8] used the similar characteristics to detect unfocus raindrops. Eigen et al. [9] removed small dirt or raindrops from a corrupt image by predicting a clean output by the convolutional neural network. You et al. [10] detected a small round area, in which the brightness did not change so much and the motion was slow, as a raindrop.

Most of conventional methods dealt with only a raindrop which has isolated spherical shape. However, when it rains hard, raindrops are connected on a windshield and show various shapes even though a windshield wiper is activated. As shown in Fig. 1, raindrops with band-like shapes turn up on a windshield after wiping. We proposed the method which can detect raindrops with various shapes on a windshield in the daytime and at night [11]. In the daytime, raindrops are detected by examining the difference of the blur between the surrounding area. False areas are removed using the assumption that raindrops on a windshield don't move so much. At night, bright areas in which a temporal change of the intensity is small are detected as raindrops. This method [11] showed good performance in the daytime, but at night, some false areas appeared around the light source in the heavy rain. To solve this problem, we change the procedure at night and show that our method is effective at various roads under various rain conditions both in the daytime and at night.

This paper is organized as follows. Section 2 describes the detail of raindrop detection on a windshield in the daytime and at night. Section 3 discusses experimental results performed to several road scenes. Conclusions are presented in Sect. 4.

## 2    Raindrop Detection Method

The appearance of a raindrop on a windshield in the daytime is quite different from that of a raindrop at night, as shown in Fig. 2. In the day time, the blurred background is seen through a raindrop on a windshield, as shown in Fig. 2(a). On the other hand, a raindrop at night appears as gray or a white region, as shown in Fig. 2(b) since a raindrop on the dark background is not visible and only a raindrop lit up by a headlight or the surrounding light source appears on a windshield. Therefore, our method uses different algorithms for raindrop detection in the daytime and at night respectively. Our method judges the day or night by the intensity level of the whole image or an in-vehicle illuminance sensor.

(a) Daytime                                    (b) Night

**Fig. 2.** Appearance of raindrop in the daytime and at night.

### 2.1    Raindrop Detection in the Daytime

We use the following assumptions for raindrop detection in the daytime.

1. When the background contains strong texture, a raindrop on a windshield blurs more than its neighbor area, as shown in the upper red rectangle of Fig. 3.
2. When the background is homogeneous, the texture of a raindrop is stronger than its neighbor area, as shown in the lower red rectangle of Fig. 3, since the intensity changes in the boundary of a raindrop.

**Fig. 3.** Raindrops on textured background and non-textured background. (Color figure online)

To detect raindrops, our method examines the degree of the blur between a raindrop area and its surrounding area. Figure 4 shows the outline of the proposed method in the daytime. At first, an input image is divided into strong textured areas and weak textured areas. Next, two edge strength images are extracted from two kinds of smoothed images and the edge ratio is calculated from these edge strength images as the degree of the blur. Finally, false areas are removed from raindrop candidates based on the assumption that the motion of a raindrop on a windshield is small. Each step is explained below in detail.

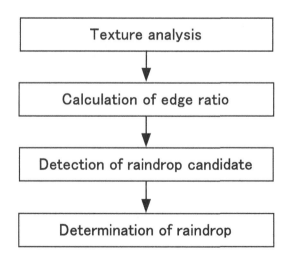

**Fig. 4.** Outline of the proposed method.

**Texture Analysis.** An input image is divided into grid blocks $B(u, v)$ ($1 \leq u \leq N, 1 \leq v \leq M$) and texture analysis is conducted by examining the edge strength in each block. In experiments, the image size is $640 \times 360$ pixels and the size of each block $B(u, v)$ is $10 \times 10$ pixels. The edge strength is detected by Sobel

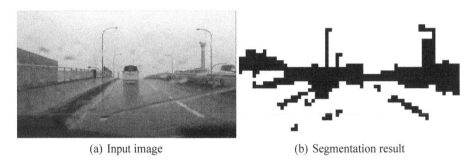

(a) Input image                    (b) Segmentation result

**Fig. 5.** Textured block and non-textured block.

operator and the total of the edge strength E(u,v) is calculated in each block $B(u,v)$. When $E(u,v)$ is larger than $T_E$, $B(u,v)$ is classified into a textured block. Otherwise, $B(u,v)$ is classified into a non-textured block. The threshold $T_E$ is determined so that a road surface may be classified into a non-textured block. Figure 5(b) shows the example of texture analysis. A black block shows a textured block and a white block shows a non-textured block.

**Detection of Raindrop Candidate.** Raindrop candidates are detected from the difference of the degree of the blur between a raindrop region and its neighbourhood. The ratio of the edge strength is calculated from two kinds of smoothed images to measure the degree of the blur without depending on the texture of the background. An input image $I(i,j)$ $(1 \leq i \leq W, 1 \leq j \leq H)$ is smoothed by the Gaussian filter. Let $I_{s1}(i,j)$ denote an image smoothed by the Gaussian filter whose variance is $\sigma_1$ and let $I_{s2}(i,j)$ denote an image smoothed by the Gaussian filter whose variance is $\sigma_2$. $\sigma_2$ is larger than $\sigma_1$. The Sobel filter is applied to $I_{s1}(i,j)$ and $I_{s2}(i,j)$ and edge strength images $I_{e1}(i,j)$ and $I_{e2}(i,j)$ are created. In clear textured area, edge strength changes greatly by smoothing. On the other hand, in blurred area, the change of the edge strength is small. The degree of the blur $D_b(i,j)$ at each pixel $(i,j)$ is defined by

$$D_b(i,j) = \frac{I_{e1}(i,j)}{I_{e2}(i,j)} \tag{1}$$

$D_b(i,j)$ is small when the degree of the blur is severe at a pixel $(i,j)$. On the other hand, it's large when the degree of the blur is light. Therefore, when the background has clear strong texture, $D_b(i,j)$ in a raindrop is smaller than $D_b(i,j)$ in the surrounding regions. When the background is homogeneous, $D_b(i,j)$ in a raindrop is larger than $D_b(i,j)$ in the surrounding regions. $D_b(i,j)$ $(1 \leq i \leq W, 1 \leq i \leq H)$ is scanned in the raster direction and the pixel $(i,j)$ satisfying following conditions is detected as the raindrop candidate.

1. The pixel $(i,j)$ is contained in the non-textured block.
   The pixel $(i,j)$ is chosen as a raindrop candidate when there are one or more neighbourhood pixels $(k,l)$ $(i-1 \leq k \leq i+1, j-1 \leq l \leq j+1)$ satisfying $D_b(i,j) - D_b(k,l) > T_n$.

2. The pixel $(i, j)$ is contained in the textured block.
   The pixel $(i, j)$ is chosen as a raindrop candidate when there are one or more neighbourhood pixels $(k, l)$ $(i - 1 \leq k \leq i + 1, j - 1 \leq l \leq j + 1)$ satisfying $D_b(k, l) - D_b(i, j) > T_t$.

In experiments, $T_n$ and $T_t$ are set to 0.78 and 2.1. In Fig. 6(b), raindrop candidates are indicated by white points. While the neighbourhood pixel $(k, l)$ satisfying $\mid D_b(i, j) - D_b(k, l) \mid < T_c$ exists, this pixel is added to raindrop candidates. In experiments, $T_c$ is set to 0.1. Figure 6(c) shows the final result of raindrop candidates.

**Determination of Raindrop.** In Fig. 6(c), raindrop candidates contain a lot of false areas in surrounding structures. Our method removes these areas by frame integration since the background moves large but a raindrop on a windshield does not move so much in an image. At first, the binary image $R_t(i, j)$ $(1 \leq i \leq W, 1 \leq j \leq H)$ in which raindrop candidates are set to 1 is created in each frame $t$. Next, an integration image $SR(i, j)$ is created by adding $R_{t-n+1}(i, j), R_{t-n+2}(i, j), \cdots, R_t(i, j)$. $SR(i, j)$ is binarized by the predetermined threshold $T_r$. In experiments, $n$ and $T_r$ are set to 3 and 2 respectively.

(a) Inputimage

(b) Raindrop candidates

(c) Raindrop candidates after neighbourhood points are added

(d) Raindrop areas

**Fig. 6.** Raindrop detection in the daytime.

(a) Optical flow in the background                    (b) Optical flow in the raindrop

**Fig. 7.** Optical flow in the background and raindrop.

When a vehicle moves along a road, some areas on the straight line toward the Focus of Expansion (FOE), such as on the guardrail, the railing of the bridge or the wall, may remain falsely since the similar texture is observed in these areas continuously. Our method gets rid of these false areas by examining the direction of the optical flow in raindrop candidates. The KLT tracker [12] is used for optical flow detection. As shown in Fig. 7(a), the flow vector detected around the lane marker converges to the FOE and shows similar direction. On the other hand, the flow vector detected in a raindrop shows various directions as shown in Fig. 7(b). To examine the variation of the flow direction, the variance of the flow direction is calculated in the block whose center is the pixel of interest for 15 frames. In experiments, the block size is $11 \times 11$. Figure 8(a)–(c) show the example of the optical flow detected in raindrop candidates. Figure 8(d) shows the histogram of the flow direction in the raindrop (blue circle) and Fig. 8(e) shows the histogram of the flow direction in the lane marker (green circle). Flow vectors detected around the lane marker show similar direction for several frames. On the other hand, flow vectors detected in the raindrop show various directions. Therefore, a pixel in the raindrop candidate is deleted when the variance of the flow direction is small. The closing process is applied to the binary image $SR_b(i,j)$ and the final result is obtained after removing small regions from $SR_b(i,j)$. Figure 6(d) shows final raindrop areas obtained from raindrop candidates of Fig. 6(c).

### 2.2 Raindrop Detection at Night

At night, a raindrop on the dark background is not visible and only a raindrop lit up by a headlight or the surrounding light source appears on the windshield as gray or a white region (Fig. 9(a)). When a raindrop is away from the light source, a change of the intensity is small even if the light source moves in an image. Therefore, a bright area in which the intensity does not change so much is detected as a raindrop except for the surrounding of the light source.

At first, a bright area in which a temporal change of the intensity is small is detected as raindrop candidates by Eq. 2. The frame differential image $FD_t(i,j)$ is created from two consecutive images $I_{t-1}(i,j)$ and $I_t(i,j)$.

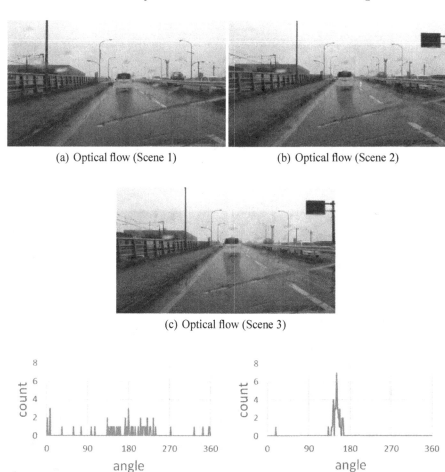

(a) Optical flow (Scene 1)                    (b) Optical flow (Scene 2)

(c) Optical flow (Scene 3)

(d) Histogram created in a raindrop       (e) Histogram created in a lane marker
(blue circle)                             (green circle)

**Fig. 8.** Detection of optical flow. (Color figure online)

$$FD_t(i,j) = \begin{cases} 1 \ if \ I_t(i,j) > T_{dark} \ and \\ \quad\ \mid I_t(i,j) - I_{t-1}(i,j) \mid < T_{dif} \\ 0 \ otherwise \end{cases} \qquad (2)$$

$T_{dark}$ is the threshold for eliminating a dark background from a processing region and $T_{dif}$ is the threshold for detecting a pixel where the frame differential value is small. In experiments, $T_{dark}$ and $T_{dif}$ are set to 30 and 20 respectively. Figure 9(b), (c) and (d) shows raindrop candidates detected at time $t$, $t-1$ and $t-2$ in white.

Next, an integration image $SFD_t(i,j)$ is created by adding the frame differential image $FD_{t-m+1}(i,j), FD_{t-m+2}(i,j), \cdots, F_t(i,j)$. $SFD_t(i,j)$ is binarized

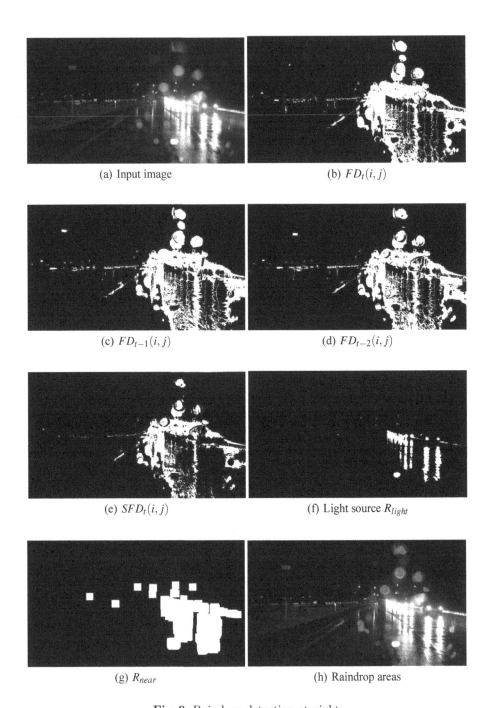

(a) Input image

(b) $FD_t(i, j)$

(c) $FD_{t-1}(i, j)$

(d) $FD_{t-2}(i, j)$

(e) $SFD_t(i, j)$

(f) Light source $R_{light}$

(g) $R_{near}$

(h) Raindrop areas

**Fig. 9.** Raindrop detection at night.

by the predetermined threshold $T_{sfd}$. In experiments, both $m$ and $T_{sfd}$ are set to 3. Figure 9(e) shows $SFD_t(i,j)$ which is created from $FD_{t-2}$, $FD_{t-1}$ and $FD_t$. A lot of false areas were removed by frame integration because bright areas in the background moves in an image. Some false areas remains around the light source near the Focus of Expansion (FOE) because these areas do not move so much when a vehicle moves along a road. To delete these areas, the light source area $R_{light}$ is estimated by simple binarization and the surrounding area $R_{near}$ is detected by applying dilation processing to $R_{light}$. Figure 9(f) and (g) show $R_{light}$ and $R_{near}$.

The labeling processing is applied to the binarized $SFD_t(i,j)$ and final raindrop areas are determined after removing labeled areas which satisfy at least one of the following conditions.

1. The circumscribed rectangle contains $R_{near}$.
2. The size is small.

Figure 9(h) shows the final result of raindrop detection at night.

## 3    Experiments

We conducted experiments to detect raindrops on a windshield by an in-vehicle video camera. The image size is $640 \times 360$ pixels. We used eight videos in the daytime and six ones at night. The length of each video is about two minutes respectively. Various backgrounds in the main road and the community road are contained in these videos. Various rain conditions from heavy rain to light rain are also contained in these videos. In each video sequences, same parameters were applied. The ground truth of the raindrop location is determined manually in each frame.

Figures 10, 11, 12, 13, 14 and 15 show examples of detection results in the daytime. In each figure, (a) shows an input image, (b) shows raindrop candidates detected by the edge ratio, (c) shows the result after removing false areas by frame integration and (d) shows the final result after correction by the variance of the optical flow direction. In Figs. 10, 11, 12, 13, 14 and 15(d), raindrops are indicated in green. In Figs. 10, 12, 13, 14 and 15(c), some false detected areas appear around the lane marker. In Fig. 11(c), some areas around a pedestrian crossing are detected as raindrops wrongly. In Figs. 12, 13, 14 and 15, some false areas appear on the wall and the fence. Since these areas are on the straight line toward the Focus of Expansion (FOE), most of them were removed by examining the variance of the flow direction in the surrounding region, as shown in Figs. 10, 11, 12, 13, 14 and 15(d).

Figures 10 and 11 show detection results in a broad main road and Figs. 12, 13, 14 and 15 show detection results in narrow community road. In both scenes, raindrops with various shapes, e.g. a band-like shape, appear on various backgrounds, such as a road surface, a building, a tree, sky and so on. Since the proposed method detects a raindrop every pixel and it uses the ratio of edge strength to detect raindrop candidates, most of raindrops with various shapes

(a) Input image

(b) Raindrop candidates

(c) Frame integration

(d) Detection result

**Fig. 10.** Experimental results in the day time (Scene 1). (Color figure online)

(a) Input image

(b) Raindrop candidates

(c) Frame integration

(d) Detection result

**Fig. 11.** Experimental results in the day time (Scene 2). (Color figure online)

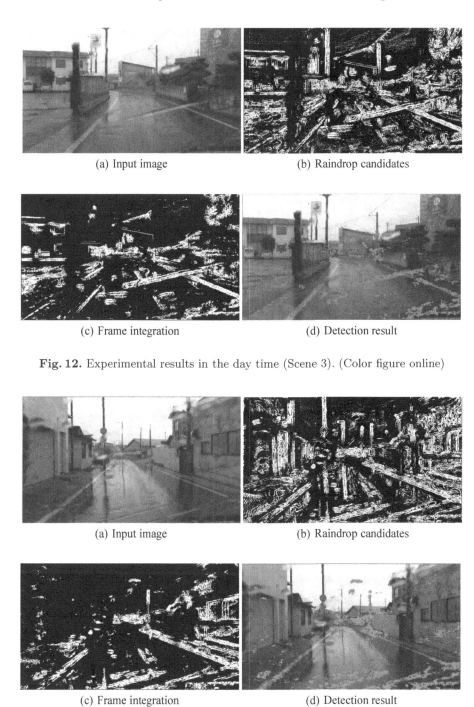

(a) Input image

(b) Raindrop candidates

(c) Frame integration

(d) Detection result

**Fig. 12.** Experimental results in the day time (Scene 3). (Color figure online)

(a) Input image

(b) Raindrop candidates

(c) Frame integration

(d) Detection result

**Fig. 13.** Experimental results in the day time (Scene 4). (Color figure online)

(a) Input image                    (b) Raindrop candidates

(c) Frame integration              (d) Detection result

**Fig. 14.** Experimental results in the day time (Scene 5). (Color figure online)

(a) Input image                    (b) Raindrop candidates

(c) Frame integration              (d) Detection result

**Fig. 15.** Experimental results in the day time (Scene 6). (Color figure online)

were detected regardless of a background. Some of raindrops in the sky were lost since our method uses fixed thresholds determined by experiments. We will develop the method to decide the threshold dynamically in the next step. The current processing time in the daytime is $10fps$ on PC with Xeon 3.2 GHz CPU. However, the video rate processing would be possible by the parallelization of the process and optimization of the software.

Figures 16, 17, 18, 19, 20 and 21 show examples of detection results at night. In each Figure, (a) shows an input image, (b) shows the light source $R_{light}$ detected by the binarization, (c) shows the mask area $R_{near}$ which is obtained by applying dilation processing to the light source $R_{light}$ and (d) shows the final result of the raindrop detection. In Figs. 16, 17, 18, 19, 20 and 21(d), raindrops are indicated in red. Figures 16, 17 and 18 show detection results in a broad main road and Figs. 19, 20 and 21 show detection results in a narrow community road. Most of raindrops except for the surrounding of the light source were detected. Since the brightness of the raindrop near the light source changes intensely as the light source moves in an image, it's difficult to distinguish them from reflected lights on the road surface. Although the present method excludes the surrounding of the light source from a detection area, we will improve the method in future so that a raindrop near the light source can be detected. The processing time at night is more than $30fps$ on the same PC.

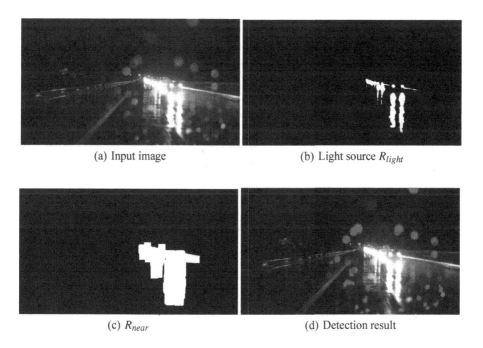

(a) Input image                    (b) Light source $R_{light}$

(c) $R_{near}$                     (d) Detection result

**Fig. 16.** Experimental results at night (Scene 7). (Color figure online)

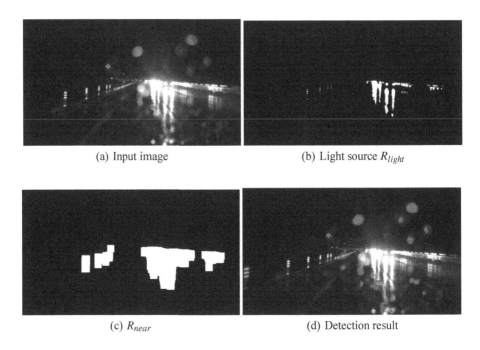

(a) Input image    (b) Light source $R_{light}$

(c) $R_{near}$    (d) Detection result

**Fig. 17.** Experimental results at night (Scene 8). (Color figure online)

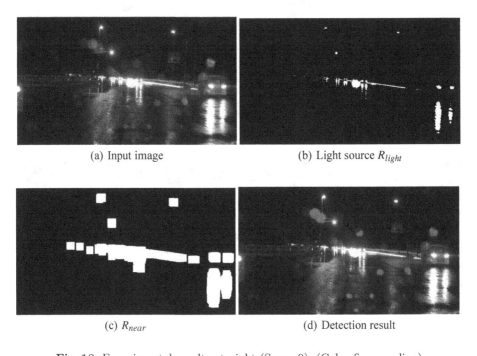

(a) Input image    (b) Light source $R_{light}$

(c) $R_{near}$    (d) Detection result

**Fig. 18.** Experimental results at night (Scene 9). (Color figure online)

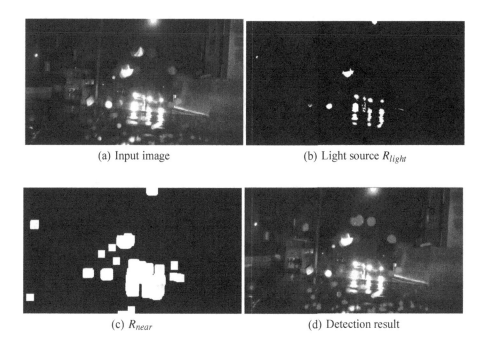

(a) Input image

(b) Light source $R_{light}$

(c) $R_{near}$

(d) Detection result

**Fig. 19.** Experimental results at night (Scene 10). (Color figure online)

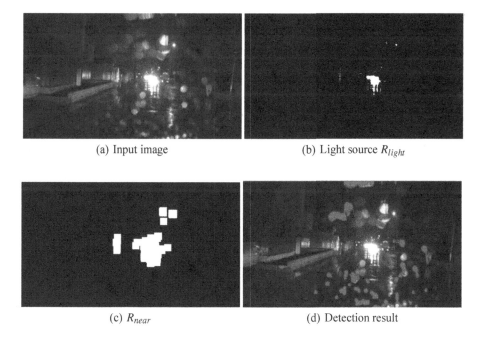

(a) Input image

(b) Light source $R_{light}$

(c) $R_{near}$

(d) Detection result

**Fig. 20.** Experimental results at night (Scene 11). (Color figure online)

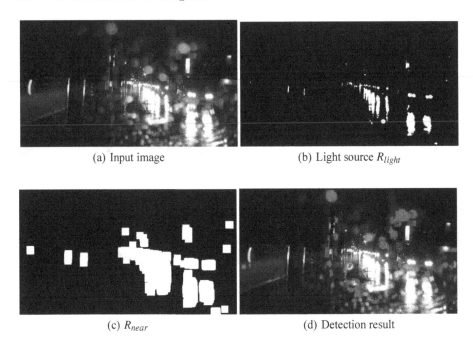

(a) Input image                          (b) Light source $R_{light}$

(c) $R_{near}$                           (d) Detection result

**Fig. 21.** Experimental results at night (Scene 12). (Color figure online)

## 4   Conclusion

In this paper, we have proposed the method for raindrop detection on a windshield from an in-vehicle monocular camera. Conventional methods are considered on isolated spherical raindrops. However, when it rains hard, raindrops are connected on a windshield and show various shapes even though a windshield wiper is activated. The proposed method can deal with raindrops with various shapes, e.g. a band-like shape. In the daytime, raindrops are detected by examining the difference of the blur between the surrounding area. The ratio of the edge strength extracted from two kinds of smoothed images is used as the degree of the blur. False areas are removed using the assumption that raindrops on a windshield don't move so much. At night, bright areas in which a temporal change of the intensity is small are detected as raindrops in an image except for the surrounding of the light source. Experimental results obtained from some real streets under various rain conditions show the effectiveness of the proposed method. In the future, we aim at reducing false detection and realizing videorate processing on the in-vehicle CPU. We are going to create the database for raindrop detection and evaluate the performance quantitatively.

# References

1. Garg, K., Nayar, S.K.: Vision and rain. Int. J. Comput. Vis. **75**(1), 3–27 (2007)
2. Barnum, P.C., Narashimhan, S., Kanade, T.: Analysis of rain and snow in frequency space. Int. J. Comput. Vis. **86**(2–3), 256–274 (2010)
3. Kurihata, H., et al.: Rainy weather recognition from in-vehicle camera images for driver assistance. In: Proceedings of IEEE Intelligent Vehicles Symposium, pp. 205–210 (2005)
4. Kurihata, H., et al.: Detection of raindrops on a windshield from an in-vehicle video camera. Int. J. Innov. Comput. Inf. Control **3**(6(B)), 1583–1591 (2007)
5. Halimeh, J.C., Roser, M.: Raindrop detection on car windshields using geometric-photometric environment of construction and intensity-based correlation. In: Proceedings of IEEE Intelligent Vehicles Symposium, pp. 610–615 (2009)
6. Sugimoto, M., Kakiuchi, N., Ozaki, N., Sugawara, R.: A novel technique for raindrop detection on a car windshield using geometric-photometric model. In: Proceedings of ITSC, pp. 740–745 (2012)
7. Liao, H., Wang, D., Yang, C., Shine, J.: Video-based water drop detection and removal method for a moving vehicle. Inf. Technol. J. **12**(4), 569–583 (2013)
8. Nashashibi, F., de Charette, R., Lia, A.: Detection of unfocused raindrops on a windscreen using low level image processing. In: Proceedings of International Conference on Control, Automation, Robotics and Vision, pp. 1410–1415 (2010)
9. Eigen, D., Krishnan, D., Fergus, R.: Restoring an image taken through a window covered with dirt or rain. In: Proceedings of ICCV, pp. 633–640 (2013)
10. You, S., Tan, R.T., Kawakami, R., Ikeuchi, K.: Adherent raindrop detection and removal in video. In: Proceedings of CVPR, pp. 1035–1042 (2013)
11. Ishizuka, J., Onoguchi, K.: Detection of raindrop with various shapes on a windshield. In: Proceedings of ICPRAM, pp. 475–483 (2016)
12. Tomasi, C., Kanade, T.: Detection and tracking of point features. CMU Technical report CMU-CS-91-132 (1991)

# Comparative Analysis of PRID Algorithms Based on Results Ambiguity Evaluation

V. Renò[1]([⊠]), A. Cardellicchio[2], T. Politi[2], C. Guaragnella[2], and T. D'Orazio[1]

[1] Institute of Intelligent Systems for Automation,
Italian National Research Council, via Amendola 122 D/O, 70126 Bari, Italy
reno@ba.issia.cnr.it
[2] Dipartimento di Ingegneria Elettrica e dellInformazione,
Politecnico di Bari, via Orabona 4, 70126 Bari, Italy

**Abstract.** The re-identification of a subject among different cameras (namely Person Re-Identification or *PRID*) is a task that implicitly defines ambiguities. Two individuals dressed in a similar manner or with a comparable body shape are likely to be misclassified by a computer vision system, especially when only poor quality images are available (i.e. the case of many surveillance systems). For this reason we introduce a method to find, exploit and classify ambiguities among the results of *PRID* algorithms. This approach is useful to analyze the results of a classical PRID pipeline on a specific dataset evaluating its effectiveness in re-identification terms with respect to the ambiguity rate (AR) value. Cumulative Matching Characteristic curves (CMC) can be consequently split according to the AR, using the proposed method to evaluate the performance of an algorithm in low, medium or high ambiguity cases. Experiments on state-of-art algorithms demonstrate that ambiguity-wise separation of results is an helpful tool in order to better understand the effective behaviour of a PRID approach.

## 1 Introduction

Person re-identification (PRID) is a crucial task in modern video surveillance systems, and concerns the retrieval of the same individual given several views acquired by a set of non-overlapping cameras. PRID is strictly related to a number of other video surveillance topics, like cross-camera tracking, event analysis, abandoned object retrieval, and so on; however, it is an extremely challenging task, and has recently drawn a lot of focus by researchers with different fields of expertise.

First of all, each camera in a surveillance system has specific hardware properties that, together with varying lighting conditions, introduce slight variations in the captured frames which, as a consequence, have to be conducted to a common baseline using proper image processing techniques. Furthermore, pose variations of the subject, along with occlusion phenomena, have to be taken into account, as they can negatively impact PRID performances hiding discriminating features that could be otherwise exploited.

© Springer International Publishing AG 2017
A. Fred et al. (Eds.): ICPRAM 2016, LNCS 10163, pp. 230–242, 2017.
DOI: 10.1007/978-3-319-53375-9_13

Another challenging issue is related to the method used to evaluate the performance of PRID algorithms. Normally, PRID techniques are tested against one or more datasets, each one with specific characteristics and challenges. In a pre-processing step, the considered dataset is split into a gallery set, which contains exactly one view per individual, and a probe set, which contains one or more views per subject. The algorithm compares each instance of the gallery set against a set of views taken by the probe set, searching for the best possible match; therefore, PRID can be seen as a multiclass classification problem, where each subset of views related to a certain individual represents a specific class. The task of determining meaningful features – and the proper classifier that should be used – is non-trivial; furthermore, there is not an *universal* dataset (i.e. a dataset that can be used to test methodologies against every specific issue) and, as a consequence, an algorithm which gives good results on a certain dataset may obtain mediocre performances on another dataset. Traditional PRID approaches deal with these problems using a recurring scheme to which we will refer to as PRID pipeline [1]. The first step in the PRID pipeline is image segmentation, where significant information are extracted using proper pre-processing techniques (i.e. background subtraction ([2–6]), human detection ([7,8]) and shadow suppression ([9])). In the second step, a discriminating signature is computed for each view, starting from robust features which can be related to appearance (i.e. color, texture, or shape [10–13]) or to other characteristics like gait [14]. In the third and last step, signatures extracted in the previous step are compared to find the most similar image pairs. Classic matching methods exploited fixed metrics, like Euclidean distance or Bhattacharyya coefficient; modern methods employs more sophisticated approaches, as distance metric learning ([15–17]) or machine learning ([18,19]). Recently, the whole pipeline has been replaced by deep learning architectures ([20–22]), which automatically extract discriminating features at different levels of abstraction, combining them into a meaningful signature, thus giving a significant boost in terms of performances.

Even the most sophisticated state of the art approaches still rely on a basic assumption: results comparison is carried out using an agnostic method based on the analysis of Cumulative Matching Characteristics (CMC) curves, as reported in Fig. 1. These curves describe re-identification results in terms of ranking, that represents the number of iterations after which the PRID algorithm is able to output the correct match. Specifically, Rank-1 represents correctly matched subjects, Rank-2 shows how many individuals are being re-identified after one iteration, and so on. However, this approach only allows us to understand the recognition percentage at a specific rank, without adding any specific information on how results have been produced. Hence, CMC curves give quantitative results, without taking into account neither the intrinsic difficulties of a given dataset, nor the qualitative meaning of the achieved results. For example, given a certain rank, it is not possible to understand if the result has been achieved comparing pair of samples which show low, medium or high ambiguity properties [24]. In case of low ambiguity, the correct match should be ideally returned

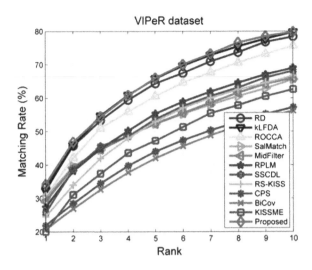

**Fig. 1.** Example of overall CMC curve taken from [23].

as a Rank-1, while worst results should be expected when the algorithm faces ambiguous observations. Normally, the most significant results should be return within Rank-5.

Furthermore, a graphical overview of a certain number of examples taken by VIPeR is depicted in Fig. 2. In this example, some images are categorized in easy, ambiguous or difficult cases. It is immediate to notice that the first are cases in which the human vision system is able to match correspondences easily, for example the textured sweater of the first subject, or red and yellow sweatshirts in the other images. Ambiguous cases are challenging situations even for an expert human operator because each subject looks like many others, especially when he is wearing dark clothes. Finally, difficult cases are always included in datasets and represent a cluster of images in which different light conditions and subject orientations make the correct association almost impossible. In these cases, a correct association would be probably due to fortuity rather than the effective recognition of features by the PRID algorithm.

With this work, we further explore the concept of Ambiguity Rate (AR) introduced in [24] – i.e. an index that compares the results given by a PRID algorithm on a specific dataset – evaluating the performances of state-of-the art PRID algorithms in predetermined ambiguity ranges. The rest of the paper is organized as follows. In Sect. 2, we will give an explanation of our method, highlighting the algorithm used to extract the AR. In Sect. 3, we will compare the results of three PRID approaches using AR, while in Sect. 4 conclusions and a perspective on future works is given.

**Fig. 2.** Example of subjects from VIPeR dataset. The first row represents some query images, while the second row contains the corresponding ground truth. In this example pictures have been manually clustered in easy, ambiguous or difficult to recognize and are respectively highlighted with a blue, black or red contour rectangle. (Color figure online)

## 2   Methodology

### 2.1   Algorithm Description

The proposed approach is related to the one presented in [24] and can be summarized as an enhancement of the PRID pipeline in terms of results' analysis.

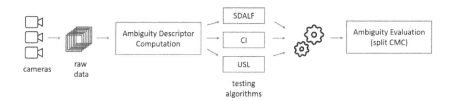

**Fig. 3.** Algorithm high level block diagram.

Looking at Fig. 3 it is immediate to notice that the testing algorithms only represent the central part of the whole approach. In fact, raw data (i.e. images coming from a video surveillance system or a dataset) are first of all pre-processed in order to compute the ambiguity descriptor while the ambiguity evaluation is done as the last step, when results in terms of iterations and ranks are available for each algorithm.

### 2.2   Pre-processing Step

An interesting aspect of this methodology is that the framework does not impose strict constraints in the definition of the ambiguity descriptor *ad*. This entity numerically describes the particular scene or the specific image patch that is going to be analyzed by a PRID algorithm using different features that can be chosen according to the phenomenon that needs to be investigated. For example, a dataset rich of people that wear textured clothes will probably be described in terms of textural features, while general purpose datasets will rely on color based descriptions. Assuming that an expert video surveillance operator evaluates the output of a semi-automatic system by observing recurrent colors in the images, ambiguities in this paper are defined in terms of color changes. Therefore, for each frame of the input dataset, we define the ambiguity descriptor as an array of color values $ad = [h_1, h_2, \ldots, h_n]^T$ where $n$ is the number of horizontal stripes used to divide the image and $h_k$ represents the modal value of the Hue coordinate of the $k$-th stripe.

## 2.3   Post-processing Steps

Ambiguity can be evaluated after executing testing algorithms on the chosen dataset. These methods basically associate a similarity score to each image pair that combines one sample from the gallery set with all the samples from the probe set. For example, the results obtained for a query image $q_i$ can be represented as an associative array

$$q_i \iff [r_1, r_2, \ldots, r_M, \ldots, r_P] \tag{1}$$

where $P$ is the number of images in the probe set and $M$ is a threshold used to consider only the best results. $[r_1, r_2, \ldots, r_M]$ can be therefore represented in terms of the ambiguity descriptor chosen in the pre-processing step, defining the Ambiguity Descriptor Matrix for the query image $q_i$

$$
\begin{aligned}
ADM_{q_i} &= [ad_{r_1}, ad_{r_2}, \ldots, ad_{r_M}] \\
&= \begin{pmatrix} h_{1r_1} & h_{1r_2} & \cdots & h_{1r_M} \\ h_{2r_1} & h_{2r_2} & \cdots & h_{2r_M} \\ \vdots & \vdots & \vdots & \vdots \\ h_{nr_1} & h_{nr_2} & \cdots & h_{nr_M} \end{pmatrix} = \begin{pmatrix} hS_1^T \\ hS_2^T \\ \vdots \\ hS_n^T \end{pmatrix}
\end{aligned} \tag{2}
$$

where $hS_j^T$ is the array in which the modal values of the best $M$ frames of the $j$-th stripe are stored. These rows are employed to compute percentage deviations of color features without losing spacial information by applying the following formula:

$$
\%_{q_i} = \begin{pmatrix} \dfrac{\max(hS_1^T) - \min(hS_1^T)}{256} \\ \vdots \\ \dfrac{\max(hS_n^T) - \min(hS_n^T)}{256} \end{pmatrix} \tag{3}
$$

Finally, the $AR$ value for $q_i$ is defined starting from the average value of the percentage deviations

$$AR_{q_i} = 1 - \frac{1}{n} \sum_{s=1}^{n} \%_{q_i}(s) \tag{4}$$

so that low variations of color percentage displacements produce high ambiguity rate. The alternation of colors in the results, instead, are related to low $AR$ values. It is worth noticing that the role of $M$ is essential for the ambiguity to be effectively computed because useful information can be extracted only within the best ranks. If all the images from the probe set were used to compute the ambiguity rate, AR would be exactly the same for each query image.

Finally, CMC curves can be split according to the ambiguity rate simply setting thresholds and filtering the results. In this paper we define three different ambiguity ranges according to the following fuzzification rule:

$$R_1 \rightarrow 0 \leq AR \leq 0.4$$
$$R_2 \rightarrow 0.4 < AR \leq 0.8 \tag{5}$$
$$R_3 \rightarrow 0.8 < AR \leq 1$$

This way CMC curves can be drawn considering multiple contributions: the first for low ambiguity rates, the second for medium ones and the third for high ambiguity results.

## 3   Experiments and Results

Ambiguity evaluation as described in the previous sections has been performed on three person re-identification algorithms known in literature:

- **Symmetry-Driven Accumulation of Local Features (SDALF)** [10], that basically exploits color (Maximally Stable Color Regions and Weighted Color Histograms) and texture (Recurrent High-Structured Patches) features around pedestrian symmetry axes for extracting image signatures;
- **Color Invariants for PRID (CI)** [25], that exploits relationships between different color patches extracted from each pedestrian image (usually two: one for the upper part and the other for the lower part);
- **Unsupervised Salience Learning for PRID (USL)** [26], that exploits salient features (unique and discriminative) to characterize each person. Both color histograms and SIFT features are used to extract signatures that are subsequently processed by a classifier (e.g. SVM or KNN).

All the algorithms have been tested on the well known VIPeR dataset [27], that contains 632 images taken from non overlapping cameras with arbitrary viewpoints. Images belonging to VIPeR have been taken under varying illumination conditions and each one is scaled to $128 \times 48$ pixels. The approach presented in this paper is mainly focused on the interpretation of results on split CMC curves, according to the fuzzification rule presented in the previous section. Different curves for different ambiguity rate values help in better understanding the algorithm capabilities and interpret if it is producing ambiguous results or not. For this reason, the experiments presented in this section will exploit the AR value to understand how a specific algorithm is working given a specific ambiguity range.

**Table 1.** Dataset separation for different values of the ambiguity rate.

| Ambiguity rate | SDALF | | CI | | USL | |
|---|---|---|---|---|---|---|
| LOW | 16 img | 5.06% | 18 img | 5.70% | 18 img | 5.70% |
| MEDIUM | 298 img | 94.30% | 287 img | 90.82% | 279 img | 88.29% |
| HIGH | 2 img | 0.63% | 11 img | 3.48% | 19 img | 6.01% |

First, the ambiguity descriptor is computed as described in Sect. 2.2 for each image of the collection using 6 horizontal stripes. Then, the rest of the PRID pipeline is executed for the chosen algorithms and finally the results are processed in order to compute the ambiguity rate. In order to obtain a visual comparison of the least ambiguous result and the most ambiguous one, examples of boxplot enriched by the corresponding frames are provided in Figs. 4 and 5. Each box refers to one of the stripes used to divide the images, as noticeable in the figure, so it is representative of ADM described in Eq. 2. A boxplot with large boxes will refer to a non ambiguous response, that should basically imply that the algorithm is operating in an *easy condition*, so the correct response should be given at the first rank. On the contrary, small boxes are related to ambiguous responses that are likely to be mistaken. In this situation, a good PRID algorithm should return the correct answer within the first ranks, but not always at rank 1. Looking at Fig. 4, the first thing to point out is that the only algorithm able to re-identify the query image is CI at rank 3. SDALF is not producing the correct answer in five ranks. Images at ranks $1, 2, 3$ depict people with beige trousers, but the upper part of the query image is not being considered by this approach. The case of USL suggests that there is a consistent amount of people that are likely to be misclassified due to extremely similar clothing. The analysis of Fig. 5 shows that images taken within the best results actually are not so ambiguous. The first 5 returned values are different one from the other: different colours of the shirt/dress (red, black, green, orange and gray) and different colours of the trousers/skirt (pink, black, red, orange). In this situation, the only algorithm that is not answering correctly is CI, while SDALF and USL achieve a rank 1 result. Both the results of CI (for the minimum AR case) and USL (for the maximum AR case) show how the features used by the algorithms can not isolate easy recognizable situations for a human eye. This is probably due to the representation of the colors in different visual systems: the human one and the digital one. For the first, peaks on different color tones can be immediately distinguishable, while in a digital color space the same peaks can generate values that are likely to be classified as similar colors even if they are different. This suggests us to investigate a methodology to quantify the global ambiguity of a dataset and associate an ambiguity level to each query image (e.g. easy, medium or difficult), as will be discussed in the future works section.

Figure 6 shows the ambiguity rate histogram for each response of the three algorithms. The background of the plot helps in the visualization of the three ranges: it is immediate to notice that a small number of responses has a

# Maximum ambiguity rate boxplot

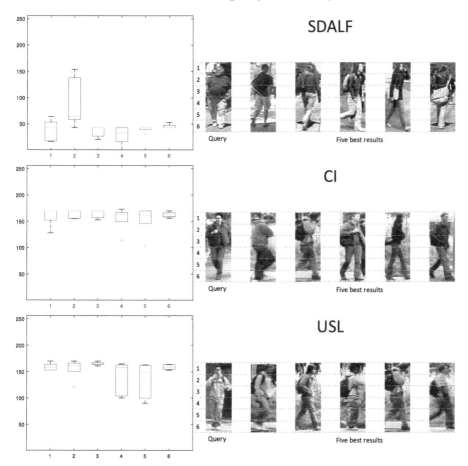

**Fig. 4.** Boxplots that represent the result with the highest ambiguity rate for each algorithm. The plot is enriched with the visual information about both query image and the first five results returned by the specific algorithm. The division stripes are reported on the x-axis, while mode values of hue coordinates are plotted on the y-axis. Small boxes are referred to high ambiguity and big boxes to low ambiguity.

corresponding low ambiguity rate ($< 0.4$) or a high one ($> 0.8$), according to the fuzzification rule presented beforehand.

Looking at Table 1, all algorithms are isolating a small percentage of the images in the tails of the distribution, namely the 5% of the results of the algorithms has low ambiguity. The behaviour for high ambiguity rates is different: only 2 images actually fall into this category for SDALF, 11 for CI and 19 for USL. This means that the algorithms tend to avoid extremely ambiguous responses. A medium level ambiguity is produced most of the time, as almost

# Minimum ambiguity rate boxplot

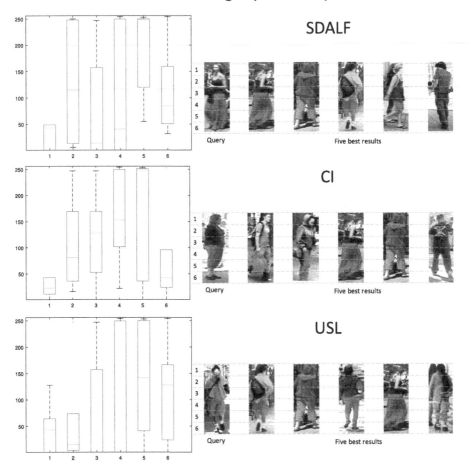

**Fig. 5.** Boxplots that represent the result with the lowest ambiguity rate for each algorithm. The plot is enriched with the visual information about both query image and the first five results returned by the specific algorithm. The division stripes are reported on the x-axis, while mode values of hue coordinates are plotted on the y-axis. Small boxes are referred to high ambiguity and big boxes to low ambiguity. (Color figure online)

90% have an AR value between 0.4 and 0.8. The corresponding CMC curves for *LOW*, *MEDIUM* and *HIGH* ambiguity rate values are reported in Fig. 7 and are called *split CMC*. For each curve, the $x$ axis reports the first 100 ranks and the $y$ axis shows the percentage of images that have been recognized at the specific rank. Due to the cumulative nature of the curve, if there is a step, it means that there are no matches at the corresponding rank. An example of this behaviour is shown in Fig. 7 (d), where there is a big step that starts

# Ambiguity Rate Histograms

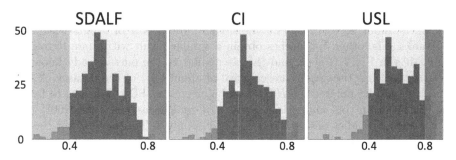

**Fig. 6.** Ambiguity rate histograms on VIPeR results. The x-axis represents the ambiguity rate while on the y-axis occurrences are counted. Low, medium and high ambiguity rate ranges are respectively orange, green and yellow highlighted. (Color figure online)

# Split CMC curves

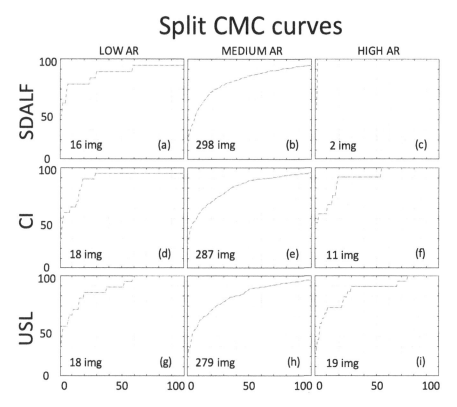

**Fig. 7.** Split cumulative matching characteristic curves obtained for the three considered algorithms.

approximatively around rank 35. The easiest operating condition for an algorithm, where the expected result would be a really high percentage at rank 1, is the *LOW* AR. Here, the best algorithm in our experiments is SDALF because it achieves about 50% of results at rank 1 and about 80% of results within the first ranks. The other approaches obtain a similar result with more iterations. The CMCs in Fig. 7 (b), (e) and (h) are similar to the ones already known in literature because they are representative of about 90% of the dataset. A final remark should be pointed for *HIGH* ambiguity rates. SDALF seems to be the best algorithm in this comparison (with 50% rank 1 responses and 100% rank 2), but the cardinality of the *HIGH* AR set is only 2. This means that the two images are immediately recognized by the algorithm, even if it is working in a challenging situation. Both CI and USL show comparable results in the middle of curves (f) and (i), where there is a step for a recognition percentage of about 90%. CI reports a good starting point, as its rank 1 accuracy is about 50%, while on the other hand USL is able to gain its performances within the first ranks, passing from 20% to 70% in a couple of iterations. Independently from a particular experiment, a generic algorithm should be able to increase the number of images that lie in the tail of its ambiguity distribution. When dealing with *LOW* AR values, the recognition percentage at rank 1 should be the highest, while the correct response can be expected within the first ranks for *HIGH* ambiguity queries.

## 4   Conclusion

In this paper, ambiguities have been exploited in order to evaluate the accuracy of a re-identification algorithm splitting well known CMC curves. The methodology basically defines an ambiguity descriptor and relies on it to compute the AR of each query performed by an algorithm on a specific dataset, actually enriching the state-of-art PRID pipeline. The definition of ambiguity evaluated in this paper can be seen as a *relative* one, because it depends on the results that the algorithm achieves on each query, as stated in Eq. 2. The AR histogram (Fig. 6) graphically explains the ambiguity distribution among the images of a specific dataset, while split CMC curves can be studied separately (ambiguous vs. non ambiguous situations), enabling us to measure the performance of different algorithms on the same dataset. However, the work presented in this paper is the first step in the exploitation of ambiguities in order to understand the capabilities of a re-identification approach. Even if *relative* ambiguity modelling is certainly useful to understand the operative conditions in which an algorithm is working, the results shown in Sect. 3 inspire future research in the direction of an *absolute* ambiguity definition. This way, each image of a dataset will be classified as *easy* (e.g. the only orange dressed man in a crowd of dark clothed subjects) or *difficult* (e.g. a black dressed man in a crowd of dark clothed people). Finally, exploiting both *relative* and *absolute* ambiguities, a generic rank of a CMC will be promoted or penalized starting from the assumption that easy cases should not be misclassified, while higher ranks can be tolerated for hard queries.

# References

1. Cardellicchio, A., D'Orazio, T., Politi, T., Renò, V.: An human perceptive model for person re-identification. In: VISAPP-International Conference on Computer Vision Theory and Applications-2015 (2015)
2. Stauffer, C., Grimson, W.E.L.: Adaptive background mixture models for real-time tracking. In: IEEE Computer Society Conference on Computer Vision and Pattern Recognition, vol. 2. IEEE (1999)
3. Zivkovic, Z.: Improved adaptive Gaussian mixture model for background subtraction. In: 17th International Conference on Pattern Recognition, ICPR 2004, vol. 2, pp. 28–31 (2004)
4. Jojic, N., Perina, A., Cristani, M., Murino, V., Frey, B.: Stel component analysis: Modeling spatial correlations in image class structure. In: IEEE Conference on Computer Vision and Pattern Recognition, CVPR 2009, pp. 2044–2051. IEEE (2009)
5. Renò, V., Marani, R., D'Orazio, T., Stella, E., Nitti, M.: An adaptive parallel background model for high-throughput video applications and smart cameras embedding. In: Proceedings of the International Conference on Distributed Smart Cameras, ICDSC 2014, pp. 30:1–30:6. ACM, New York (2014)
6. Spagnolo, P., Leo, M., D'Orazio, T., Distante, A.: Robust moving objects segmentation by background subtraction. In: The International Workshop on Image Analysis for Multimedia Interactive Services (WIAMIS) (2004)
7. Dalal, N., Triggs, B.: Histograms of oriented gradients for human detection. In: IEEE Computer Society Conference on Computer Vision and Pattern Recognition, CVPR 2005, vol. 1, pp. 886–893. IEEE (2005)
8. Corvee, E., Bak, S., Bremond, F., et al.: People detection and re-identification for multi surveillance cameras. In: International Conference on Computer Vision Theory and Applications, VISAPP-2012 (2012)
9. Lu, J., Zhang, E.: Gait recognition for human identification based on ICA and fuzzy SVM through multiple views fusion. Pattern Recognit. Lett. **28**, 2401–2411 (2007)
10. Farenzena, M., Bazzani, L., Perina, A., Murino, V., Cristani, M.: Person re-identification by symmetry-driven accumulation of local features. In: 2010 IEEE Conference on Computer Vision and Pattern Recognition (CVPR), pp. 2360–2367. IEEE (2010)
11. Gheissari, N., Sebastian, T.B., Hartley, R.: Person re-identification using spatiotemporal appearance. In: 2006 IEEE Computer Society Conference on Computer Vision and Pattern Recognition, vol. 2, pp. 1528–1535. IEEE (2006)
12. Roy, A., Sural, S., Mukherjee, J.: A hierarchical method combining gait and phase of motion with spatiotemporal model for person re-identification. Pattern Recognit. Lett. **33**, 1891–1901 (2012)
13. D'Orazio, T., Guaragnella, C.: A graph-based signature generation for people re-identification in a multi-camera surveillance system. In: VISAPP, vol. 1, pp. 414–417 (2012)
14. Bauml, M., Stiefelhagen, R.: Evaluation of local features for person re-identification in image sequences. In: 2011 8th IEEE International Conference on Advanced Video and Signal-Based Surveillence (AVSS), pp. 291–296. IEEE (2011)
15. Zheng, W.S., Gong, S., Xiang, T.: Person re-identification by probabilistic relative distance comparison. In: 2011 IEEE Conference on Computer Vision and Pattern Recognition (CVPR), pp. 649–656. IEEE (2011)

16. Hirzer, M., Roth, P.M., Köstinger, M., Bischof, H.: Relaxed pairwise learned metric for person re-identification. In: Fitzgibbon, A., Lazebnik, S., Perona, P., Sato, Y., Schmid, C. (eds.) ECCV 2012. LNCS, vol. 7577, pp. 780–793. Springer, Heidelberg (2012). doi:10.1007/978-3-642-33783-3_56

17. Köstinger, M., Hirzer, M., Wohlhart, P., Roth, P.M., Bischof, H.: Large scale metric learning from equivalence constraints. In: 2012 IEEE Conference on Computer Vision and Pattern Recognition (CVPR), pp. 2288–2295. IEEE (2012)

18. Prosser, B., Zheng, W.S., Gong, S., Xiang, T., Mary, Q.: Person re-identification by support vector ranking. In: BMVC, vol. 2, p. 6 (2010)

19. Layne, R., Hospedales, T.M., Gong, S., Mary, Q.: Person re-identification by attributes. In: BMVC, vol. 2, p. 8 (2012)

20. Li, W., Zhao, R., Xiao, T., Wang, X.: Deepreid: deep filter pairing neural network for person re-identification. In: Proceedings of the IEEE Conference on Computer Vision and Pattern Recognition, pp. 152–159 (2014)

21. Ahmed, E., Jones, M., Marks, T.K.: An improved deep learning architecture for person re-identification. In: Proceedings of the IEEE Conference on Computer Vision and Pattern Recognition, pp. 3908–3916 (2015)

22. Ding, S., Lin, L., Wang, G., Chao, H.: Deep feature learning with relative distance comparison for person re-identification. Pattern Recogn. **48**, 2993–3003 (2015)

23. An, L., Chen, X., Yang, S.: Person re-identification via hypergraph-based matching. Neurocomputing **182**, 247–254 (2016)

24. Renò, V., Cardellicchio, A., Politi, T., Guaragnella, C., D'Orazio, T.: Exploiting ambiguities in the analysis of cumulative matching curves for person re-identification. In: ICPRAM 2016 - Proceedings of the 5th International Conference on Pattern Recognition Applications and Methods, pp. 484–494 (2016)

25. Kviatkovsky, I., Adam, A., Rivlin, E.: Color invariants for person reidentification. IEEE Trans. Pattern Anal. Mach. Intell. **35**, 1622–1634 (2013)

26. Zhao, R., Ouyang, W., Wang, X.: Unsupervised salience learning for person re-identification. In: Proceedings of the IEEE Conference on Computer Vision and Pattern Recognition, pp. 3586–3593 (2013)

27. Gray, D., Brennan, S., Tao, H.: Evaluating appearance models for recognition, reacquisition, and tracking. In: Proceedings of IEEE International Workshop on Performance Evaluation for Tracking and Surveillance (PETS), vol. 3. Citeseer (2007)

# Author Index

Printed in the United States
By Bookmasters